Contents

STUDENT SOLUTIONS MANUAL AND STUDY GUIDE TO ACCOMPANY

Understanding Elementary Algebra

THIRD EDITION

and

Understanding Elementary Algebra with Geometry

ARTHUR GOODMAN
The Queens College of the City University of New York

LEWIS HIRSCH
Rutgers University

Prepared by
STEVEN KAHAN
The Queens College of the City University of New York

WEST PUBLISHING COMPANY

Minneapolis/St. Paul ▩ New York ▩ Los Angeles ▩ San Francisco

WEST'S COMMITMENT TO THE ENVIRONMENT

In 1906, West Publishing Company began recycling materials left over from the production of books. This began a tradition of efficient and responsible use of resources. Today, up to 95% of our legal books and 70% of our college texts and school texts are printed on recycled, acid-free stock. West also recycles nearly 22 million pounds of scrap paper annually—the equivalent of 181,717 trees. Since the 1960s, West has devised ways to capture and recycle waste inks, solvents, oils, and vapors created in the printing process. We also recycle plastics of all kinds, wood, glass, corrugated cardboard, and batteries, and have eliminated the use of Styrofoam book packaging. We at West are proud of the longevity and the scope of our commitment to the environment.

Production, Prepress, Printing and Binding by West Publishing Company.

Printed in the United States of America
01 00 99 98 97 96 95 94 8 7 6 5 4 3 2 1 0

ISBN 0–314–03995–3

CHAPTER 1
THE INTEGERS

Exercises 1.1

1. True 3. True 5. True 7. True

9. False

11. {1, 2, 3, 4, 5, 6, 7} 13. {0, 2, 4, 6, 8, 10, 12, 14, 16, 18}

15. {0, 1, 2, 3, 4, 5, 6} 17. {0, 1, 2, 3, 4, 5}

19. {6, 7, 8, 9,...} 21. {7, 8, 9, 10,...}

23. {3, 4, 5} 25. {1, 2, 3, 4, 6, 8, 12, 24}

27. {1, 2, 3, 4, 5, 6, 12, 15} 29. {0, 4, 8, 12,...}

31. {0, 12, 24, 36,...} 33. $\varnothing$ 35. $>$ 37. $=$

39. $>$ 41. $=$ 43. $=$ 45. $<, \leq, \neq$

47. $\leq, \geq, =$ 49. $>, \geq, \neq$ 51. $<, \leq, \neq$ 53. $>, \geq, \neq$

55. $2 \cdot 7$ 57. $3 \cdot 11$ 59. $2 \cdot 3 \cdot 5$ 61. Prime

63. $2 \cdot 2 \cdot 2 \cdot 2 \cdot 2 \cdot 2$ 65. $2 \cdot 2 \cdot 2 \cdot 2 \cdot 2 \cdot 3$ 67. $3 \cdot 29$

69.
0 1 2 3 4 5 6 7

71. 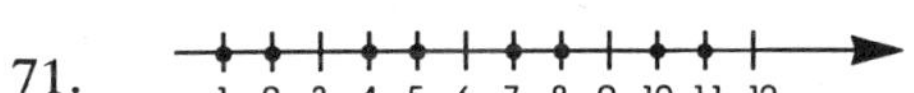

73. W contains the number 0; N does not.

74. In a sum, the numbers to be added are called terms; in a product, the numbers to be multiplied are called factors. In 2 + 5, 2 is a term; in $2 \cdot 3$, 2 is a factor.

75. A factor of n is a number that divides exactly into n; a multiple of n is a number that is exactly divisible by n. 3 is a factor of 12; 12 is a multiple of 3.

76. C is the set containing the first three multiples of 6.

77. C is the set containing the first three multiples of 12.

78. (a) Sum of three terms : x, y, and z

(b) Product of three factors : x, y, and z

(c) Sum of two terms: xy and z

(d) Product of two factors : x and (y + z)

Exercises 1.2

1. True (commutative law of addition)
3. True (commutative law of addition)
5. False
7. False
9. True (commutative law of multiplication)
11. True (commutative law of multiplication)
13. False
15. True (associative law of addition)
17. True (associative law of addition)
19. True (associative law of multiplication)
21. True (associative law of multiplication)
23. False
25. False
27. Associative law of addition
29. Associative law of multiplication
31. Commutative law of addition, followed by associative law of addition.
33. Commutative law of multiplication, followed by associative law of multiplication.

35. -4 37. $+4$ 39. 4 41. 4

43. -4 45. -4 47. 5 49. 5

51. 2 53. 11 55. 5

57. $7+3\cdot 4=7+12=19$

59. $4\cdot 3+6=12+6=18$

61. $11-4\cdot 2=11-8=3$

63. $15-5\cdot 3=15-15=0$

65. $6+3\cdot 5-2=6+15-2=19$

67. $12-4\cdot 3+6=12-12+6=6$

69. $6+4(3+2)=6+4\cdot 5$
$=6+20=26$

71. $6+(4\cdot 3+2)=6+(12+2)$
$=6+14=20$

73. $6+(4\cdot 3)+2=6+12+2=20$

75. $(6+4)(3+2)=10\cdot 5=50$

77. $\dfrac{15+2\cdot 3}{3+2\cdot 2}=\dfrac{15+6}{3+4}=\dfrac{21}{7}=3$

79. $3+2[3+2(3+2)]=3+2[3+2\cdot 5]$
$=3+2[3+10]$
$=3+2\cdot 13$
$=3+26=29$

81. $9-4[6-2(3-1)]=9-4[6-2\cdot 2]$
$=9-4[6-4]$
$=9-4\cdot 2=9-8=1$

83. $4+[3+5[2+2(3+1)]]$
$=4+[3+5[2+2\cdot 4]]$
$=4+[3+5[2+8]]=4+[3+5\cdot 10]$
$=4+[3+50]=4+53=57$

5. $\dfrac{10+2(5+3)}{2\cdot 5+3} = \dfrac{10+2\cdot 8}{10+3}$

$= \dfrac{10+16}{10+3} = \dfrac{26}{13} = 2$

87. Jane has a checking account with overdraft privileges. If she has \$283 in her account and she writes a check for \$300, by how much is Jane's account overdrawn?

(\$283 – \$300 = –\$17)

88. A B

Exercises 1.3

1. $+5 + (-7) = -(7-5) = -2$

3. $-9 + (-3) = -(9+3) = -12$

5. $-4 + (+11) = +(11-4) = +7$

7. $+8 + (-3) = +(8-3) = +5$

9. $+6 + (-2) = +(6-2) = +4$

11. $-6 + (-2) = -(6+2) = -8$

13. $-6 + (+2) = -(6-2) = -4$

15. $+6 + (+2) = 6 + 2 = 8$

17. $-7 + (-8) = -(7+8) = -15$

19. $10 + (-14) = -(14-10) = -4$

21. $8 + (-8) = 0$

23. $9 + (-16) = -(16-9) = -7$

25. $-30 + (-14) = -(30+14) = -44$

27. $-6 + 12 = +(12-6) = +6$

29. $-5+(-4)+(-6) = -(5+4)+(-6)$
$= -9+(-6) = -(9+6)$
$= -15$

31. $5+(-4)+(-6) = +(5-4)+(-6)$
$= +1+(-6)$
$= -(6-1) = -5$

33. $-5+4+(-6) = -(5-4)+(-6)$
$= -1+(-6)$
$= -(1+6) = -7$

35. $-5+(-4)+6 = -(5+4)+6$
$= -9+6$
$= -(9-6) = -3$

37. $7+(-10)+(-2) = -(10-7)+(-2)$
$= -3+(-2)$
$= -(3+2) = -5$

39. $-3+9+(-5) = +(9-3)+(-5)$
$= +6+(-5)$
$= +(6-5) = 1$

41. $-4+(-3)+(-8) = -(4+3)+(-8)$
$= -7+(-8)$
$= -(7+8) = -15$

43. $15+(-3)+(-6) = +(15-3)+(-6)$
$= +12+(-6)$
$= +(12-6) = 6$

45. $-8+(-2)+6=-(8+2)+6$
$=-10+6$
$=-(10-6)=-4$

47. $16+(-5)+(-7)+2$
$=+(16-5)+(-7)+2$
$=+11+(-7)+2=+(11-7)+2$
$=+4+2=6$

49. $-8+6+(-5)+1$
$=-(8-6)+(-5)+1$
$=-2+(-5)+1=-(2+5)+1$
$=-7+1=-(7-1)=-6$

51. $2+(-9)+(-3)+(-1)+6$
$=-(9-2)+(-3)+(-1)+6$
$=-7+(-3)+(-1)+6$
$=-(7+3)+(-1)+6$
$=-10+(-1)+6=-(10+1)+6$
$=-11+6=-(11-6)=-5$

53. $27+(-56)=-(56-27)=-29$

55. $-22+(-45)=-(22+45)=-67$

57. $-31+(-26)+48=-(31+26)+48$
$=-57+48$
$=-(57-48)=-9$

59. $-5+[7+(-3)]=-5+[+(7-3)]$
$=-5+(+4)$
$=-(5-4)=-1$

61. $-3+5\cdot 2+(-6)=-3+10+(-6)$
$=+(10-3)+(-6)$
$=+7+(-6)$
$=+(7-6)=+1$

63. $|2+(-6)|+|2|+|-6|$
$=|-(6-2)|+|2|+|-6|$
$=|-4|+|2|+|-6|$
$=4+2+6=12$

65. $|-7|+(-7)=7+(-7)=0$

67. $+48+[(-18)+(-22)+(-15)]$
$+(+50)+[(-28)+(-12)]$
$+(+20)+(-17)+(+27)$
$=+48+(-55)+(+50)+(-40)$
$+(+20)+(-17)+(+27)=+33$

So the final balance in Carla's checking account is $33.

69. $+25+(+8)+(-14)+(-5)$
$=+(25+8)+(-14)+(-5)$
$=+33+(-14)+(-5)$
$=+(33-14)+(-5)$
$=+19+(-5)$
$=+(19-5)=+14$

So on fourth down, the team is located on its own 14 yard line.

71. $+\$10,432+(-\$1,678)+(-\$2,046)+\$7,488$
$=\$8,754+(-\$2,046)+\$7,488$
$=\$6,708+\$7,488=\$14,196$ (profit)

73. $10,000+380+540+(-275)+(-600)+(-72)$
$=10,920+(-947)=9,973$

75. When we add an integer to its opposite, we get 0 as out result. This occurs because an integer and its opposite have the same absolute value.

76. If the sum of two integers is zero, the integers must be opposites. In all other cases, the sum of the integers would either be positive or negative.

77. $4 - 7 = -3$, because $-3 + 7 = 4$

78. $4 - (-7) = 11$, because $11 + (-7) = 4$

Exercises 1.4

1. $6 - (+10) = 6 + (-10) = -4$

3. $-7 - (+4) = -7 + (-4) = -11$

5. $3 - (-6) = 3 + (+6) = 9$

7. $-8 - (-2) = -8 + (+2) = -6$

9. $-5 + (+8) = 3$

11. $5 - (+8) = 5 + (-8) = -3$

13. $5 - (-8) = 5 + (+8) = 13$

15. $-5 - (+8) = -5 + (-8) = -13$

17. $-5 - (-8) = -5 + (+8) = 3$

19. $5 + (-8) = -3$

21. $-5 + (-8) = -13$

23. $6 - (-7) = 6 + (+7) = 13$

25. $-6 - (-7) = -6 + (+7) = 1$

27. $2 + (-6) - (+7) = 2 + (-6) + (-7)$
$= -4 + (-7) = -11$

29. $2 - 6 - 7 = 2 + (-6) + (-7)$
$= -4 + (-7) = -11$

31. $2 - (6 - 7) = 2 - (-1)$
$= 2 + (+1) = 3$

33. $7 - 9 - 3 + 2 = 7 + (-9) + (-3) + 2$
$= -2 + (-3) + 2$
$= -5 + 2 = -3$

35. $11 - 5 + 4 - 7 = 11 + (-5) + 4 + (-7)$
$= 6 + 4 + (-7)$
$= 10 + (-7) = 3$

37. $2 - 3 - 6 - 2 = 2 + (-3) + (-6) + (-2)$
$= -1 + (-6) + (-2)$
$= -7 + (-2) = -9$

39. $-10 + 4 - 9 - (-3)$
$= -10 + 4 + (-9) + (+3)$
$= -6 + (-9) + (+3)$
$= -15 + (+3) = -12$

41. $3 - 6 + 1 - (-4) = 3 + (-6) + 1 + (+4)$
$= -3 + 1 + (+4)$
$= -2 + (+4) = 2$

43. $-1 - 4 - 2 - (-5)$
$= -1 + (-4) + (-2) + (+5)$
$= -5 + (-2) + (+5)$
$= -7 + (+5) = -2$

45. $4 - 8 - 6 + 3 = 4 + (-8) + (-6) + 3$
$= -4 + (-6) + 3$
$= -10 + 3 = -7$

47. $4 - 8 - (6 + 3) = 4 - 8 - 9$
$= 4 + (-8) + (-9)$
$= -4 + (-9) = -13$

49. $4 - (8 - 6) + 3 = 4 - 2 + 3$
$= 4 + (-2) + 3$
$= 2 + 3 = 5$

51. $4-(8-6+3)=4-(8+(-6)+3)$
$=4-(2+3)=4-5$
$=4+(-5)=-1$

53. $-8+8=0$

55. $-8-8=-8+(-8)=-16$

57. $-8-(-8)=-8+(+8)=0$

59. $9-5\cdot 4-2=9-20-2$
$=9+(-20)+(-2)$
$=-11+(-2)=-13$

61. $9-5(4-2)=9-5\cdot 2=9-10$
$=9+(-10)=-1$

63. $9-(5\cdot 4-2)=9-(20-2)=9-18$
$=9+(-18)=-9$

65. $|6-2|-|2-6|=|4|-|-4|$
$=4-4=0$

67. $|2-6|-(2-6)=|-4|-(-4)=4-(-4)$
$=4+(+4)=8$

69. $|-4-3+2|-4-3+2$
$=|-4+(-3)+2|+(-4)+(-3)+2$
$=|-7+2|+(-4)+(-3)+2$
$=|-5|+(-4)+(-3)+2$
$=5+(-4)+(-3)+2$
$=1+(-3)+2=-2+2=0$

71. $29{,}028-(-1{,}290)=29{,}028+1{,}290$
$=30{,}318$ feet

73. $\$643.47-(-\$82.94)=\$643.47+\$82.94$
$=\$726.41$

75. $-8-2-7=(-8)+(-2)+(-7)$.

The terms are –8, –2, and –7.

77. $-3-m-n=(-3)+(-m)+(-n)$.

The terms are –3, –m, and –n.

79. The left side is equal to 4, while the right side is equal to –4. In general, the absolute value of a difference of two integers is not the same as the difference of the absolute values of those integers.

80. 4 times –3 should mean "add –3 four times," giving $(-3)+(-3)+(-3)+(-3)=-12$.

81. We can interpret –4 times 3 to mean "subtract 3 four times," giving $-3-3-3-3=-12$. Similarly, we can interpret –4 times –3 to mean "subtract –3 four times," giving

$-(-3)-(-3)-(-3)-(-3)$
$=+3+3+3+3=+12$.

Exercises 1.5

1. $(+5)(-3)=-(5\cdot 3)=-15$

3. $(-5)(+3)=-(5\cdot 3)=-15$

5. $(-5)(-3)=+(5\cdot 3)=15$

7. $-5-3=-8$

9. $-(5-3)=-2$

11. $\frac{20}{-5} = -\left(\frac{20}{5}\right) = -4$

13. $\frac{-20}{5} = -\left(\frac{20}{5}\right) = -4$

15. $\frac{-20}{-5} = +\left(\frac{20}{5}\right) = 4$

17. $\frac{0}{7} = 0$

19. No answer (division by 0 is not allowed)

21. $4(-2) - 6 = -8 - 6 = -14$

23. $4 - 2 - 6 = 2 - 6 = -4$

25. $4 - (2 - 6) = 4 - (-4) = 8$

27. $4(-2 - 6) = 4(-8) = -32$

29. $-6(2)(-5)(-1) = -12(-5)(-1)$
$= 60(-1) = -60$

31. $9 - 3(1 - 3) = 9 - 3(-2)$
$= 9 + 6 = 15$

33. $8 - 3(2 - 5) = 8 - 3(-3)$
$= 8 + 9 = 17$

35. $8 - 3 \cdot 2 - 5 = 8 - 6 - 5$
$= 2 - 5 = -3$

37. $8 - (3 \cdot 2 - 5) = 8 - (6 - 5)$
$= 8 - 1 = 7$

39. $(8 - 3)(2 - 5) = (5)(-3) = -15$

41. $\frac{-12 - 3 + 1}{-7} = \frac{-15 + 1}{-7} = \frac{-14}{-7} = 2$

43. $\frac{-8 - 2 - 4}{-2} = \frac{-10 - 4}{-2} = \frac{-14}{-2} = 7$

45. $\frac{-8 - (2 - 4)}{-2} = \frac{-8 - (-2)}{-2} = \frac{-8 + 2}{-2}$
$= \frac{-6}{-2} = 3$

47. $\frac{-8 - 2(-4)}{-2} = \frac{-8 + 8}{-2} = \frac{0}{-2} = 0$

49. $\frac{8}{-2} - \frac{12}{-3} = -4 - (-4) = -4 + 4 = 0$

51. $\frac{8(-6)}{8 - 6} = \frac{-48}{2} = -24$

53. $\frac{-4(-5)(-4)}{-4(-5) - 4} = \frac{20(-4)}{20 - 4} = \frac{-80}{16} = -5$

55. $\frac{3(-2) - 4}{-4 - 1} = \frac{-6 - 4}{-4 - 1} = \frac{-10}{-5} = 2$

57. $\frac{6 - 4}{4 - 4} = \frac{2}{0}$, which has no answer, since division by 0 is not allowed.

59. $\frac{4(-6)}{-2 - 1} - \frac{4 - 6}{-2(-1)} = \frac{-24}{-3} - \frac{-2}{2}$
$= 8 - (-1) = 9$

61. $-6[-3 - 2(5 - 3)] = -6[-3 - 2 \cdot 2]$
$= -6[-3 - 4]$
$= -6(-7) = 42$

63. $5 - 2[3 - (4 + 6)] = 5 - 2[3 - 10]$
$= 5 - 2(-7)$
$= 5 + 14 = 19$

65. $5 - \frac{3 - 5}{-2} = 5 - \frac{-2}{-2} = 5 - 1 = 4$

67. $6 - \frac{8 - 2(-3)}{-7} = 6 - \frac{8 + 6}{-7} = 6 - \frac{14}{-7}$
$= 6 - (-2) = 6 + 2 = 8$

69. – .209

71. – 24.57

73. 1.1

75. .302

77. 2.26

79. Even though the symbols are the same in both examples, the parentheses lead to different results. Without the parentheses, we have a subtraction example: $5 - 2 = 3$; with the parentheses, we have a multiplication example:

$5\,(-2) = -10$.

80. Again, the same symbols are used in both examples, but the location of the parentheses causes the difference this time. Our order of operations requires that we multiply before we subtract. Therefore,

$6(-3) - 2 = -18 - 2 = -20$, while $6 - 3\,(-2) = 6 + 6 = 12$.

81. When we see $-7 - 8$, it must be interpreted as a subtraction, since no parentheses are present. Therefore, (a) and (b) are both wrong, since $-7 - 8 = -15$. (It would be correct to write $-7\,(-8) = +56$ or $(-7)(-8) = +56$.) Part (c) is wrong also. Since $2 - 4 = -2$, it follows that $-3 - (2 - 4) = -3 - (-2) = -3 + 2 = -1$. In part (d), the order of operations has been followed. Here, $-4 - 2(5 - 6) = -4 - 2(-1) = -4 + 2 = -2$. ($-4 - 2 = -6$ never legitimately enters into the calculation.)

82. The statement "two negatives make a positive" is accurate for multiplication and division but **not** for addition and subtraction. For instance, $(-3) + (-2) = -5$ and $(-3) - (-2) = -1$, and neither of these answers is positive.

Exercises 1.6

1. $4.2 + 5.9 = 10.1$ 3. $4(5.1) = 20.4$ 5. $\frac{12.8}{3.2} = 4$ 7. $\frac{36.8}{1.5} = 24.53$

9. $\frac{8}{.2} + \frac{12}{.4} = 40 + 30 = 70$ 11. $-2 - 5.3 = -7.3$ 13. $-2(5.3) = -10.6$

15. $-2(-5.3) = 10.6$ 17. $\frac{-8.4}{1.2} = -7$ 19. $\frac{-6}{1.5} + \frac{21.6}{-1.2} = -4 - 18 = -22$

21 Between 4 and 5 23. Between 2 and 3 25. Between 7 and 8 27. Between −1 and 0

29.
-4 -2 0 2 4

31.
-2 0 2 4 6

33.
-4 -2 0 2 4

35.
0 1 2 3 4 5 6 7 8 9 10

37.
-4 -2 0 2 4

39. (a) Use the commutative law of addition to rewrite the problem as 12.9 – 1.7 – 8.3. Then this equals 12.9 – 10 = 2.9.

(b) Use the commutative law of multiplication to rewrite the problem as 7(10)(–2.3). Then this becomes 7(–23) = –161.

(c) Use the associative law of multiplication to rewrite the problem as $-10\left(\frac{3.2}{-1.6}\right)$. Then this equals –10(–2) = 20.

(d) First, use the commutative law of multiplication to rewrite the problem as $-30(-10)\left(\frac{1.6}{3}\right)$, which becomes $300\left(\frac{1.6}{3}\right)$. Then use the associative law of multiplication to write this as $\frac{300}{3}(1.6) = 100(1.6) = 160$.

40. Since (1)(1) = 1 and (2)(2) = 4, the square root of 3, to the nearest whole number, is 2. Since (1.7)(1.7) = 2.89 and (1.8)(1.8) = 3.24, the square root of 3, to the nearest tenth, is 1.7. Since (1.73)(1.73) = 2.9929 and (1.74)(1.74) = 3.0276, the square root of 3, to the nearest hundredth, is 1.73. Since (1.732)(1.732) = 2.999824 and (1.733)(1.733) = 3.003289, the square root of 3, to the nearest thousandth, is 1.732.

Chapter 1 Review Exercises

1. {2, 4, 6, 8, 10, 12, 14, 16, 18}

3. {2, 3, 5, 7, 11, 13, 17, 19}

5. {23, 29}

7. $2 \cdot 2 \cdot 5$

9. Prime

11. $2 \cdot 2 \cdot 5 \cdot 5$

13. 3 – 7 = –4

15. –7 –5 = – 12

17. –2 – (–6) = –2 + 6 = 4

19. –4 –5 – 6 = –9 – 6 = –15

21. –7 + 12 – 5 = + 5 –5 = 0

23. 7 – 4 + 3 – 9 = 3 + 3 – 9 = 6 – 9 = – 3

25. 8 – 5 – 6 = 3 – 6 = – 3

27. 8 – (5 – 6) = 8 – (–1) = 8 + 1 = 9

29. 8(5 – 6) = 8(–1) = –8

31. 8 – 3 – 6 = 5 – 6 = –1

33. 8 – 3(–6) = 8 + 18 = 26

35. 8 – (3 – 6) = 8 – (–3) = 8 + 3 = 11

37. 8(–3)(–6) = –24(–6) = 144

39. 9 – 4(3 – 7) = 9 – 4(–4) = 9 + 16 = 25

41. $9 - 4 \cdot 3 - 7 = 9 - 12 - 7$
$= -3 - 7 = -10$

43. $9 - (4 \cdot 3 - 7) = 9 - (12 - 7)$
$= 9 - 5 = 4$

45. $(9-4)(3-7) = 5(-4) = -20$

47. $|4-9| - |3-7| = |-5| - |-4| = 5 - 4 = 1$

49. $\frac{-7-3}{-2(-5)} = \frac{-10}{+10} = -1$

51. $\frac{-4(-2)(-8)}{-4(2)-8} = \frac{8(-8)}{-8-8} = \frac{-64}{-16} = 4$

Chapter 1 Practice Test

1. (a) True (b) False (c) False (d) True (both have eight elements)

   (e) C = ∅

3. $|3-8|-|1-6| = |-5|-|-5| = 5-5 = 0$

5. $-4(-3)(-2) = 12(-2) = -24$

7. $-3(-5) - 2(-1) = 15 - (-2) = 15 + 2 = 17$

9. $\frac{4(-8)}{4-8} - \frac{3-6}{-2-1} = \frac{-32}{-4} - \frac{-3}{-3} = 8 - 1 = 7$

11. $\frac{(-2)(-3)(-4)}{(-2)(-3)-4} = \frac{6(-4)}{6-4} = \frac{-24}{2} = -12$

13. $8 - 5 \cdot 4 - 7 = 8 - 20 - 7 = -12 - 7 = -19$

15. $8 - 3[8 - 3(8 - 3)] = 8 - 3[8 - 3(5)]$

$= 8 - 3[8 - 15] = 8 - 3(-7)$

$= 8 - (-21) = 8 + 21 = 29$

CHAPTER 2
ALGEBRAIC EXPRESSIONS

Exercises 2.1

1. $x \cdot x \cdot x \cdot x \cdot x \cdot x$

3. $(-x)(-x)(-x)(-x)$

5. $-(x \cdot x \cdot x \cdot x)$

7. $x \cdot x \cdot y \cdot y \cdot y$

9. $x \cdot x + y \cdot y \cdot y$

11. $x \cdot y \cdot y \cdot y$

13. a^4

15. x^2y^3

17. $-r^2s^3$

19. $-x^2(-y)^3$

21. $x^3x^5 = x^{3+5} = x^8$

23. $3^5 = 3 \cdot 3 \cdot 3 \cdot 3 \cdot 3 = 9 \cdot 3 \cdot 3 \cdot 3$
$= 27 \cdot 3 \cdot 3 = 81 \cdot 3 = 243$

25. $-2^3 = -(2 \cdot 2 \cdot 2) = -(4 \cdot 2) = -8$

27. $(-2)^4 = (-2)(-2)(-2)(-2)$
$= 4(-2)(-2) = -8(-2) = 16$

29. $3^2 + 3^3 - 3^4 = 3 \cdot 3 + 3 \cdot 3 \cdot 3 - 3 \cdot 3 \cdot 3 \cdot 3$
$= 9 + 27 - 81 = -45$

31. $4 \cdot 3^2 - 2 \cdot 5^2 = 4 \cdot 3 \cdot 3 - 2 \cdot 5 \cdot 5$
$= 36 - 50 = -14$

33. $(3-7)^2 - (4-5)^3$
$= (-4)^2 - (-1)^3$
$= (-4)(-4) - (-1)(-1)(-1)$
$= 16 - (-1) = 17$

35. $3^2 4^3 = 3 \cdot 3 \cdot 4 \cdot 4 \cdot 4 = 576$

37. $3^2 3^3 = 3^{2+3} = 3^5 = 3 \cdot 3 \cdot 3 \cdot 3 \cdot 3 = 243$

39. $x^3x^5 = x^{3+5} = x^8$

41. $aa^2a^4 = a^{1+2+4} = a^7$

43. $(3x)(5x)(4x) = (3 \cdot 5 \cdot 4)(x \cdot x \cdot x)$
$= 60x^3$

45. $(3r^2)(2r^3) = (3 \cdot 2)(r^2r^3)$
$= 6r^{2+3} = 6r^5$

47. $(-3x^3)(5x^2) = (-3 \cdot 5)(x^3x^2)$
$= -15x^{3+2} = -15x^5$

49. $(-c^4)(2^3)(-5c) = (-2^3(-5))(c^4c)$
$= 40c^{4+1} = 40c^5$

51. $(8x^3y^2)(4xy^5)$
$= (8 \cdot 4)(x^3x)(y^2y^5)$
$= 32x^{3+1}y^{2+5} = 32x^4y^7$

53. $(-2xy)(x^2y^2)(-3xy)$
$= (-2(-3))(xx^2x)(yy^2y)$
$= 6x^{1+2+1}y^{1+2+1} = 6x^4y^4$

55. $(3a^4)^2 = (3a^4)(3a^4) = (3 \cdot 3)(a^4a^4)$
$= 9a^{4+4} = 9a^8$

57. $(-4n^2)^3 = (-4n^2)(-4n^2)(-4n^2)$
$= ((-4)(-4)(-4))(n^2n^2n^2)$
$= -64n^{2+2+2} = -64n^6$

59. $(x^2)^3(x^4)^2 = (x^2x^2x^2)(x^4x^4)$
$= x^{2+2+2+4+4} = x^{14}$

61. .073 63. 14.758 65. 107.916 67. 48.337

69. In $3\cdot 2^4$, the exponent of 4 applies only to the 2, not to the 3; in $(3\cdot 2)^4$, the exponent of 4 applies to both the 2 and the 3. Put another way, we compute $3\cdot 2^4$, by first raising 2 to the fourth power and then multiplying the result by 3. We compute $(3\cdot 2)^4$ by first multiplying 2 by 3 and then raising the result to the fourth power.

70. (a) The exponent of 2 applies only to the 2, not to the 3.

(b) Only x should be raised to the fourth power.

(c) 5^4 is not the same as $5\cdot 4$.

(d) When we square 3, we get 9, not 6. Addition is performed in the exponents.

(e) -3^4 is the opposite of 3^4, so is equal to –81.

(f) We add exponents when we multiply powers of x, not when we add powers of x.

(g) When multiplying powers of x, we add exponents. So $x^2x^7 = x^{2+7} = x^9$.

(h) Here, we do not add exponents, since $(x^2)^4 = x^2x^2x^2x^2 = x^{2+2+2+2} = x^8$.

Exercises 2.2

1. $-y = -(-3) = 3$

3. $-|y| = -|(-3)| = -3$

5. $x + y = (2) + (-3) = -1$

7. $x - y = (2) - (-3) = 5$

9. $|x - y| = |(2) - (-3)| = |5| = 5$

11. $|x| - y = |(2)| - (-3) = 2 + 3 = 5$

13. $|x| - |y| = |(2)| - |(-3)| = 2 - 3 = -1$

15. $x + y + z = (2) + (-3) + (-4) = -5$

17. $x + y - z = (2) + (-3) - (-4) = 3$

19. $xy - z = (2)(-3) - (-4) = -6 + 4 = -2$

21. $x(y - z) = (2)((-3) - (-4)) = 2(1) = 2$

23. $x - (y - z) = (2) - ((-3) - (-4))$
$= 2 - 1 = 1$

25. $xy^2 = (2)(-3)^2 = 2(9) = 18$

27. $x + y^2 = (2) + (-3)^2 = 2 + 9 = 11$

29. $x^2 + y^2 = (2)^2 + (-3)^2 = 4 + 9 = 13$

31. $(x + y + z)^2 = ((2) + (-3) + (-4))^2$
$= (-5)^2 = 25$

33. $-y^2 = -(-3)^2 = -9$

35. $-x^2z = -(2)^2(-4) = -4(-4) = 16$

37. $xyz^2 = (2)(-3)(-4)^2 = -6(16) = -96$

39. $x(y-z)^2 = (2)((-3)-(-4))^2$
$= (2)(1)^2 = 2(1) = 2$

41. $xy^2 - (xy)^2 = (2)(-3)^2 - ((2)(-3))^2$
$= 2(9) - (-6)^2$
$= 18 - 36 = -18$

43. $(z - 3x)^2 = ((-4) - 3(2))^2$
$= (-4-6)^2 = (-10)^2 = 100$

45. $(5x + y)(3x - y)$
$= (5(2) + (-3)((3(2) - (-3))$
$= (10 - 3)(6 + 3)$
$= (7)(9) = 63$

47. $y^2 - 3y + 2 = (-3)^2 - 3(-3) + 2$
$= 9 + 9 + 2 = 20$

49. $3x^2 + 4x + 1 = 3(2)^2 + 4(2) + 1$
$= 3(4) + 4(2) + 1$
$= 12 + 8 + 1 = 21$

51. $x^2 + 3x^2y - 3xy^2 + y^3$
$= (2)^2 + 3(2)^2(-3) - 3(2)(-3)^2$
$+(-3)^2$
$= 4 + 3(4)(-3) - 3(2)(9) + (-27)$
$= 4 - 36 - 54 - 27 = -113$

53. $\frac{xy}{x+y} = \frac{(2)(-3)}{(2)+(-3)} = \frac{-6}{-1} = 6$

55. $\frac{2}{x} - \frac{y}{3} + \frac{z}{2} = \frac{2}{(2)} - \frac{(-3)}{3} + \frac{(-4)}{2}$
$= 1 - (1) + (-2) = 0$

57. $3x - 4y$ has two terms. The first has a coefficient of 3 and a literal part of x; the second has a coefficient of –4 and a literal part of y.

59. $3x(-4y) = -12xy$ has one term, with a coefficient of –12 and a literal part of xy.

61. $3x(z - y)$ has one term with a coefficient of 3 and a literal part of $x(z - y)$.

63. $4x^2 - 3x + 2$ has three terms. The first has a coefficient of 4 and a literal part of x^2; the second has a coefficient of –3 and literal part of x; the third has no literal part—it is just the constant 2.

65. $-x^2 + y - 13$ has three terms. The first has a coefficient of –1 and a literal part of x^2; the second has a coefficient of 1 and a literal part of y; the third has no literal part—it is just the constant –13.

67. $3x(2x) + 4y(5y) = 6x^2 + 20y^2$ has two terms. The first has a coefficient of 6 and a literal part of x^2; the second has a coefficient of 20 and a literal part of y^2.

69. 2.031 71. –9.862 73. .046 75. 25.667

77. A term is an algebraic expression that is connected by multiplication (and/or division). If a term is formed by multiplying two or more expressions, each is called a factor of that term.

78. x + xy + xyz is one of many possible answers.

Exercises 2.3

1. Essential

3. Non-essential

5. Essential

7. Non-essential

9. Essential

11. Essential (both)

13. The first is non-essential, the second is essential.

15. x, 2x: literal part is x, coefficients are 1 and 2.
y, 3y: literal part is y, coefficients are 1 and 3.

17. $2x^2, -x^2$: literal part is x^2, coefficients are 2 and –1.
–3x, –x: literal part is x, coefficients are –3 and –1.
$4x^3$: literal part is x^3, coefficient is 4.

19. 4, 5: constant terms.
4u, 5u: literal part is u, coefficients are 4 and 5.
$4u^2$: literal part is u^2, coefficient is 4.

21. $5x^2$: literal part is x^2, coefficient is 5.
$5x^2y$: literal part is x^2y, coefficient is 5.
$5y^2$: literal part is y^2, coefficient is 5.
$5xy^2$: literal part is xy^2, coefficient is 5.

23. $-x^2y$, $-2x^2y$: literal part is x^2y, coefficients are –1 and –2.
$2xy^2$, $3xy^2$: literal part is xy^2, coefficients are 2 and 3.
x^2y^2: literal part is x^2y^2, coefficient is 1.

25. $2x + 5x = (2 + 5)x = 7x$

27. $2x^2 + 5x^2 = (2 + 5)x^2 = 7x^2$

29. $3a - 8a + 2a = (3 - 8 + 2)a = -3a$

31. $-3y + y - 2y = (-3 + 1 - 2)y = -4y$

33. $-x - 2x - 3x = (-1 - 2 - 3)x = -6x$

35. $$\begin{aligned} &2x - 3y - 7x + 5y \\ &= 2x - 7x - 3y + 5y \\ &= (2 - 7)x + (-3 + 5)y \\ &= -5x + 2y \end{aligned}$$

37. 3x + 5y + 2z cannot be simplified

39. $$\begin{aligned} &3x^2 + 7x + x^2 + 3x \\ &= 3x^2 + x^2 + 7x + 3x \\ &= (3 + 1)x^2 + (7 + 3)x \\ &= 4x^2 + 10x \end{aligned}$$

41. $x^2 - 2x + x^2 - x$
$= x^2 + x^2 - 2x - x$
$= (1+1)x^2 + (-2-1)x$
$= 2x^2 - 3x$

43. $5x^2y - 3x^2 + x^2y - x^2$
$= 5x^2y + x^2y - 3x^2 - x^2$
$= (5+1)x^2y + (-3-1)x^2$
$= 6x^2y - 4x^2$

45. $2x + 5x - 3y + y - 7x$
$= 2x + 5x - 7x - 3y + y$
$= (2+5-7)x + (-3+1)y$
$= 0x - 2y = -2y$

47. $-5s^2 + 3st - s^2 + 6s^2$
$= -5s^2 - s^2 + 6s^2 + 3st$
$= (-5-1+6)s^2 + 3st$
$= 0s^2 + 3st = 3st$

49. $3a^2b + ab^2 - ab^2 - 2a^2b - ab^2$
$= 3a^2b - 2a^2b + ab^2 - ab^2 - ab^2$
$= (3-2)a^2b + (1-1-1)ab^2$
$= 1a^2b - 1ab^2 = a^2b - ab^2$

51. $2x + 10 = 2x + 2\cdot 5 = 2(x+5)$

53. $5y - 20 = 5y - 5\cdot 4 = 5(y-4)$

55. $9x + 3y - 6 = 3(3x) + 3y - 3\cdot 2$
$= 3(3x + y - 2)$

57. $x^2 + xy = xx + xy = x(x+y)$

59. $3(x + 4) = 3x + 12$

61. $5(y - 2) = 5y - 10$

63. $-2(x + 7) = -2x - 14$

65. $3(5x + 2) = 15x + 6$

67. $-4(3x + 1) = -12x - 4$

69. $x(x+3) = x^2 + 3x$

71. $x(x^2 + 3x) = x^3 + 3x^2$

73. $5x(2x-4) = 10x^2 - 20x$

75. $3(x+y) + 4x - y = 3x + 3y + 4x - y$
$= 3x + 4x + 3y - y$
$= (3+4)x + (3-1)y$
$= 7x + 2y$

77. $3(x+y) + 4(x-y)$
$= 3x + 3y + 4x - 4y$
$= 3x + 4x + 3y - 4y$
$= (3+4)x + (3-4)y$
$= 7x + (-1)y = 7x - y$

79. $3(x + y) + 4x(-y) = 3x + 3y - 4xy$

81. $5x(x^2+3) + 2x(3+x^2)$
$= 5x^3 + 15x + 6x + 2x^3$
$= 5x^3 + 2x^3 + 15x + 6x$
$= (5+2)x^3 + (15+6)x$
$= 7x^3 + 21x$

83. $5x(x^2+3) + 2x(3x^2)$
$= 5x^3 + 15x + 6x^3$
$= 5x^3 + 6x^3 + 15x$
$= (5+6)x^3 + 15x$
$= 11x^3 + 15x$

85. The distributive law allows us to combine like terms.

86. (a) Our order of operations requires that multiplication be done before addition. Therefore, we should not add 5 and 3 first.

(b) The distributive law requires that we multiply each term of $x - 4$ by 3.

(c) We cannot add 5 and $3x$ to get $8x$, since these are not like terms.

(d) This is correct as written.

Exercises 2.4

1. $$\begin{aligned} 4x + y + 4(x + y) &= 4x + y + 4x + 4y \\ &= 4x + 4x + y + 4y \\ &= 8x + 5y \end{aligned}$$

3. $$\begin{aligned} &5(m + 2n) + 3(m - n) \\ &= 5m + 10n + 3m - 3n \\ &= 5m + 3m + 10n - 3n \\ &= 8m + 7n \end{aligned}$$

5. $$\begin{aligned} &-2(x - 3y) + 5(y - x) \\ &= -2x + 6y + 5y - 5x \\ &= -2x - 5x + 6y + 5y \\ &= -7x + 11y \end{aligned}$$

7. $$\begin{aligned} 5 + 3(x - 2) &= 5 + 3x - 6 \\ &= 3x + 5 - 6 \\ &= 3x - 1 \end{aligned}$$

9. $$\begin{aligned} 5 - 3(x - 2) &= 5 - 3x + 6 \\ &= -3x + 5 + 6 \\ &= -3x + 11 \end{aligned}$$

11. $$\begin{aligned} (5 - 3)(x - 2) &= 2(x - 2) \\ &= 2x - 4 \end{aligned}$$

13. $$\begin{aligned} &4(m - 3n) + 2(5m + 6n) \\ &= 4m - 12n + 10m + 12n \\ &= 4m + 10m - 12n + 12n \\ &= 14m \end{aligned}$$

15. $$\begin{aligned} &2(a - 2b) - 4(b - 2a) \\ &= 2a - 4b - 4b + 8a \\ &= 2a + 8a - 4b - 4b \\ &= 10a - 8b \end{aligned}$$

17. $$\begin{aligned} &3(2x^2 - 4y) + 4(5y - 3x^2) \\ &= 6x^2 - 12y + 20y - 12x^2 \\ &= 6x^2 - 12x^2 - 12y + 20y \\ &= -6x^2 + 8y \end{aligned}$$

19. $$\begin{aligned} 8 - (3x - 4) &= 8 - 3x + 4 \\ &= -3x + 8 + 4 \\ &= -3x + 12 \end{aligned}$$

21. $$\begin{aligned} 5y - (1 - 2y) &= 5y - 1 + 2y \\ &= 5y + 2y - 1 \\ &= 7y - 1 \end{aligned}$$

23. $$\begin{aligned} 5(x - 3y) - x - 3y &= 5x - 15y - x - 3y \\ &= 5x - x - 15y - 3y \\ &= 4x - 18y \end{aligned}$$

25. $$\begin{aligned} &5(x - 3y) - (x - 3y) \\ &= 5x - 15y - x + 3y \\ &= 5x - x - 15y + 3y \\ &= 4x - 12y \end{aligned}$$

27. $5(x - 3y) - x(-3y) = 5x - 15y + 3xy$

29. $5x(-3y) - x(-3y) = -15xy + 3xy$
$= -12xy$

31. $5x(-3y)(-x)(-3y)$
$= ((5)(-3)(-1)(-3))(x \cdot x)(y \cdot y)$
$= -45x^2y^2$

33. $2x^2(x-2) + x(3x^2 - 4x)$
$= 2x^3 - 4x^2 + 3x^3 - 4x^2$
$= 2x^3 + 3x^3 - 4x^2 - 4x^2$
$= 5x^3 - 8x^2$

35. $3a(4a-1) - a(4-a)$
$= 12a^2 - 3a - 4a + a^2$
$= 12a^2 + a^2 - 3a - 4a$
$= 13a^2 - 7a$

37. $4(x^2 + 7x) - (x^2 - 7x)$
$= 4x^2 + 28x - x^2 - 7x$
$= 4x^2 - x^2 + 28x - 7x$
$= 3x^2 + 21x$

39. $3a(a^2 + 3b) + 4b^2(a^2 - b)$
$= 3a^3 + 9ab + 4a^2b^2 - 4b^3$

41. $3x^2 - 7x + 4 - 8x^2 - 3 - x$
$= 3x^2 - 8x^2 - 7x - x + 4 - 3$
$= -5x^2 - 8x + 1$

43. $x^2y(xy - x) - 5xy(x^2y - x^2)$
$= x^3y^2 - x^3y - 5x^3y^2 + 5x^3y$
$= x^3y^2 - 5x^3y^2 - x^3y + 5x^3y$
$= -4x^3y^2 + 4x^3y$

45. $4u^2v(u-v) - (uv^3 + u^2v^2)$
$= 4u^3v - 4u^2v^2 - uv^3 - u^2v^2$
$= 4u^3v - 4u^2v^2 - u^2v^2 - uv^3$
$= 4u^3v - 5u^2v^2 - uv^3$

47. $4(x + 3y) + (4x + 3y) + 4x(3y)$
$= 4x + 12y + 4x + 3y + 12xy$
$= 4x + 4x + 12y + 3y + 12xy$
$= 8x + 15y + 12xy$

49. $6(m - 2n) + (6m - 2n) + 6m(-2n)$
$= 6m - 12n + 6m - 2n - 12mn$
$= 6m + 6m - 12n - 2n - 12mn$
$= 12m - 14n - 12mn$

51. $3t^5(t^4 - 4) - (t^5 + t^4) - 2t^3(3t)(-t^5)$
$= 3t^9 - 12t^5 - t^5 - t^4 + 6t^9$
$= 3t^9 + 6t^9 - 12t^5 - t^5 - t^4$
$= 9t^9 - 13t^5 - t^4$

53. $-3(-x + 2) + (8 - 5x) - (2 - 2x)$
$= 3x - 6 + 8 - 5x - 2 + 2x$
$= 3x - 5x + 2x - 6 + 8 - 2$
$= 0x + 0 = 0$

55. $a - 2[a - 2(a - 2)] = a - 2[a - 2a + 4]$
$= a - 2[-a + 4]$
$= a + 2a - 8$
$= 3a - 8$

57. $x\{x - 4[x - (x - 4)]\}$
$= x\{x - 4[x - x + 4]\}$
$= x\{x - 4(4)\}$
$= x(x - 16)$
$= x^2 - 16x$

59. $3(x + 2) + 4[x - 3(2 - x)]$
$= 3(x + 2) + 4[x - 6 + 3x]$
$= 3(x + 2) + 4(4x - 6)$
$= 3x + 6 + 16x - 24$
$= 3x + 16x + 6 - 24$
$= 19x - 18$

61. $4(y-3)-2[3y-5(y-1)]$

$$= 4(y-3)-2[3y-5y+5]$$
$$= 4(y-3)-2(-2y+5)$$
$$= 4y-12+4y-10$$
$$= 8y-22$$

63. (a) The solution is correct up to $3+2[5x+12]$. At this point, it is wrong to add 3 + 2 = 5. Instead, use the Distributive Law: $3+2[5x+12] = 3+10x+24$

$$= 10x+3+24$$
$$= 10x+27$$

(b) Within the brackets, we must add x and x + 3, not multiply them. As a result, we get $3x+5[x+(x+3)]$

$$= 3x+5[2x+3]$$
$$= 3x+10x+15$$
$$= 13x+15$$

64. (a) Do not subtract 5 – 3; distribute –3 instead.

(b) When –3 is distributed, it must multiply –4 as well as x.

(c) When we multiply –3 by –4, we get 12, not –12.

(d) This is done correctly.

Exercises 2.5

1. n + 4   3. n – 4   5. n – 4   7. 5n + 6

9. 2n – 9   11. n(n + 7)   13. (n + 2)(n – 6)   15. 2n – 8 = 14

17. 5n + 4 = n – 2   19. r + s = rs

21. 2(r + s) = rs – 3   23. x + (x + 1) = 2x + 1, where x = smaller integer

25. x + (x + 2) = 2x + 2, where x = smaller even integer

27. x(x + 2)(x + 4), where x = smallest of the odd integers

29. 8x = 7(x + 2) – 4, where x = smaller even integer

31. $x^2+(x+1)^2+(x+2)^2=5$, where x = smallest of the integers

33. (a) 8 nickels   (b) 40¢ (= 8 × 5¢)   (c) 12 dimes
(d) 120¢ (= 12 × 10¢) or $1.20   (e) 9 quarters
(f) 225¢ (= 9 × 25¢) or $2.25   (g) 29 coins
(h) 40¢ + 120¢ + 225¢ = 385¢ or $3.85

35. (a) A reads 200 cards per minute.   (b) A reads cards for 15 minutes.

(c) A reads 3,000 cards (= 200×15). (d) B reads 160 cards per minute.

(e) B reads cards for 20 minutes. (f) B reads 3,200 cards (= 160×20).

(g) A and B read 6,200 cards altogether (= 3,000 + 3,200).

37. (a) She walks at the rate of 100 meters per minute.

(b) She walks for 25 minutes.

(c) She walks 2,500 meters (= 100×25).

(d) She jogs at the rate of 220 meters per minute.

(e) She jogs for 35 minutes.

(f) She jogs 7,700 meters (= 220×35).

(g) She covers a distance of 10,200 meters altogether (= 2,500 + 7,700).

Chapter 2 Review Exercises

1. $x \cdot y \cdot y \cdot y$

3. $-(x \cdot x \cdot x \cdot x)$

5. $3 \cdot x \cdot x$

7. x^2y^3

9. $a^2 - b^3$

11. $z^4 = -(-2)^4 = -(-2)(-2)(-2)(-2)$
$= -16$

13. $xy^2 = (-3)(4)^2 = -3(16) = -48$

15. $xyz - (x + y + z)$
$= (-3)(4)(-2) - ((-3)) + (4) + (-2))$
$= 24 - (-1) = 25$

17. $|xy| + z - |z|$
$= |(-3)(4)| + (-2) - |(-2)|$
$= |-12| + (-2) - |-2|$
$= 12 - 2 - 2 = 8$

19. $2x^2 - (x + y)^2$
$= 2(-3)^2 - ((-3) + (4))^2$
$= 2(9) - (1)^2$
$= 18 - 1 = 17$

21. $x^3x^4x = x^{3+4+1} = x^8$

23. $a^7a^2 + a^3a^6 = a^{7+2} + a^{3+6}$
$= a^9 + a^9 = 2a^9$

25. $4x^3x^2 + 3x^2x^4 = 4x^{3+2} + 3x^{2+4}$
$= 4x^5 + 3x^6$

27. $3x^2 - 7x + 7 - 5x^2 - x - 3$
$= 3x^2 - 5x^2 - 7x - x + 7 - 3$
$= -2x^2 - 8x + 4$

29. $2a^2b(3ab^4) = (2 \cdot 3)(a^2a)(bb^4)$
$= 6a^{2+1}b^{1+4} = 6a^3b^5$

31. $2a^2b(3a + b^4)$
$= 2a^2b(3a) + 2a^2b(b^4)$
$= 6a^3b + 2a^2b^5$

33. $3x(2x+4)+5(x^2-3)$
$= 6x^2+12x+5x^2-15$
$= 6x^2+5x^2+12x-15$
$= 11x^2+12x-15$

35. $4y^2(y-2)-y(y^2-5y)$
$= 4y^3-8y^2-y^3+5y^2$
$= 4y^3-y^3-8y^2+5y^2$
$= 3y^3-3y^2$

37. $3xy(x^2-2y)+4xy^2(y-x)$
$= 3x^3y-6xy^2+4xy^3-4x^2y^2$

39. $3(2x-4y)-(x+2y)-(x-2y)$
$= 6x-12y-x-2y-x+2y$
$= 6x-x-x-12y-2y+2y$
$= 4x-12y$

41. $3x^4(x^3-2y^2)-4x(x^2)(x^4)$
$= 3x^7-6x^4y^2-4x^7$
$= 3x^7-4x^7-6x^4y^2$
$= -x^7-6x^4y^2$

43. $3-[x-3-(x-3)]$
$= 3-[x-3-x+3]$
$= 3-[x-x-3+3]$
$= 3-0=3$

45. $5x^2+10=5x^2+5\cdot 2=5(x^2+2)$

47. $3y-6z+9=3y-3(2z)+3(3)$
$= 3(y-2z+3)$

49. $n+7=3n-4$

51. $n+(n+2)=n-5$

53. (a) He sells 12 newspaper subscriptions.

(b) He earns \$2 for each newspaper subscription.

(c) He sells 9 magazine subscriptions.

(d) He earns \$5 for each magazine subscription.

(e) He earns \$24 (= 12×2) for the newspaper subscriptions.

(f) He earns \$45 (= 9×5) for the magazine subscriptions.

(g) He earns \$69 (= 24 + 45) altogether.

Chapter 2 Practice Test

1. $-3^4=-(3\cdot 3\cdot 3\cdot 3)=-81$

3. $(-3-4+6)^5=(-7+6)^5=(-1)^5$
$= (-1)(-1)(-1)(-1)(-1)$
$= -1$

5. $x-y-z=(-2)-(3)-(-4)$
$= -2-3+4=-5+4=-1$

7. $x^3-z^2=(-2)^3-(-4)^2$
$= (-2)(-2)(-2)-(-4)(-4)$
$= -8-(+16)$
$= -8-16=-24$

$4x^2y - 5xy + y^2 - 3xy - 2y^2 - x^2y$
$= 4x^2y - x^2y - 5xy - 3xy + y^2 - 2y^2$
$= 3x^2y - 8xy - y^2$

11. $2x(x^2 - y) - 3(x - xy) - (2x^3 - 3x)$
$= 2x^3 - 2xy - 3x + 3xy - 2x^3 + 3x$
$= 2x^3 - 2x^3 - 2xy + 3xy - 3x + 3x$
$= xy$

13. $4 - [x - 4(x - 4)] = 4 - [x - 4x + 16]$
$= 4 - [-3x + 16]$
$= 4 + 3x - 16$
$= 3x - 12$

15. (a) \$3 (b) x (c) \$2 (d) $2x - 5$
(e) $3(2x - 5)$ dollars (f) $2x$ dollars
(g) $2x + 3(2x - 5) = 2x + 6x - 15 = 8x - 15$ dollars

CHAPTER 3
FIRST DEGREE EQUATIONS AND INEQUALITIES

Exercises 3.1

1. $2(x-3) = 2x - 6$
   $2x - 6 = 2x - 6$
   identity

3. $5(x+2) = 5x + 2$
   $5x + 10 = 5x + 2$
   $10 = 2$
   contradiction

5. $2a + 4 + 3a = 6a + 4 - a$
   $5a + 4 = 5a + 4$
   identity

7. $2z^2 + 3z - 2z^2 - z = z + z + 1$
   $2z^2 - 2z^2 + 3z - z = z + z + 1$
   $2z = 2z + 1$
   $0 = 1$
   contradiction

9. $5u - 4(u-1) - u = u - 4 - (u-2)$
   $5u - 4u + 4 - u = u - 4 - u + 2$
   $5u - 4u - u + 4 = u - u - 4 + 2$
   $4 = -2$
   contradiction

11. $7 - 3(y-2) = y + 4(5-y) - 7$
    $7 - 3y + 6 = y + 20 - 4y - 7$
    $7 + 6 - 3y = 20 - 7 + y - 4y$
    $13 - 3y = 13 - 3y$
    identity

13. $w(w-2) - w^2 + 2w$
    $= 3(w+1) - (3w-1)$
    $w^2 - 2w - w^2 + 2w$
    $= 3w + 3 - 3w + 1$
    $w^2 - w^2 - 2w + 2w$
    $= 3w - 3w + 3 + 1$
    $0 = 4$
    contradiction

15. $2(x^2-3) - x(2x-1) + x$
    $= 2 - x - (x+8) + 4x$
    $2x^2 - 6 - 2x^2 + x + x$
    $= 2 - x - x - 8 + 4x$
    $2x^2 - 2x^2 + x + x - 6$
    $= 4x - x - x + 2 - 8$
    $2x - 6 = 2x - 6$
    identity

17. CHECK x = –3:
    $x + 5 = -2$
    $(-3) + 5 \stackrel{?}{=} -2$
    $2 \neq -2$
    Therefore x = –3 does not satisfy the equation.
    CHECK x = –7:
    $x + 5 = -2$
    $(-7) + 5 \stackrel{?}{=} -2$
    $-2 \stackrel{\checkmark}{=} -2$
    Therefore x = –7 does satisfy the equation.

19. CHECK a = –5:
    $2 - a = 3$
    $2 - (-5) \stackrel{?}{=} 3$
    $7 \neq 3$
    Therefore a = –5 does not satisfy the equation.
    CHECK a = 5:
    $2 - a = 3$
    $2 - (5) \stackrel{?}{=} 3$
    $-3 \neq 3$
    Therefore a = 5 does not satisfy the equation.
    CHECK a = 1:
    $2 - a = 3$
    $2 - (1) \stackrel{?}{=} 3$
    $1 \neq 3$
    Therefore a = 1 does not satisfy the equation.

21. CHECK y = 2:

$$5y - 6 = y - 3$$
$$5(2) - 6 \stackrel{?}{=} 2 - 3$$
$$10 - 6 \stackrel{?}{=} 2 - 3$$
$$4 \neq -1$$

Therefore y = 2 does not satisfy the equation.

CHECK $y = \frac{3}{4}$:

$$5y - 6 = y - 3$$
$$5\left(\frac{3}{4}\right) - 6 \stackrel{?}{=} \frac{3}{4} - 3$$
$$\frac{15}{4} - 6 \stackrel{?}{=} \frac{3}{4} - 3$$
$$\frac{15}{4} - \frac{24}{4} \stackrel{?}{=} \frac{3}{4} - \frac{12}{4}$$
$$\frac{-9}{4} \stackrel{\checkmark}{=} \frac{-9}{4}$$

Therefore $y = \frac{3}{4}$ does satisfy the equation.

23. CHECK w = –4:

$$6 - 2w = 10 - 3w$$
$$6 - 2(-4) \stackrel{?}{=} 10 - 3(-4)$$
$$6 + 8 \stackrel{?}{=} 10 + 12$$
$$14 \neq 22$$

Therefore w = –4 does not satisfy the equation.

CHECK w = 1:

$$6 - 2w = 10 - 3w$$
$$6 - 2(1) \stackrel{?}{=} 10 - 3(1)$$
$$6 - 2 \stackrel{?}{=} 10 - 3$$
$$4 \neq 7$$

Therefore w = 1 does not satisfy the equation.

25. CHECK $x = \frac{2}{5}$:

$$4(x - 7) - (x + 1) = 15 - x$$
$$4\left(\left(\frac{2}{5}\right) - 7\right) - \left(\left(\frac{2}{5}\right) + 1\right) \stackrel{?}{=} 15 - \left(\frac{2}{5}\right)$$
$$4\left(\frac{2}{5} - \frac{35}{5}\right) - \left(\frac{2}{5} + \frac{5}{5}\right) \stackrel{?}{=} \frac{75}{5} - \frac{2}{5}$$
$$4\left(\frac{-33}{5}\right) - \frac{7}{5} \stackrel{?}{=} \frac{73}{5}$$
$$\frac{-132}{5} - \frac{7}{5} \stackrel{?}{=} \frac{73}{5}$$
$$\frac{-139}{5} \neq \frac{73}{5}$$

Therefore $x = \frac{2}{5}$ does not satisfy the equation.

CHECK x = 7:

$$4(x - 7) - (x + 1) = 15 - x$$
$$4((7) - 7) - ((7) + 1) \stackrel{?}{=} 15 - 7$$
$$4(0) - 8 \stackrel{?}{=} 15 - 7$$
$$0 - 8 \stackrel{?}{=} 15 - 7$$
$$-8 \neq 8$$

Therefore x = 7 does not satisfy the equation.

27. CHECK z = –2:

$$3z + 2(z - 1) = 4(z + 2) - (z + 5)$$
$$3(-2) + 2((-2) - 1) \stackrel{?}{=} 4((-2) + 2) - ((-2) + 5)$$
$$-6 + 2(-3) \stackrel{?}{=} 4(0) - 3$$
$$-6 - 6 \stackrel{?}{=} 0 - 3$$
$$-12 \neq -3$$

Therefore z = –2 does not satisfy the equation.

CHECK z = 1:

$$3z + 2(z - 1) = 4(z + 2) - (z + 5)$$
$$3(1) + 2((1) - 1) \stackrel{?}{=} 4((1) + 2) - ((1) + 5)$$
$$3 + 2(0) \stackrel{?}{=} 4(3) - 6$$
$$3 + 0 \stackrel{?}{=} 12 - 6$$
$$3 \neq 6$$

Therefore z = 1 does not satisfy the equation.

29. CHECK x = –2:

$$x^2 - 3x = 2x - 6$$
$$(-2)^2 - 3(-2) \stackrel{?}{=} 2(-2) - 6$$
$$4 + 6 \stackrel{?}{=} -4 - 6$$
$$10 \neq -10$$

Therefore x = –2 does not satisfy the equation.

CHECK $x = 2$:

$$x^2 - 3x = 2x - 6$$
$$(2)^2 - 3(2) \stackrel{?}{=} 2(2) - 6$$
$$4 - 6 \stackrel{?}{=} 4 - 6$$
$$-2 \stackrel{\checkmark}{=} -2$$

Therefore $x = 2$ does satisfy the equation.

31. CHECK $a = -1$:

$$a^2 - 4a = 4 - a$$
$$(-1)^2 - 4(-1) \stackrel{?}{=} 4 - (-1)$$
$$1 + 4 \stackrel{?}{=} 4 + 1$$
$$5 \stackrel{\checkmark}{=} 5$$

Therefore $a = -1$ does satisfy the equation.

CHECK $a = 4$:

$$a^2 - 4a = 4 - a$$
$$(4)^2 - 4(4) \stackrel{?}{=} 4 - (4)$$
$$16 - 16 \stackrel{?}{=} 4 \quad - 4$$
$$0 \stackrel{\checkmark}{=} 0$$

Therefore $a = 4$ does satisfy the equation.

33. CHECK $y = -2$:

$$y(y + 6) = (y + 2)^2$$
$$-2((-2) + 6) \stackrel{?}{=} ((-2) + 2)^2$$
$$-2(4) \stackrel{?}{=} 0^2$$
$$-8 \neq 0$$

Therefore $y = -2$ does not satisfy the equation.

CHECK $y = 2$:

$$y(y + 6) = (y + 2)^2$$
$$2((2) + 6) \stackrel{?}{=} ((2) + 2)^2$$
$$2(8) \stackrel{?}{=} 4^2$$
$$16 \stackrel{\checkmark}{=} 16$$

Therefore $y = 2$ does satisfy the equation.

35. A value satisfies or is a solution to an equation if both sides of the equation are equal when the value is substituted for the variable.

36. An identity is an equation that is always true, no matter what value is chosen for the variable. A contradiction is an equation that is never true, no matter what value is chosen for the variable. A conditional equation is one that is true for some values of the variable and false for the others.

37. If two quantities are equal, then we will not disturb this equality if we change each quantity in exactly the same way.

38. Two equations are called equivalent if their solution sets are exactly the same.

39. $x + 1 = 16$
$x - 3 = 12$

$x = 15$

$2x = 30$

$x = -1$

$x + 7 = 6$
$2x = 18$

$x = 9$

$x + 2 = 11$

$x - 2 = 7$
$2x = 10$

$5x = 25$

41. -5

43. 4

45. 64

47. -11

49. $10 + 4(-6) = 10 - 24 = -14$

51. $-5 - 2(-8) = -5 + 16 = 11$

Exercises 3.2

1.
$$\begin{array}{rcr} x + 3 &=& 8 \\ -3 && -3 \\ \hline x &=& 5 \end{array}$$

CHECK $x = 5$:

$$5 + 3 \stackrel{?}{=} 8$$
$$8 \stackrel{\checkmark}{=} 8$$

3.
$$\begin{array}{rcr} y - 4 &=& 7 \\ +4 && +4 \\ \hline y &=& 11 \end{array}$$

CHECK y = 11:

$11 - 4 \stackrel{?}{=} 7$

$7 \stackrel{\checkmark}{=} 7$

5. $a + 3 = 1$

$\underline{\quad -3 \quad -3}$

$a = -2$

CHECK a = –2:

$-2 + 3 \stackrel{?}{=} 1$

$1 \stackrel{\checkmark}{=} 1$

7. $a - 5 = -8$

$\underline{\quad +5 \quad +5}$

$a = -3$

CHECK a = –3:

$-3 - 5 \stackrel{?}{=} -8$

$-8 \stackrel{\checkmark}{=} -8$

9. $3x = 21$

$\frac{\cancel{3}x}{\cancel{3}} = \frac{21}{3}$

$x = 7$

CHECK x = 7:

$3(7) \stackrel{?}{=} 21$

$21 \stackrel{\checkmark}{=} 21$

11. $4x = 15$

$\frac{\cancel{4}x}{\cancel{4}} = \frac{15}{4}$

$x = \frac{15}{4}$

CHECK $x = \frac{15}{4}$:

$4\left(\frac{15}{4}\right) \stackrel{?}{=} 15$

$15 = 15$

13. $-4x = -12$

$\frac{-\cancel{4}x}{\cancel{4}} = \frac{-12}{-4}$

$x = 3$

CHECK x = 3:

$-4(3) \stackrel{?}{=} -12$

$-12 \stackrel{\checkmark}{=} -12$

15. $3x + x = 8 - 12$

$4x = -4$

$\frac{\cancel{4}x}{\cancel{4}} = \frac{-4}{4}$

$x = -1$

CHECK x = –1:

$3(-1) + (-1) \stackrel{?}{=} 8 - 12$

$-3 - 1 \stackrel{?}{=} 8 - 12$

$-4 \stackrel{\checkmark}{=} -4$

17. $4x - x = 2 - 8$

$3x = -6$

$\frac{\cancel{3}x}{\cancel{3}} = \frac{-6}{3}$

$x = -2$

CHECK x = –2:

$4(-2) - (-2) \stackrel{?}{=} 2 - 8$

$-8 + 2 \stackrel{?}{=} 2 - 8$

$-6 \stackrel{\checkmark}{=} -6$

19. $2z - 3z - 11z = -4(6)$

$-12z = -24$

$\frac{-\cancel{12}z}{-\cancel{12}z} = \frac{-24}{-12}$

$z = 2$

CHECK z = 2:

$2(2) - 3(2) - 11(2) \stackrel{?}{=} -4(6)$

$4 - 6 - 22 \stackrel{?}{=} -24$

$-24 \stackrel{\checkmark}{=} -24$

21. $3w - 7w + 4w = 8 - 3$

$0w = 5$

$0 = 5$

Contradiction

23. $2x + 1 = 7$

$\underline{\quad -1 \quad -1}$

$2x = 6$

$\frac{\cancel{2}x}{\cancel{2}} = \frac{6}{2}$

$x = 3$

CHECK x = 3:

$2(3)+1 \stackrel{?}{=} 7$

$6+1 \stackrel{?}{=} 7$

$7 \stackrel{\surd}{=} 7$

25. $10x - 4 = 22$

$\underline{\quad +4 \quad +4}$

$10x = 26$

$\frac{\cancel{10}x}{\cancel{10}} = \frac{26}{10}$

$x = \frac{13}{5}$

CHECK $x = \frac{13}{5}$:

$10\left(\frac{13}{5}\right) - 4 \stackrel{?}{=} 22$

$26 - 4 \stackrel{?}{=} 22$

$22 \stackrel{\surd}{=} 22$

27. $2t = 3t + 5$

$\underline{-3t \quad -3t}$

$-t = 5$

$\frac{-t}{-1} = \frac{5}{-1}$

$t = -5$

CHECK t = –5:

$2(-5) \stackrel{?}{=} 3(-5) + 5$

$-10 \stackrel{?}{=} -15 + 5$

$-10 \stackrel{\surd}{=} -10$

29. $9 = 6 - 3a$

$\underline{-6 \quad -6}$

$3 = -3a$

$\frac{3}{-3} = \frac{\cancel{-3}a}{\cancel{-3}}$

$-1 = a$

CHECK a = –1:

$9 \stackrel{?}{=} 6 - 3(-1)$

$9 \stackrel{?}{=} 6 + 3$

$9 \stackrel{\surd}{=} 9$

31. $20 = 3w - 1$

$\underline{+1 \qquad +1}$

$21 = 3w$

$\frac{21}{3} = \frac{\cancel{3}w}{\cancel{3}}$

$7 = w$

CHECK w = 7:

$20 \stackrel{?}{=} 3(7) - 1$

$20 \stackrel{?}{=} 21 - 1$

$20 \stackrel{\surd}{=} 20$

33. $8y + 4 = 5y + 19$

$\underline{-5y \qquad -5y}$

$3y + 4 = 19$

$\underline{-4 \qquad -4}$

$3y = 15$

$\frac{\cancel{3}y}{\cancel{3}} = \frac{15}{3}$

$y = 5$

CHECK y = 5:

$8(5) + 4 \stackrel{?}{=} 5(5) + 19$

$40 + 4 \stackrel{?}{=} 25 + 19$

$44 \stackrel{\surd}{=} 44$

35. $2a + 5 = 4a + 13$

$\underline{-2a \qquad -2a}$

$5 = 2a + 13$

$\underline{-13 \qquad -13}$

$-8 = 2a$

$\frac{-8}{2} = \frac{\cancel{2}a}{\cancel{2}}$

$-4 = a$

CHECK a = –4:

$2(-4) + 5 \stackrel{?}{=} 4(-4) + 13$

$-8 + 5 \stackrel{?}{=} -16 + 13$

$-3 \stackrel{\surd}{=} -3$

37. $5r - 8 = 3r - 20$

$\underline{-3r \qquad -3r}$

$2r - 8 = -20$

$\underline{+8 \qquad +8}$

$2r = -12$

$$\frac{\not{2}r}{\not{2}} = \frac{-12}{2}$$

$$r = -6$$

CHECK r = –6:

$$5(-6) - 8 \stackrel{?}{=} 3(-6) - 20$$

$$-30 - 8 \stackrel{?}{=} -18 - 20$$

$$-38 \stackrel{\checkmark}{=} -38$$

39. $$\begin{array}{rcl} 10 - x &=& 4 - 3x \\ \underline{+3x} && \underline{+3x} \\ 10 + 2x &=& 4 \\ \underline{-10} && \underline{-10} \\ 2x &=& -6 \end{array}$$

$$\frac{2x}{2} = \frac{-6}{2}$$

$$x = -3$$

CHECK x = –3:

$$10 - (-3) \stackrel{?}{=} 4 - 3(-3)$$

$$10 + 3 \stackrel{?}{=} 4 + 9$$

$$13 \stackrel{\checkmark}{=} 13$$

41. $$\begin{array}{rcl} -4 - 3u &=& -2 - u \\ \underline{+3u} && \underline{+3u} \\ -4 &=& -2 + 2u \\ \underline{+2} && \underline{+2} \\ -2 &=& 2u \end{array}$$

$$\frac{-2}{2} = \frac{\not{2}u}{\not{2}}$$

$$-1 = u$$

CHECK u = –1:

$$-4 - 3(-1) \stackrel{?}{=} -2 - (-1)$$

$$-4 + 3 \stackrel{?}{=} -2 + 1$$

$$-1 \stackrel{\checkmark}{=} -1$$

43. $$\begin{array}{rcl} x + 7 &=& 7 - x \\ \underline{+x} && \underline{+x} \\ 2x + 7 &=& 7 \\ \underline{-7} && \underline{-7} \\ 2x &=& 0 \end{array}$$

$$\frac{\not{2}x}{\not{2}} = \frac{0}{2}$$

$$x = 0$$

CHECK x = 0:

$$0 + 7 \stackrel{?}{=} 7 - 0$$

$$7 \stackrel{\checkmark}{=} 7$$

45. $$\begin{array}{rcl} x + 7 &=& 7 + x \\ \underline{-x} && \underline{-x} \\ 7 &=& 7 \end{array}$$

Identity

47. $$\begin{array}{rcl} x - 7 &=& 7 + x \\ \underline{-x} && \underline{-x} \\ -7 &=& 7 \end{array}$$

Contradiction

49. $$\begin{array}{rcl} x - 7 &=& 7 - x \\ \underline{-x} && \underline{+x} \\ 2x - 7 &=& 7 \\ \underline{+7} && \underline{+7} \\ 2x &=& 14 \end{array}$$

$$\frac{\not{2}x}{\not{2}} = \frac{14}{2}$$

$$x = 7$$

CHECK x = 7:

$$7 - 7 \stackrel{?}{=} 7 - 7$$

$$0 \stackrel{\checkmark}{=} 0$$

51. $$\begin{array}{rcl} 2(t + 1) + 4t &=& 27 \\ 2t + 2 + 4t &=& 27 \\ 6t + 2 &=& 27 \\ \underline{-2} && \underline{-2} \\ 6t &=& 25 \end{array}$$

$$\frac{\not{6}t}{\not{6}} = \frac{25}{6}$$

$$t = \frac{25}{6}$$

CHECK $t = \frac{25}{6}$:

$$2\left(\frac{25}{6} + 1\right) + 4\left(\frac{25}{6}\right) \stackrel{?}{=} 27$$

$$2\left(\frac{31}{6}\right) + 4\left(\frac{25}{6}\right) \stackrel{?}{=} 27$$

$$\frac{31}{3}+\frac{50}{3}\overset{?}{=}27$$
$$\frac{81}{3}\overset{?}{=}27$$
$$27\overset{\checkmark}{=}27$$

53. $2(y+3)+4(y-2)=22$
$$2y+6+4y-8=22$$
$$6y-2=22$$
$$\underline{\quad +2 \quad +2}$$
$$6y \quad = 24$$
$$\frac{\cancel{6}y}{\cancel{6}}=\frac{24}{6}$$
$$y=4$$

CHECK y = 4:
$$2(4+3)+4(4-2)\overset{?}{=}22$$
$$2(7)+4(2)\overset{?}{=}22$$
$$14+8\overset{?}{=}22$$
$$22\overset{\checkmark}{=}22$$

55. $4+3(3y-5)=2y-11+y$
$$4+9y-15=2y-11+y$$
$$9y-11=3y-11$$
$$\underline{-3y \qquad -3y}$$
$$6y-11= \quad -11$$
$$\underline{\quad +11 \qquad +11}$$
$$6y \quad = \quad 0$$
$$\frac{\cancel{6}y}{\cancel{6}}=\frac{0}{6}$$
$$y=0$$

CHECK y = 0:
$$4+3(3(0)-5)\overset{?}{=}2(0)-11+0$$
$$4+3(0-5)\overset{?}{=}0-11+0$$
$$4+3(-5)\overset{?}{=}11$$
$$4-15\overset{?}{=}-11$$
$$-11\overset{\checkmark}{=}-11$$

57. $3(a-2)+4(2-a)=a+2(a+1)$
$$3a-6+8-4a=a+2a+2$$
$$-a+2=3a+2$$
$$\underline{\quad +a \qquad +a}$$
$$2=4a+2$$
$$\underline{-2 \qquad -2}$$
$$\frac{0}{4}=\frac{\cancel{4}a}{\cancel{4}}$$
$$0=a$$

CHECK a = 0:
$$3(0-2)+4(2-0)\overset{?}{=}0+2(0+1)$$
$$3(-2)+4(2)\overset{?}{=}0+2(1)$$
$$-6+8\overset{?}{=}0+2$$
$$2\overset{\checkmark}{=}2$$

59. $8z-3(z-3)=-9$
$$8z-3z+9=-9$$
$$5z+9=-9$$
$$\underline{\quad -9 \quad -9}$$
$$5z \quad =-18$$
$$\frac{\cancel{5}z}{\cancel{5}}=\frac{-18}{5}$$
$$z=\frac{-18}{5}$$

CHECK $z=\frac{-18}{5}$:
$$8\left(\frac{-18}{5}\right)-3\left(\frac{-18}{5}-3\right)\overset{?}{=}-9$$
$$8\left(\frac{-18}{5}\right)-3\left(\frac{-33}{5}\right)\overset{?}{=}-9$$
$$\frac{-144}{5}+\frac{99}{5}\overset{?}{=}-9$$
$$\frac{-45}{5}\overset{?}{=}-9$$
$$-9\overset{\checkmark}{=}-9$$

61. $20t-5(t-1)=23$
$$20t-5t+5=23$$
$$15t+5=23$$
$$\underline{\quad -5 \quad -5}$$
$$15t \quad =18$$

$$\frac{\not{15}t}{\not{15}} = \frac{18}{15}$$

$$t = \frac{6}{5}$$

CHECK $t = \frac{6}{5}$:

$$20\left(\frac{6}{5}\right) - 5\left(\frac{6}{5} - 1\right) \stackrel{?}{=} 23$$

$$20\left(\frac{6}{5}\right) - 5\left(\frac{1}{5}\right) \stackrel{?}{=} 23$$

$$24 - 1 \stackrel{?}{=} 23$$

$$23 \stackrel{\checkmark}{=} 23$$

63. $20 - 5(t - 1) = 23$

$$20 - 5t + 5 = 23$$

$$-5t + 25 = 23$$

$$\underline{\quad -25 \quad -25}$$

$$-5t = -2$$

$$\frac{-\not{5}t}{-\not{5}} = \frac{-2}{-5}$$

$$t = \frac{2}{5}$$

CHECK $t = \frac{2}{5}$:

$$20 - 5\left(\frac{2}{5} - 1\right) \stackrel{?}{=} 23$$

$$20 - 5\left(-\frac{3}{5}\right) \stackrel{?}{=} 23$$

$$20 + 3 \stackrel{?}{=} 23$$

$$23 \stackrel{\checkmark}{=} 23$$

65. $2(y - 3) - 3(y - 5) = 5y - 5(y - 2)$

$$2y - 6 - 3y + 15 = 5y - 5y + 10$$

$$-y + 9 = 10$$

$$\underline{\quad -9 = -9}$$

$$-y = 1$$

$$\frac{-y}{-1} = \frac{1}{-1}$$

$$y = -1$$

CHECK $y = -1$:

$$2(-1 - 3) - 3(-1 - 5) \stackrel{?}{=} 5(-1) - 5(-1 - 2)$$

$$2(-4) - 3(-6) \stackrel{?}{=} 5(-1) - 5(-3)$$

$$-8 + 18 \stackrel{?}{=} -5 + 15$$

$$10 \stackrel{\checkmark}{=} 10$$

67. $4x - 3(x + 8) = 5x - 2(x - 12) - 2x$

$$4x - 3x - 24 = 5x - 2x + 24 - 2x$$

$$x - 24 = x + 24$$

$$\underline{\quad -x \qquad -x}$$

$$-24 = 24$$

Contradiction

69. $a - (5 - 3a) = 7a - (a - 3) - 8$

$$a - 5 + 3a = 7a - a + 3 - 8$$

$$4a - 5 = 6a - 5$$

$$\underline{-4a \qquad -4a}$$

$$-5 = 2a - 5$$

$$\underline{+5 \qquad +5}$$

$$0 = 2a$$

$$\frac{0}{2} = \frac{\not{2}a}{\not{2}}$$

$$0 = a$$

CHECK $a = 0$:

$$0 - (5 - 3(0)) \stackrel{?}{=} 7(0) - (0 - 3) - 8$$

$$0 - (5 - 0) \stackrel{?}{=} 0 - (-3) - 8$$

$$0 - 5 \stackrel{?}{=} 0 + 3 - 8$$

$$-5 \stackrel{\checkmark}{=} -5$$

71. $x^2 + 3x - 7 = x^2 - 5x + 1$

$$\underline{-x^2 \qquad -x^2}$$

$$3x - 7 = -5x + 1$$

$$\underline{+5x \qquad +5x}$$

$$8x - 7 = 1$$

$$\underline{\quad +7 \qquad +7}$$

$$8x = 8$$

$$\frac{\not{8}x}{\not{8}} = \frac{8}{8}$$

$$x = 1$$

CHECK $x = 1$:

$$(1)^2 + 3(1) - 7 \stackrel{?}{=} (1)^2 - 5(1) + 1$$

$$1 + 3 - 7 \stackrel{?}{=} 1 - 5 + 1$$

$$-3 \stackrel{\checkmark}{=} -3$$

73. $x(x+2)+3x = x(x-1)-12$

$$x^2+2x+3x = x^2-x-12$$
$$\underline{-x^2 \qquad\qquad -x^2}$$
$$2x+3x = -x-12$$
$$5x = -x-12$$
$$\underline{+x \qquad +x}$$
$$6x = -12$$
$$\frac{\not{6}x}{\not{6}} = \frac{-12}{6}$$
$$x = -2$$

CHECK $x = -2$:

$$-2(-2+2)+3(-2) \stackrel{?}{=} (-2)(-2-1)-12$$
$$-2(0)+3(-2) \stackrel{?}{=} (-2)(-3)-12$$
$$0-6 \stackrel{?}{=} 6-12$$
$$-6 \stackrel{\checkmark}{=} -6$$

75. $2z(z+1)+3(z+2) = 3z(z+2)-z^2$

$$2z^2+2z+3z+6 = 3z^2+6z-z^2$$
$$2z^2+5z+6 = 2z^2+6z$$
$$\underline{-2z^2 \qquad\qquad -2z^2}$$
$$5z+6 = 6z$$
$$\underline{-5z \qquad\qquad -5z}$$
$$6 = z$$

CHECK $z = 6$:

$$2(6)(6+1)+3(6+2) \stackrel{?}{=} 3(6)(6+2)-(6)^2$$
$$2(6)(7)+3(8) \stackrel{?}{=} 3(6)(8)-36$$
$$84+24 \stackrel{?}{=} 144-36$$
$$108 \stackrel{\checkmark}{=} 108$$

77. $x = 6.5$

79. $t = .03$

81. $t = -8.19$

83. To check your answer after you have solved an equation, write your answer in place of the variable in the original equation. If the result is a true statement, then your answer satisfies the equation.

84. (a) After subtracting 4 from both sides of the equation, we should get $2x = 4$, not $2x = 12$.

(b) Subtracting 3 from $3x$ will not give x. (These are not like terms.)

(c) When we divide both sides of the equation by 3, we must remember to divide both terms on the left. We would then get $x - 2 = 4$, not $x - 6 = 4$.

(d) We cannot divide both sides by x, since x might be equal to 0. (In this case, it is!)

(e) In the last step, we should divide both sides by -3. Then we get $\frac{-\not{3}x}{-\not{3}} = \frac{-6}{-3}$, which gives $x = 2$.

(f) We need to divide both sides of the final equation by -1 to obtain $x = -2$.

(g) This is correct as written.

85. $|5-9|-|-3-8| = |-4|-|-11|$
$$= 4-11 = -7$$

87. $\frac{-4(-2)(-6)}{-4(-2)-6} = \frac{8(-6)}{8-6} = \frac{-48}{2} = -24$

89. $72 = 2 \cdot 2 \cdot 2 \cdot 3 \cdot 3$

Exercises 3.3

1. Let W = width of the rectangle.

$$6W = 40$$

$$\frac{\not{6}W}{\not{6}} = \frac{40}{6}$$

$$W = \frac{40}{6} = \frac{20}{3}$$

Thus, the width of the rectangle is $\frac{20}{3}$ or $6\frac{2}{3}$ in.

CHECK $6\left(\frac{20}{3}\right) \stackrel{\checkmark}{=} 40$

3. Let L = length of the rectangle

$$3.5L = 73.5$$

$$\frac{\not{3.5}L}{\not{3.5}} = \frac{73.5}{3.5}$$

$$L = 21$$

Thus, the length of the rectangle is 21 in.

CHECK: $(3.5)(21) \stackrel{\checkmark}{=} 73.5$

5. Let T = time required.

$$52T = 234$$

$$\frac{\not{52}T}{\not{52}} = \frac{234}{52}$$

$$T = \frac{234}{52} = \frac{9}{2} \text{ or } 4\frac{1}{2}$$

Thus, it takes $4\frac{1}{2}$ hours.

CHECK: $52\left(\frac{9}{2}\right) \stackrel{\checkmark}{=} 234$

7. Let m = number of miles driven.

$$3(29.95) + .17m = 123.17$$

$$89.85 + .17m = 123.17$$

$$\underline{-89.85 \qquad\qquad -89.85}$$

$$.17m = 33.32$$

$$\frac{\not{.17}m}{\not{.17}} = \frac{33.32}{.17}$$

$$m = \frac{33.32}{.17} = 196$$

Thus, 196 miles were driven.

CHECK: To rent the car for 3 days at \$29.95 per day costs 3(29.95) = \$89.85. To drive 196 miles at \$.17 per mile costs 196(.17) = \$33.32.

\$89.85 + \$33.32 = \$123.17

9. Let x = length of the shorter piece (in feet).

x + 8 = length of the longer piece (in feet).

$$x + (x + 8) = 30$$

$$2x + 8 = 30$$

$$\underline{-8 \quad -8}$$

$$2x = 22$$

$$\frac{\not{2}x}{\not{2}} = \frac{22}{2}$$

$$x = 11$$

Then x + 8 = 11 + 8 = 19. Thus, the two pieces are 11 ft. and 19 ft.

CHECK: 19 is 8 more than 11, and 11 + 19 = 30.

11. Let x = one of the numbers.

3x + 4 = the other number.

$$x + (3x + 4) = 24$$

$$4x + 4 = 24$$

$$\underline{-4 \quad -4}$$

$$4x = 20$$

$$\frac{\not{4}x}{\not{4}} = \frac{20}{4}$$

$$x = 5$$

Then 3x + 4 = 3(5) + 4 = 19. Thus, the numbers are 5 and 19.

CHECK: 19 is 4 more than 3 times 5 and the sum of 5 and 19 is 24.

13. Let x = one of the numbers. 4x – 5 = the other number.

$$x + (4x - 5) = 11$$

$$5x - 5 = 11$$

$$\underline{+5 \quad +5}$$

$$5x = 16$$

$$\frac{\not{5}x}{\not{5}} = \frac{16}{5}$$

$$x = \frac{16}{5}$$

Then $4x - 5 = 4\left(\frac{16}{5}\right) - 5 = \frac{39}{5}$. Thus, the numbers are $\frac{16}{5}$ and $\frac{39}{5}$.

CHECK:
$\frac{39}{5}$ is 5 less than 4 times $\frac{16}{5}$ and the sum of $\frac{16}{5}$ and $\frac{39}{5}$ is $\frac{55}{5} = 11$.

15. Let x = the number.

$$x + (5x - 4) = 27$$
$$6x - 4 = 27$$
$$\underline{\quad +4 \quad +4}$$
$$6x = 31$$
$$\frac{\not{6}x}{\not{6}} = \frac{31}{6}$$
$$x = \frac{31}{6}$$

CHECK: 4 less than 5 times $\frac{31}{6}$ is $\frac{131}{6}$,

and $\frac{31}{6} + \frac{131}{6} = \frac{162}{6} = 27$.

17. Let x = the number.

$$(2x + 4) - x = 12$$
$$x + 4 = 12$$
$$\underline{\quad -4 \quad -4}$$
$$x = 8$$

Thus, the number is 8.

CHECK: 4 more than twice 8 is 20, and 20 – 8 = 12.

19. Let x = the smallest number.

2x – 5 = the middle number.

2x + 10 = the largest number.

$$x + (2x - 5) + (2x + 10) = 80$$
$$5x + 5 = 80$$
$$\underline{\quad -5 \quad -5}$$
$$5x = 75$$
$$\frac{\not{5}x}{\not{5}} = \frac{75}{5}$$
$$x = 15$$

Then 2x – 5 = 2(15) – 5 = 25 and 2x + 10 = 2(15) + 10 = 40. Thus, the numbers are 15, 25, and 40.

CHECK: 25 is 5 less than twice 15, and 40 is 10 more than twice 15. The sum of 15, 25, and 40 is 80.

21. Let n = first integer.

n + 1 = second integer.

n + 2 = third integer.

$$n + (n + 1) + (n + 2) = 45$$
$$3n + 3 = 45$$
$$\underline{\quad -3 \quad -3}$$
$$3n = 42$$
$$\frac{\not{3}n}{\not{3}} = \frac{42}{3}$$
$$n = 14$$

Then n + 1 = 14 + 1 = 15 and n + 2 = 14 + 2 = 16. Thus, the consecutive integers are 14, 15, and 16.

CHECK: 14, 15, and 16 are three consecutive integers. The sum of 14, 15, and 16 is 45.

23. Let n = first odd integer.

n + 2 = second odd integer.

n + 4 = third odd integer.

$$n + (n + 2) + (n + 4) = 2(n + 4) + 29$$
$$3n + 6 = 2n + 8 + 29$$
$$3n + 6 = 2n + 37$$
$$\underline{2n \quad -2n}$$
$$n + 6 = 37$$
$$\underline{\quad -6 \quad -6}$$
$$n = 31$$

Then n + 2 = 31 + 2 = 33 and n + 4 = 31 + 4 = 35. Thus, the consecutive odd integers are 31, 33, and 35.

CHECK: 31, 33, and 35 are three consecutive integers. Their sum is 99, which is 29 more than 2(35) = 70.

25. Let W = width of the rectangle.

2W + 1 = length of the rectangle.

$$2W + 2(2W + 1) = 26$$
$$2W + 4W + 2 = 26$$
$$6W + 2 = 26$$
$$\underline{\quad -2 \quad -2}$$
$$6W = 24$$
$$\frac{\not{6}W}{\not{6}} = \frac{24}{6}$$
$$W = 4$$

Then 2W + 1 = 2(4) + 1 = 9 is the length of the rectangle. Thus, the rectangle has a width of 4 cm and a length of 9 cm.

CHECK: 9 is one more than twice 4, and 2(4) + 2(9) = 8 + 18 = 26.

27. Let x = length of shortest side.

x + 2 = length of "middle" side.

x + 4 = length of longest side.

$$x + (x + 2) + (x + 4) = 24$$
$$3x + 6 = 24$$
$$\underline{\quad -6 \quad -6}$$
$$3x = 18$$
$$\frac{\not{3}x}{\not{3}} = \frac{18}{3}$$
$$x = 6$$

Then x + 2 = 6 + 2 = 8 and x + 4 = 6 + 4 = 10. Thus, the sides of the triangle have lengths of 6 cm, 8 cm, and 10 cm.

CHECK: 6, 8, and 10 are consecutive even integers. A triangle whose sides are 6 cm, 8 cm, and 10 cm has a perimeter of 6 + 8 + 10 = 24 cm.

29. Let W = width of the rectangle.

W + 6 = length of the rectangle.

$$2(W + 10) + 2(3(W + 6)) = 2W + 2(W + 6) + 56$$
$$2W + 20 + 6W + 36 = 2W + 2W + 12 + 56$$
$$8W + 56 = 4W + 68$$
$$\underline{-4W \qquad -4W}$$
$$4W + 56 = 68$$
$$\underline{-56 \qquad -56}$$
$$\underline{4W = 12}$$
$$\frac{\not{4}W}{\not{4}} = \frac{12}{4}$$
$$W = 3$$

Then W + 6 = 9, so that the original rectangle has a width of 3 inches and a length of 9 inches.

CHECK: 9 is 6 more than 3. The new rectangle will have a width of 3 + 10 = 13 and a length of 3(9) = 27, so that its perimeter is 2(13) + 2(27) = 80. This is 56 more than 2(3) + 2(9) = 24, the original perimeter.

31. Let m = number of miles driven.

$$2(45) + .40m = 170$$
$$90 + .40m = 170$$
$$\underline{-90 \qquad -90}$$
$$.40m = 80$$
$$\frac{.\not{40}m}{.\not{40}} = \frac{80}{.40}$$
$$m = 200$$

The truck was driven 200 miles.

CHECK: To rent the truck for two days at $45 per day costs 2(45) = $90. To drive 200 miles at $.40 per mile costs .40(200) = $80. $90 + $80 = $170.

33. Let d = daily rental charge.

$$125 + 5d = 275$$
$$\underline{-125 \qquad -125}$$
$$5d = 150$$

$$\frac{\not{5}d}{\not{5}} = \frac{150}{5}$$

$$d = 30$$

The daily rental charge is $30.

CHECK: To rent a computer for 5 days at $30 per day costs 5(30) = $150. With an installation charge of $125, the total five-day rental costs $150 + $125 = $275.

35. Let q = number of quarters.

20 – q = number of dimes.

$$25q + 10(20 - q) = 425$$
$$25q + 200 - 10q = 425$$
$$15q + 200 = 425$$
$$\underline{\quad -200 \quad -200}$$
$$15q = 225$$
$$\frac{\not{15}q}{\not{15}} = \frac{225}{15}$$
$$q = 15$$

Then 20 – q = 20 – 15 = 5. Thus, there are 15 quarters and 5 dimes in the collection.

CHECK: 15 + 5 = 20. The value of 15 quarters is 15(.25) = $3.75. The value of 5 dimes is 5(.10) = $.50.

$3.75 + $.50 = $4.25.

37. Let x = number of advanced-purchase tickets sold. 150 – x = number of tickets sold at the door.

$$10x + 12(150 - x) = 1580$$
$$10x + 1800 - 12x = 1580$$
$$-2x + 1800 = 1580$$
$$\underline{\quad -1800 \quad -1800}$$
$$-2x = -220$$
$$\frac{\not{-2}x}{\not{-2}} = \frac{-220}{-2}$$
$$x = 110$$

110 advanced-purchase tickets were sold.

CHECK: If 110 tickets were sold in advance, then 40 were sold at the door. 110(10) = $1100 was collected from the advance sale and 40(12) = $480 was collected at the door. $1100 + $480 = $1580.

39. Let x = number of half-dollars.
x + 11 = number of quarters.
45 – x – (x + 11) = number of dimes.

$$50x + 25(x + 11) + 10(45 - x - (x + 11)) = 1110$$
$$50x + 25x + 275 + 10(34 - 2x) = 1110$$
$$75x + 275 + 340 - 20x = 1110$$
$$55x + 615 = 1110$$
$$\underline{\quad -615 \quad -615}$$
$$55x = 495$$
$$\frac{\not{55}x}{\not{55}} = \frac{495}{55}$$
$$x = 9$$

Then x + 11 = 9 + 11 = 20 and 45 – x – (x + 11) = 45 – 9 – (9 + 11) = 16. So there are 9 half-dollars, 20 quarters, and 16 dimes in the collection.

CHECK: 9 + 20 + 16 = 45. The value of 9 half-dollars is $4.50. The value of 20 quarters is $5.00. The value of 16 dimes is $1.60. Total value is $4.50 + $5.00 + $1.60 = $11.10.

41. Let t = number of hours the electrician worked.

t + 3 = number of hours her assistant worked.

$$24t + 12(t + 3) = 228$$
$$24t + 12t + 36 = 228$$
$$36t + 36 = 228$$
$$\underline{\quad -36 \quad -36}$$
$$36t = 192$$
$$\frac{\not{36}t}{\not{36}} = \frac{192}{36}$$
$$t = \frac{16}{3} \text{ or } 5\frac{1}{3}$$

The electrician worked for $5\frac{1}{3}$ hours.

CHECK: The electrician earns $\frac{16}{3}$($24) = $128 for her time. Her assistant earns

$\frac{25}{3}(\$12) = \$100$. Together, they earn \$128 + \$100 = \$228.

43. Let t = the time (in hours) that each train travels up to the point that they pass by each other.

Using d = rt, 20t = distance traveled by slower train and 40t = distance traveled by faster train. Then,

$$20t + 40t = 300$$
$$60t = 300$$
$$\frac{\cancel{60}t}{\cancel{60}} = \frac{300}{60}$$
$$t = 5$$

Thus, the trains pass by each other five hours later than 10:00 a.m., or at 3:00 p.m.

CHECK: In 5 hours, the slower train travels 20(5) = 100 miles and the faster one travels 40(5) = 200 miles. 100 + 200 = 300 miles.

45. Let t = the time (in hours) traveled by the person who left earlier.

t – 1 = the time (in hours) traveled by the person who left 1 hour later.

$$55t + 45(t-1) = 280$$
$$55t + 45t - 45 = 280$$
$$100t - 45 = 280$$
$$\underline{\qquad +45 \quad +45}$$
$$100t \qquad = 325$$
$$\frac{\cancel{100}t}{\cancel{100}} = \frac{325}{100}$$
$$t = \frac{13}{4} \text{ or } 3\frac{1}{4}$$

Thus, the people will be 280 km apart $3\frac{1}{4}$ hours later than 2:00 p.m., or at 5:15 p.m.

CHECK: The person who left earlier travels $\frac{13}{4}(55) = \frac{715}{4}$ km. The person who left later travels $\frac{9}{4}(45) = \frac{405}{4}$ km. The distance between them is $\frac{715}{4} + \frac{405}{4} = \frac{1120}{4} = 280$ km.

47. Let t = number of hours needed to complete the running section. 6 – t = number of hours needed to complete the bicycling section.

$$18t + 50(6-t) = 172$$
$$18t + 300 - 50t = 172$$
$$-32t + 300 = 172$$
$$\underline{\qquad -300 \quad -300}$$
$$-32t \qquad = -128$$
$$\frac{\cancel{-32}t}{\cancel{-32}} = \frac{-128}{-32}$$
$$t = 4$$

Thus, it takes 4 hours to complete the running section of the course. Since the running rate is 18 kph, the running section of the course is 18(4) = 72 km.

CHECK: The first person runs a distance of 18(4) = 72 km. The second person bicycles a distance of 50(6 – 4) = 50(2) = 100 km. The total distance covered is 72 + 100 = 172 km.

49. Let r = slower rate.

r + 15 = faster rate.

$$5r = 4(r + 15)$$
$$5r = 4r + 60$$
$$\underline{-4r \quad -4r}$$
$$r = 60$$

A car that travels for 5 hours at the rate of 60 kph covers 5(60) = 300 km, which is the distance between town A and town B.

CHECK: At the rate of 60 kph, the car travels 5(60) = 300 km in 5 hours. At the rate of 60 + 15 = 75 kph, the car travels 4(75) = 300 km in 4 hours, which is the same distance.

51. Let t = number of hours that the trainee works.

t – 2 = number of hours that the secretary works.

$$7t + 15(t-2) = 124$$
$$7t + 15t - 30 = 124$$
$$22t - 30 = 124$$
$$\underline{\qquad +30 \quad +30}$$
$$22t \qquad = 154$$

$$\frac{\cancel{22}t}{\cancel{22}} = \frac{154}{22}$$
$$t = 7$$

Thus, the pile of forms will be finished 7 hours after 9:00 a.m., or at 4:00 p.m.

CHECK: The trainee processes 7(7) = 49 forms. The secretary processes 15(7 – 2) = 15(5) = 75 forms. The total number of forms is 49 + 75 = 124.

53. Let x = number of shares Margaret buys at a lower price.

200 – x = number of shares Margaret buys at a higher price.

$$8.125x + 9.375(200 - x) = 1725$$
$$8.125x + 1875 - 9.375x = 1725$$
$$-1.25x + 1875 = 1725$$
$$-1875 \quad -1875$$
$$\frac{\cancel{-1.25}x}{\cancel{-1.25}} = \frac{-150}{-1.25}$$
$$x = 120$$

Then 200 – x = 200 – 120 = 80. Thus, Margaret buys 120 shares at the lower price and 80 shares at the higher price

CHECK: 120 shares at \$8.125 per share costs \$975. 80 shares at \$9.375 per share costs \$750. Total cost of all shares = \$1725.

55. Let x = number of lighter boxes.

x + 89 = number of heavier boxes.

$$6.58x + 9.32(x + 89) = 1974.28$$
$$6.58x + 9.32x + 829.48 = 1974.28$$
$$15.9x + 829.48 = 1974.28$$
$$-829.48 = -829.48$$
$$\frac{\cancel{15.9}x}{\cancel{15.9}} = \frac{1144.80}{15.9}$$
$$x = 72$$

Then x + 89 = 72 + 89 = 161. Thus, there are 72 lighter boxes and 161 heavier boxes on the truck.

CHECK: 72 boxes that weigh 6.58 kg each weigh 473.76 kg altogether; 161 boxes that weigh 9.32 kg each weigh 1500.52 kg altogether. The total weight of the boxes is 473.76 + 1500.52 = 1974.28 kg.

Exercises 3.4

1. $$x + 4 < 3$$
$$-2 + 4 \overset{?}{<} 3$$
$$2 < 3$$

Therefore, – 2 satisfies the inequality.

3. $$a - 2 > -1$$
$$-3 - 2 \overset{?}{>} -1$$
$$-5 \not> -1$$

Therefore, –3 does not satisfy the inequality.

5. $$-y + 3 \le 5$$
$$-(-2) + 3 \overset{?}{\le} 5$$
$$2 + 3 \overset{?}{\le} 5$$
$$5 \overset{\checkmark}{\le} 5$$

Therefore, –2 satisfies the inequality.

7. $$-8 \le 2 - x$$
$$-8 \overset{?}{\le} 2 - 6$$
$$-8 \overset{\checkmark}{\le} -4$$

Therefore, 6 satisfies the inequality.

9. $2z - 5 < 3$

$2(1) - 5 \overset{?}{<} -3$

$2 - 5 \overset{?}{<} -3$

$-3 \not< -3$

Therefore, 1 does not satisfy the inequality.

11. $12 < 5 + 2u$

$12 \overset{?}{<} 5 + 2(3)$

$12 \overset{?}{<} 5 + 6$

$12 \not< 11$

Therefore, 3 does not satisfy the inequality.

13. $7 - 4x < 8$

$7 - 4(-4) \overset{?}{<} 8$

$7 + 16 \overset{?}{<} 8$

$23 \not< 8$

Therefore, –4 does not satisfy the inequality.

15. $-2 < 8 - x < 3$

$-2 \overset{?}{<} 8 - 6 \overset{?}{<} 3$

$-2 \overset{\checkmark}{<} 2 \overset{\checkmark}{<} 3$

Therefore, 6 satisfies the inequality.

17. $6 + 2(a - 3) < 1$

$6 + 2(-2 - 3) \overset{?}{<} 1$

$6 + 2(-5) \overset{?}{<} 1$

$6 - 10 \overset{?}{<} 1$

$-4 \overset{\checkmark}{<} 1$

Therefore, –2 satisfies the inequality.

19. $-12 < 9 - 5(x + 1) < -5$

$-12 \overset{?}{<} 9 - 5(3 + 1) \overset{?}{<} -5$

$-12 \overset{?}{<} 9 - 5(4) \overset{?}{<} -5$

$-12 \overset{?}{<} 9 - 20 \overset{?}{<} -5$

$-12 \overset{\checkmark}{<} -11 \overset{\checkmark}{<} -5$

Therefore, 3 satisfies the inequality.

21. $x - 3 < 2$

$\underline{\quad +3 \quad +3}$

$x \quad < 5$

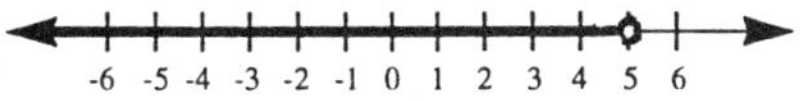

23. $a + 7 > 4$

$\underline{\quad -7 \quad -7}$

$a \quad > -3$

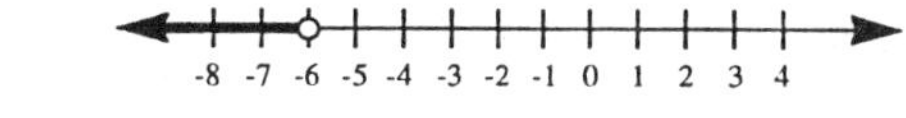

25. $-4 > w + 2$

$\underline{-2 \qquad -2}$

$-6 > w$

$w < -6$

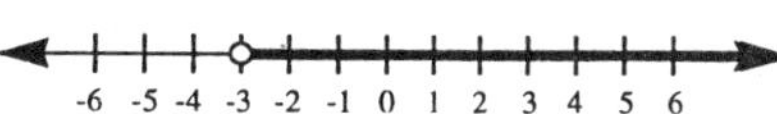

27. $z + 3 > 0$

$\underline{\quad -3 \quad -3}$

$z \quad > -3$

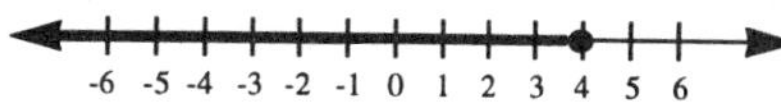

29. $3x \le 12$

$\frac{\cancel{3}x}{\cancel{3}} \le \frac{12}{3}$

$x \le 4$

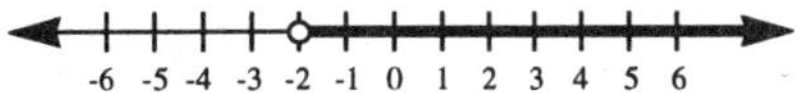

31. $4y > -8$

$\frac{\cancel{4}y}{\cancel{4}} > \frac{-8}{4}$

$y > -2$

33. $-3x < 6$

$\frac{-3x}{-3} > \frac{6}{-3}$

$x > -2$

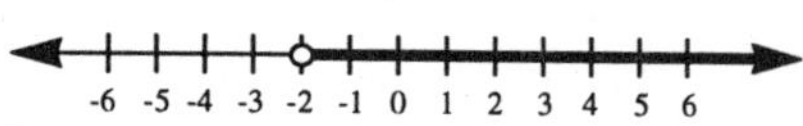

35. $-3x < -6$

$\frac{-3x}{-3} > \frac{-6}{-3}$

$x > 2$

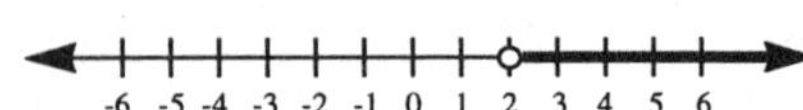

37. $7a > 0$

$\frac{7a}{7} > \frac{0}{7}$

$a > 0$

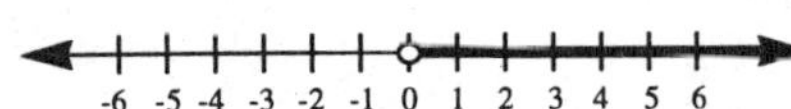

39. $-7a \geq 0$

$\frac{-7a}{-7} \leq \frac{0}{-7}$

$a \leq 0$

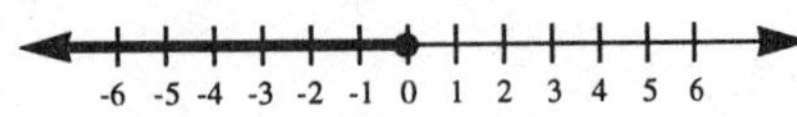

41. $-x < 3$

$(-1)(-x) > (-1)(3)$

$x > -3$

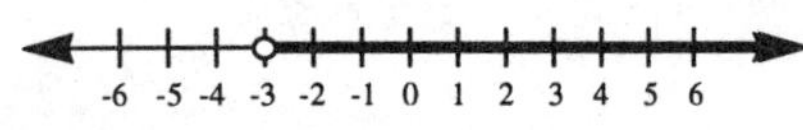

43. $-x < -3$

$(-1)(-x) > (-1)(-3)$

$x > 3$

45. $-5 < a - 4 \leq 2$

$+4 \quad +4 \quad +4$

$-1 < a \leq 6$

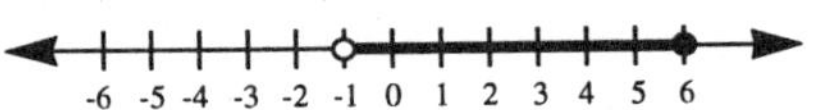

47. $1 \leq x + 3 \leq 5$

$-3 \quad -3 \quad -3$

$-2 \leq x \leq 2$

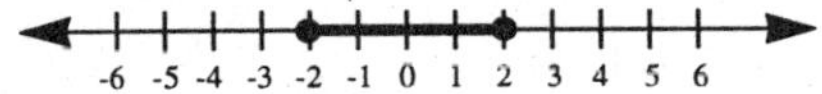

49. $-6 < 3y < 3$

$\frac{-6}{3} < \frac{3y}{3} < \frac{3}{3}$

$-2 < y < 1$

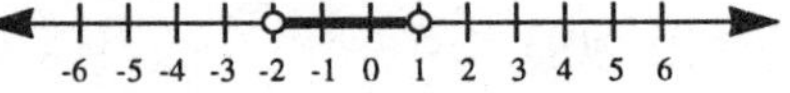

51. $0 \leq -2x < 2$

$\frac{0}{-2} \geq \frac{-2x}{-2} > \frac{2}{-2}$

$0 \geq x > -1$

or

$-1 < x \leq 0$

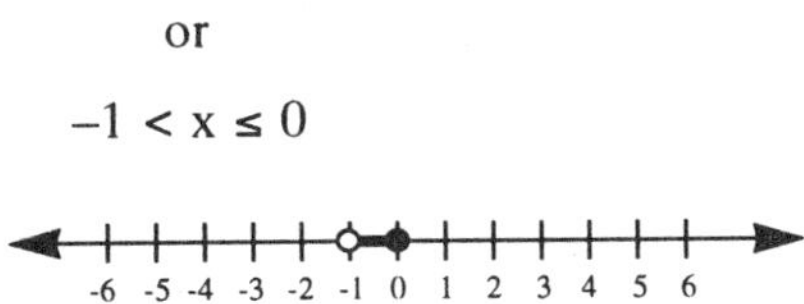

53. $$-5 < -x < -1$$
$$(-1)(-5) > (-1)(-x) > (-1)(-1)$$
$$5 > x > 1$$
or
$$1 < x < 5$$

55. One number is less than a second number if the first number is located to the left of the second on the number line. (Equivalently, we could say that the second number is located to the right of the first on the number line.)

56. If we begin with an inequality, we may add the same quantity to both sides, subtract the same quantity from both sides, or multiply or divide both sides by the same positive quantity and get another inequality with the same symbol. If we multiply or divide both sides by the same negative quantity, we get another inequality with the reversed symbol.

57. When we multiply or divide both sides of an equality by the same quantity, it does not matter whether the quantity is positive or negative. This is not the case for inequalities (see (56)).

58. $2 < x \leq 5$ means that x is between 2 and 5 on the number line, possibly equal to 5, but not equal to 2. That is, x is greater than 2 <u>and</u> less than or equal to 5.

59. $(2x^2)(3x^3) = (2 \cdot 3)(x^2x^3) = 6x^5$

61. $$\begin{aligned} 4(x-3y) + 2(3x-y) &= 4x - 12y + 6x - 2y \\ &= 4x + 6x - 12y - 2y \\ &= 10x - 14y \end{aligned}$$

63. $$\begin{aligned} x^2(x^3 - 4x) - (2x^5 - 4x^3) &= x^5 - 4x^3 - 2x^5 + 4x^3 \\ &= x^5 - 2x^5 - 4x^3 + 4x^3 \\ &= -x^5 \end{aligned}$$

Exercises 3.5

1. $$\begin{array}{rl} x + 5 & < 3 \\ -5 & \;\; -5 \\ \hline x & < -2 \end{array}$$

3. $$\begin{array}{rl} a - 2 & > -3 \\ +2 & \;\; +2 \\ \hline a & > -1 \end{array}$$

5. $$2y < 8$$
$$\frac{\not{2}y}{\not{2}} < \frac{8}{2}$$
$$y < 4$$

7. $$2y > -8$$
$$\frac{\not{2}y}{\not{2}} > \frac{-8}{2}$$
$$y > -4$$

9. $$-2y < 8$$
$$\frac{-2y}{-2} > \frac{8}{-2}$$
$$y > -4$$

11. $$-2y > -8$$
$$\frac{\not{-2}y}{\not{-2}} < \frac{-8}{-2}$$
$$y < 4$$

13. $-x < 4$

$$\frac{-x}{-1} > \frac{4}{-1}$$

$$x > -4$$

or

$$-x < 4$$

$$(-1)(-x) > (-1)(4)$$

$$x > -4$$

15. $-1 > -y$

$$\frac{-1}{-1} < \frac{-y}{-1}$$

$$1 < y$$

or

$$-1 > -y$$

$$(-1)(-1) < (-1)(-y)$$

$$1 < y$$

17. $5x + 3 \le 8$

$$\underline{\quad -3 \;\; -3}$$

$$5x \le 5$$

$$\frac{\cancel{5}x}{\cancel{5}} \le \frac{5}{5}$$

$$x \le 1$$

19. $2x - 9 \ge 16$

$$\underline{\quad +9 \;\; +9}$$

$$2x \ge 25$$

$$\frac{\cancel{2}x}{\cancel{2}} \ge \frac{25}{2}$$

$$x \ge \frac{25}{2}$$

21. $2(z-3) + 4 > -6$

$$2z - 6 + 4 > -6$$

$$2z - 2 > -6$$

$$\underline{\quad +2 \;\; +2}$$

$$2z > -4$$

$$\frac{\cancel{2}z}{\cancel{2}} > \frac{-4}{2}$$

$$z > -2$$

23. $3(x+4) + 2(x-1) < 20$

$$3x + 12 + 2x - 2 < 20$$

$$5x + 10 < 20$$

$$\underline{\quad -10 \;\; -10}$$

$$5x < 10$$

$$\frac{\cancel{5}x}{\cancel{5}} < \frac{10}{5}$$

$$x < 2$$

25. $5(w+3) - 7w \le 7$

$$5w + 15 - 7w \le 7$$

$$-2w + 15 \le 7$$

$$\underline{\quad -15 \; -15}$$

$$-2w \le -8$$

$$\frac{-\cancel{2}w}{-\cancel{2}} \ge \frac{-8}{-2}$$

$$w \ge 4$$

27. $3(a+4) - 4(a-1) < 10$

$$3a + 12 - 4a + 4 < 10$$

$$-a + 16 < 10$$

$$\underline{\quad -16 \;\; -16}$$

$$-a < -6$$

$$\frac{-a}{-1} > \frac{-6}{-1}$$

$$a > 6$$

29. $4(y-3) - (3y-12) \ge 2$

$$4y - 12 - 3y + 12 \ge 2$$

$$y \ge 2$$

31. $2(u+2) - 2(u-1) < 5$

$$2u + 4 - 2u + 2 < 5$$

$$6 < 5$$

Contradiction

33. $4(x-2) - (4x-3) < 6$

$$4x - 8 - 4x + 3 < 6$$

$$-5 < 6$$

Identity

35. $x + 3 < 2x + 7$

$\underline{-x \qquad -x}$

$3 < x + 7$

$\underline{-7 \qquad -7}$

$-4 < x$

37. $5t - 3 \geq 3t + 10$

$\underline{-3t \qquad -3t}$

$2t - 3 \geq 10$

$\underline{+3 \qquad +3}$

$2t \geq 13$

$\frac{\cancel{2}t}{\cancel{2}} \geq \frac{13}{2}$

$t \geq \frac{13}{2}$

39. $2(a-5) + 3a > 6a - 6$

$2a - 10 + 3a > 6a - 6$

$5a - 10 > 6a - 6$

$\underline{-5a \qquad -5a}$

$-10 > a - 6$

$\underline{+6 \qquad +6}$

$-4 > a$

41. $4(w+2) - 3(w-1) > 5(w-1) - 5w$

$4w + 8 - 3w + 3 > 5w - 5 - 5w$

$w + 11 > -5$

$\underline{-11 \quad -11}$

$w > -16$

43. $2y - 4(y+1) \leq 8 - (y+2)$

$2y - 4y - 4 \leq 8 - y - 2$

$-2y - 4 \leq 6 - y$

$\underline{+2y \qquad +2y}$

$-4 \leq 6 + y$

$\underline{-6 \quad -6}$

$-10 \leq y$

45. $2 < x + 7 < 10$

$\underline{-7 \quad -7 \quad -7}$

$-5 < x < 3$

47. $3 < 2a + 5 < 7$

$\underline{-5 \quad -5 \quad -5}$

$-2 < 2a < 2$

$\frac{-2}{2} < \frac{\cancel{2}a}{\cancel{2}} < \frac{2}{2}$

$-1 < a < 1$

49. $-5 \leq -4y + 3 < 7$

$\underline{-3 \quad -3 \quad -3}$

$-8 \leq -4y < 4$

$\frac{-8}{-4} \geq \frac{\cancel{-4}y}{\cancel{-4}} > \frac{4}{-4}$

$2 \geq y \geq -1$

51. $1 \leq 6 - x < 3$

$\underline{-6 \quad -6 \quad -6}$

$-5 \leq -x < -3$

$\frac{-5}{-1} \geq \frac{-x}{-1} > \frac{-3}{-1}$

$5 \geq x > 3$

53. $x + 4 < 2x - 1$

$\underline{-x \qquad -x}$

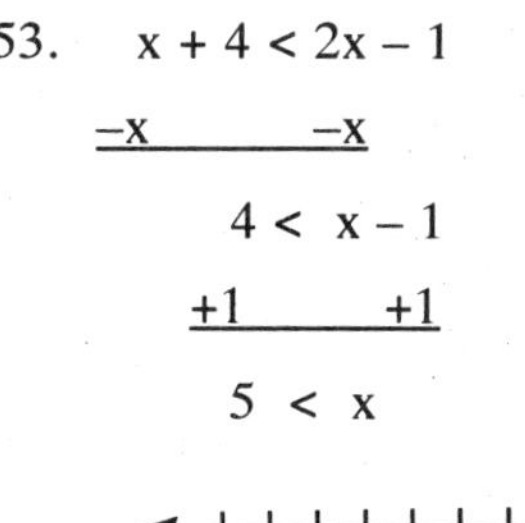

$4 < x - 1$

$\underline{+1 \qquad +1}$

$5 < x$

-6 -5 -4 -3 -2 -1 0 1 2 3 4 5 6

55. $3(a+2) - 5a \geq 2 - a$

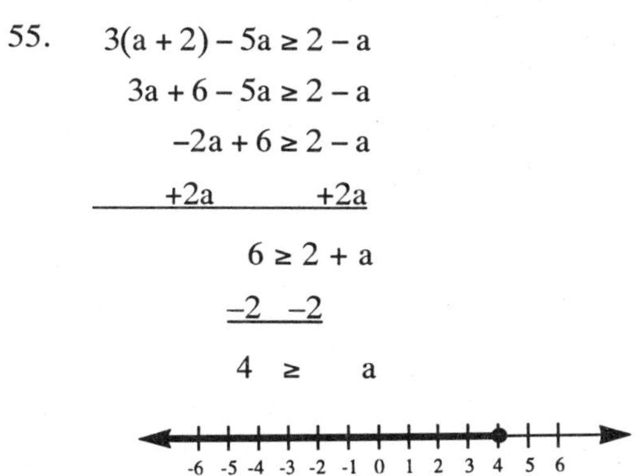

$3a + 6 - 5a \geq 2 - a$

$-2a + 6 \geq 2 - a$

$\underline{+2a \qquad +2a}$

$6 \geq 2 + a$

$\underline{-2 \quad -2}$

$4 \geq a$

-6 -5 -4 -3 -2 -1 0 1 2 3 4 5 6

57. $-1 < x + 3 < 2$

$\underline{-3 \qquad -3 \quad -3}$

$-4 < x \qquad < -1$

59. $-1 \le y - 3 \le 2$

$\underline{+3 \qquad +3 \quad +3}$

$2 \le y \qquad \le 5$

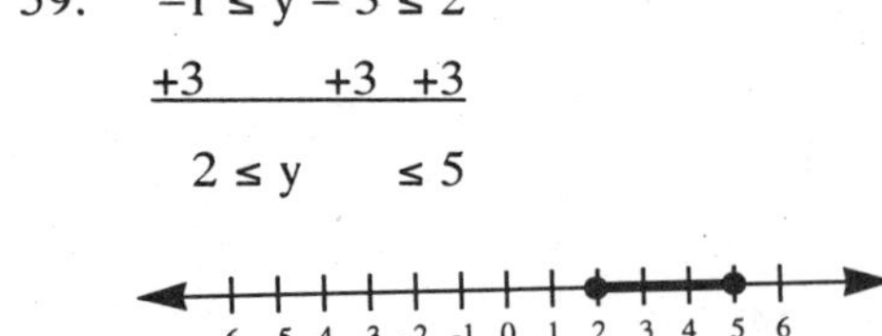

61. $-3 \le 4t + 5 < 9$

$\underline{-5 \qquad -5 \quad -5}$

$-8 \le 4t \qquad < 4$

$\frac{-8}{4} \le \frac{\cancel{4}t}{\cancel{4}} < \frac{4}{4}$

$-2 \le \ t \qquad < 1$

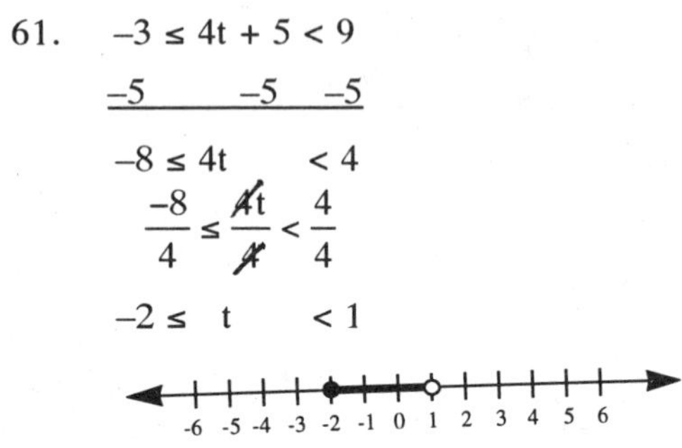

63. $-5 < 3 - 2x \le 9$

$\underline{-3 \ -3 \qquad\quad -3}$

$-8 < \ -2x \le 6$

$\frac{-8}{-2} > \frac{-\cancel{2}x}{-\cancel{2}} \ge \frac{6}{-2}$

$4 > \qquad x \ge -3$

65. $x \ge 3.2$

67. $6.32 < t \le 11.51$

69. (a) Starting with $2 \le -2x$, we must divide both sides of this inequality by -2, obtaining $\frac{2}{-2} \ge \frac{-2x}{-2}$ or $-1 \ge x$.

(b) When we subtract 4 from both sides of the original inequality, the resulting inequality should read $2x < -2$, not $2x > -2$. (We only reverse the inequality sign when we multiply or divide by a negative quantity.)

(c) We must divide both sides of the inequality $-9 > 3x$ by 3, giving $\frac{-9}{3} > \frac{\cancel{3}x}{\cancel{3}}$ or $-3 > x$.

70. Only part (a) makes sense, since it requires that x be a number between -3 and 2. Parts (b), (c), (e) and (f) all lead to contradictions. For instance, the inequality $-5 < x < -8$ implies that $-5 < -8$, which is false. (The others are similar.) Part (d) makes no sense, since we cannot write double inequalities in which the inequality symbols point in opposite directions.

Chapter 3 Review Exercises

1.
$$\begin{aligned} 5(x-4) - 3(x-3) &= 3 - (14 - 2x) \\ 5x - 20 - 3x + 9 &= 3 - 14 + 2x \\ 2x - 11 &= 2x - 11 \\ \underline{-2x \qquad} & \underline{\quad\ -2x} \\ -11 &= \ -11 \end{aligned}$$

Identity

3.
$$\begin{aligned} 3a(a+3) - a(2a+4) &= a^2 + 10 \\ 3a^2 + 9a - 2a^2 - 4a &= a^2 + 10 \\ a^2 + 5a &= a^2 + 10 \\ \underline{-a^2 \qquad} & \underline{\ -a^2 \qquad} \\ 5a &= \qquad 10 \\ \frac{\cancel{5}a}{\cancel{5}} &= \frac{10}{5} \\ a &= 2 \end{aligned}$$

Conditional equation

5. $2x - 5 = -7$

CHECK $x = -6$:

$$2(-6) - 5 \stackrel{?}{=} -7$$
$$-12 - 5 \stackrel{?}{=} -7$$
$$-17 \neq -7$$

So –6 does not satisfy the equation.

CHECK $x = -1$:

$$2(-1) - 5 \stackrel{?}{=} -7$$
$$-2 - 5 \stackrel{?}{=} -7$$
$$-7 \stackrel{\checkmark}{=} -7$$

So –1 satisfies the equation.

7. $4y + 3 \leq 10 + 2y$

CHECK $y = 4$:

$$4(4) + 3 \stackrel{?}{\leq} 10 + 2(4)$$
$$16 + 3 \stackrel{?}{\leq} 10 + 8$$
$$19 \not\leq 18$$

So 4 does not satisfy the inequality.

CHECK $y = \frac{5}{2}$:

$$4\left(\frac{5}{2}\right) + 3 \stackrel{?}{\leq} 10 + 2\left(\frac{5}{2}\right)$$
$$10 + 3 \stackrel{?}{\leq} 10 + 5$$
$$13 \stackrel{\checkmark}{\leq} 15$$

So $\frac{5}{2}$ satisfies the inequality.

9. $3t + 2(t + 1) = 3t + 3$

CHECK $t = -2$:

$$3(-2) + 2(-2 + 1) \stackrel{?}{=} 3(-2) + 3$$
$$3(-2) + 2(-1) \stackrel{?}{=} 3(-2) + 3$$
$$-6 - 2 \stackrel{?}{=} -6 + 3$$
$$-8 \neq -3$$

So –2 does not satisfy the equation.

CHECK $t = \frac{1}{2}$:

$$3\left(\frac{1}{2}\right) + 2\left(\frac{1}{2} + 1\right) \stackrel{?}{=} 3\left(\frac{1}{2}\right) + 3$$
$$3\left(\frac{1}{2}\right) + 2\left(\frac{3}{2}\right) \stackrel{?}{=} 3\left(\frac{1}{2}\right) + 3$$
$$\frac{3}{2} + 3 \stackrel{?}{=} \frac{3}{2} + 3$$
$$\frac{9}{2} \stackrel{\checkmark}{=} \frac{9}{2}$$

So $\frac{1}{2}$ satisfies the equation.

11. $8 - 3(x - 2) > x - 4$

CHECK $x = -5$:

$$8 - 3(-5 - 2) \stackrel{?}{>} -5 - 4$$
$$8 - 3(-7) \stackrel{?}{>} -5 - 4$$
$$8 + 21 \stackrel{?}{>} -5 - 4$$
$$29 \stackrel{\checkmark}{>} -9$$

So –5 satisfies the inequality.

CHECK $x = 5$:

$$8 - 3(5 - 2) \stackrel{?}{>} 5 - 4$$
$$8 - 3(3) \stackrel{?}{>} 5 - 4$$
$$8 - 9 \stackrel{?}{>} 5 - 4$$
$$-1 \not> 1$$

So 5 does not satisfy the inequality.

13. $a^2 + (a - 2)^2 = 20$

CHECK $a = -2$:

$$(-2)^2 + (-2 - 2)^2 \stackrel{?}{=} 20$$
$$(-2)^2 + (-4)^2 \stackrel{?}{=} 20$$
$$4 + 16 \stackrel{?}{=} 20$$
$$20 \stackrel{\checkmark}{=} 20$$

So –2 satisfies the equation.

CHECK $a = 2$:

$$(2)^2 + (2 - 2)^2 \stackrel{?}{=} 20$$
$$(2)^2 + (0)^2 \stackrel{?}{=} 20$$
$$4 + 0 \stackrel{?}{=} 20$$
$$4 \neq 20$$

So 2 does not satisfy the equation.

15. $5x + 8 = 2x - 7$

$\underline{-2x \quad\quad -2x}$

$3x + 8 = -7$

$\underline{-8 \quad\quad -8}$

$3x = -15$

$\frac{3x}{3} = \frac{-15}{3}$

$x = -5$

17. $2(y + 4) - 2y = 8$

$2y + 8 - 2y = 8$

$8 = 8$

Identity

19. $2(3a + 4) + 8 = 5(3a - 1)$

$6a + 8 + 8 = 15a - 5$

$6a + 16 = 15a - 5$

$\underline{-6a \quad\quad -6a}$

$16 = 9a - 5$

$\underline{+5 \quad\quad +5}$

$21 = 9a$

$\frac{21}{9} = \frac{9a}{9}$

$\frac{7}{3} = a$

21. $8x - 3(x - 4) = 4(x + 3) + 28$

$8x - 3x + 12 = 4x + 12 + 28$

$5x + 12 = 4x + 40$

$\underline{-4x \quad\quad -4x}$

$x + 12 = 40$

$\underline{-12 \quad\quad -12}$

$x = 28$

23. $a(a + 3) - 2(a - 1) = a(a - 1) + 7$

$a^2 + 3a - 2a + 2 = a^2 - a + 7$

$a^2 + a + 2 = a^2 - a + 7$

$\underline{-a^2 \quad\quad -a^2}$

$a + 2 = -a + 7$

$\underline{+a \quad\quad +a}$

$2a + 2 = 7$

$\underline{-2 \quad\quad -2}$

$2a = 5$

$\frac{2a}{2} = \frac{5}{2}$

$a = \frac{5}{2}$

25. $8 - 3(x - 1) < 2$

$8 - 3x + 3 < 2$

$-3x + 11 < 2$

$\underline{-11 \quad -11}$

$-3x < -9$

$\frac{-3x}{-3} > \frac{-9}{-3}$

$x > 3$

27. $2(x - 3) - 4(x - 1) \geq 7 - x$

$2x - 6 - 4x + 4 \geq 7 - x$

$-2x - 2 \geq 7 - x$

$\underline{+2x \quad\quad +2x}$

$-2 \geq 7 + x$

$\underline{-7 \quad -7}$

$-9 \geq x$

-11 -10 -9 -8 -7 -6 -5 -4 -3 -2 -1 0 1

29. $2 \leq 3a + 8 < 20$

$\underline{-8 \quad\quad -8 \quad -8}$

$-6 \leq 3a < 12$

$\frac{-6}{3} \leq \frac{3a}{3} < \frac{12}{3}$

$-2 \leq a < 4$

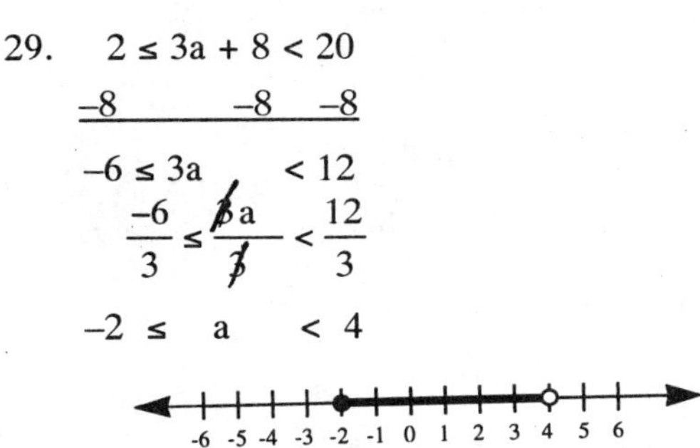

-6 -5 -4 -3 -2 -1 0 1 2 3 4 5 6

31. Let x = one of the numbers.

2x – 3 = the other number.

$x + (2x - 3) = 18$

$3x - 3 = 18$

$\underline{+3 \quad +3}$

$3x = 21$

$\frac{3x}{3} = \frac{21}{3}$

$x = 7$

Then $2x - 3 = 2(7) - 3 = 11$. Thus, the numbers are 7 and 11.

CHECK: 11 is 3 less than twice 7. The sum of 7 and 11 is 18.

33. Let W = width of the rectangle.

$5W + 4$ = length of the rectangle.

$$\begin{aligned} 2W + 2(5W + 4) &= 80 \\ 2W + 10W + 8 &= 80 \\ 12W + 8 &= 80 \\ -8 \quad & \quad -8 \\ \hline 12W &= 72 \\ \frac{\cancel{12}W}{\cancel{12}} &= \frac{72}{12} \\ W &= 6 \end{aligned}$$

Then $5W + 4 = 5(6) + 4 = 34$. Thus, the dimensions of the rectangle are 6 cm and 34 cm.

CHECK: 34 is 4 more than 5 times 6. $2(6) + 2(34) = 80$.

35. Let x = number of first quality skirts bought.

$150 - x$ = number of irregular skirts bought.

$$\begin{aligned} 12x + 7(150 - x) &= 1500 \\ 12x + 1050 - 7x &= 1500 \\ 5x + 1050 &= 1500 \\ -1050 \quad & \quad -1050 \\ \hline 5x &= 450 \\ \frac{\cancel{5}x}{\cancel{5}} &= \frac{450}{5} \\ x &= 90 \end{aligned}$$

Then $150 - x = 150 - 90 = 60$. Thus, the wholesaler bought 90 first quality skirts and 60 irregular skirts.

CHECK: 90 first quality skirts cost 12(90) = \$1080. 60 irregular skirts cost 7(60) = \$420. The total cost of the skirts is \$1080 + \$420 = \$1500.

37. Let x = number of overtime hours the laborer must work.

$$\begin{aligned} 6(40) + 9x &\geq 348 \\ 240 + 9x &\geq 348 \\ -240 \quad & \quad -240 \\ \hline 9x &\geq 108 \\ \frac{\cancel{9}x}{\cancel{9}} &\geq \frac{108}{9} \\ x &\geq 12 \end{aligned}$$

Thus, the minimum number of overtime hours that laborer must work is 12.

Chapter 3 Practice Test

1. (a)
$$\begin{aligned} 3x - 5(x - 2) &= -2x + 8 \\ 3x - 5x + 10 &= -2x + 8 \\ -2x + 10 &= -2x + 8 \\ +2x \quad & \quad +2x \\ \hline 10 &= 8 \end{aligned}$$

Contradiction

(b)
$$\begin{aligned} 3x - 5(x - 2) &= -2x + 10 \\ 3x - 5x + 10 &= -2x + 10 \\ -2x + 10 &= -2x + 10 \end{aligned}$$

Identity

(c) $3x - 5(x - 2) = 2x + 10$

$$
\begin{aligned}
3x - 5x + 10 &= 2x + 10 \\
-2x + 10 &= 2x + 10 \\
+2x \quad &\quad +2x \\
\hline
10 &= 4x - 10 \\
+10 \quad &\quad +10 \\
\hline
20 &= 4x \\
\frac{20}{4} &= \frac{\cancel{4}x}{\cancel{4}} \\
5 &= x
\end{aligned}
$$

Conditional

3. (a) $6 - 3x = 3x - 10$

$$
\begin{aligned}
+3x \quad &\quad +3x \\
\hline
6 &= 6x - 10 \\
+10 \quad &\quad +10 \\
\hline
16 &= 6x \\
\frac{16}{6} &= \frac{\cancel{6}x}{\cancel{6}} \\
\frac{8}{3} &= x
\end{aligned}
$$

(b) $2(3y - 5) - 4y = 2 - (y + 12)$

$$
\begin{aligned}
6y - 10 - 4y &= 2 - y - 12 \\
2y - 10 &= -y - 10 \\
+y \quad &\quad +y \\
\hline
3y - 10 &= -10 \\
+10 \quad &\quad +10 \\
\hline
3y &= 0 \\
\frac{3y}{3} &= \frac{0}{3} \\
y &= 0
\end{aligned}
$$

(c) $2a^2 - 3(a - 4) = 2a(a - 6) + 3a$

$$
\begin{aligned}
2a^2 - 3a + 12 &= 2a^2 - 12a + 3a \\
2a^2 - 3a + 12 &= 2a^2 - 9a \\
-2a^2 \quad &\quad -2a^2 \\
\hline
-3a + 12 &= -9a \\
+3a \quad &\quad +3a \\
\hline
12 &= -6a \\
\frac{12}{-6} &= \frac{\cancel{-6}a}{\cancel{-6}} \\
-2 &= a
\end{aligned}
$$

(d) $9 - 5(x - 2) \geq 4$

$$
\begin{aligned}
9 - 5x + 10 &\geq 4 \\
-5x + 19 &\geq 4 \\
-19 \quad &\quad -19 \\
\hline
-5x &\geq -15 \\
\frac{\cancel{-5}x}{\cancel{-5}} &\leq \frac{-15}{-5} \\
x &\leq 3
\end{aligned}
$$

-6 -5 -4 -3 -2 -1 0 1 2 3 4 5 6

(e) $1 < 3 - x \leq 5$

$$
\begin{aligned}
-3 \quad -3 \quad & \quad -3 \\
\hline
-2 < -x &\leq 2 \\
\frac{-2}{-1} > \frac{-x}{-1} &\geq \frac{2}{-1} \\
2 > x &\geq -2
\end{aligned}
$$

or

$$-2 \leq x < 2$$

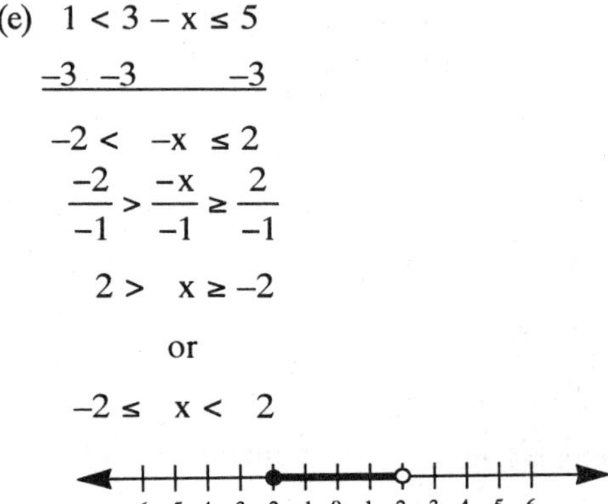

-6 -5 -4 -3 -2 -1 0 1 2 3 4 5 6

5. Let x = number of \$1 cassettes.

$20 - x$ = number of \$3 cassettes.

$$
\begin{aligned}
1x + 3(20 - x) &= 46 \\
x + 60 - 3x &= 46 \\
-2x + 60 &= 46 \\
-60 \quad &\quad -60 \\
\hline
-2x &= -14 \\
\frac{\cancel{-2}x}{\cancel{-2}} &= \frac{-14}{-2} \\
x &= 7
\end{aligned}
$$

Then $20 - x = 20 - 7 = 13$. Thus, the person bought 7 one-dollar cassettes and 13 three-dollar cassettes.

CHECK: $7 + 13 \stackrel{\checkmark}{=} 20$. 7 cassettes at \$1 each cost \$7.

13 cassettes at \$3 each cost 13(3) = \$39.

$\$7 + \$39 \stackrel{\checkmark}{=} \$46$

CUMULATIVE REVIEW CHAPTERS 1–3

1 $-8-5-7=-13-7=-20$

3. $12-4(3-5)=12-4(-2)$
$=12+8$
$=20$

5. $-5^2=-(5\cdot 5)=-25$

7. $xx^2x^3=x^{1+2+3}=x^6$

9. $x^2y-2xy^2-xy^2-3x^2y$
$=x^2y-3x^2y-2xy^2-xy^2$
$=(1-3)x^2y+(-2-1)xy^2$
$=-2x^2y-3xy^2$

11. $2x(3x^2-4y)=2x(3x^2)-2x(4y)$
$=6x^3-8xy$

13. $-3u^2(u^3)(-5v)=((-3)(-5))(u^2u^3)v$
$=15u^5v$

15. $4(m-3n)+3(2m-n)$
$=4m-12n+6m-3n$
$=4m+6m-12n-3n$
$=10m-15n$

17. $2ab(a^2-ab)-4a^2(ab-b^2)$
$=2a^3b-2a^2b^2-4a^3b+4a^2b^2$
$=2a^3b-4a^3b-2a^2b^2+4a^2b^2$
$=(2-4)a^3b+(-2+4)a^2b^2$
$=-2a^3b+2a^2b^2$

19. $x^2y-xy^2-(xy^2-x^2y)$
$=x^2y-xy^2-xy^2+x^2y$
$=x^2y+x^2y-xy^2-xy^2$
$=(1+1)x^2y+(-1-1)xy^2$
$=2x^2y-2xy^2$

21. $x-3\{x-4(x-5)\}$
$=x-3\{x-4x+20\}$
$=x-3\{-3x+20\}$
$=x+9x-60$
$=10x-60$

23. $3xy(4x^3y-2y)-2x(3y^2)(2x^3)$
$=12x^4y^2-6xy^2-12x^4y^2$
$=12x^4y^2-12x^4y^2-6xy^2$
$=(12-12)x^4y^2-6xy^2$
$=-6xy^2$

25. $x^2=(-2)^2=4$

27. $xy^2-(xy)^2=(-2)(-3)^2-((-2)(-3))^2$
$=-2(9)-(6)^2$
$=-18-36$
$=-54$

29. $|x-y-z|=|(-2)-(-3)-(5)|$
$=|-2+3-5|$
$=|-4|$
$=4$

31. $2x-4y^2=2(-2)-4(-3)^2$
$=2(-2)-4(9)$
$=-4-36$
$=-40$

33. $2x+11=5x+10$
$\underline{-2x\quad\;\; -2x}$
$11=3x+10$
$\underline{-10\quad\;\; -10}$
$1=3x$
$\frac{1}{3}=\frac{\cancel{3}x}{\cancel{3}}$
$\frac{1}{3}=x$

35. $9(a+1)-3(2a-2)=12$

$$9a+9-6a+6=12$$
$$3a+15=12$$
$$-15 \quad -15$$
$$3a = -3$$
$$\frac{\cancel{3}a}{\cancel{3}} = \frac{-3}{3}$$
$$a = -1$$

37. $4(5-x)-2(6-2x)=8$

$$20-4x-12+4x=8$$
$$8=8$$

Identity

39. $2(s+4)+3(2s+2)>6s$

$$2s+8+6s+6>6s$$
$$8s+14>6s$$
$$-8s \quad -8s$$
$$14>-2s$$
$$\frac{14}{-2} < \frac{\cancel{-2}s}{\cancel{-2}}$$
$$-7<s$$

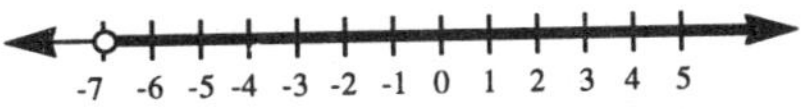

41. $1<2y-5\le 3$

$$+5 \quad +5+5$$
$$6<2y \le 8$$
$$\frac{6}{2} < \frac{\cancel{2}y}{\cancel{2}} \le \frac{8}{2}$$
$$3<y \le 4$$

43. $4(3d-2)+6(8-d)=9d-4(d-10)$

$$12d-8+48-6d=9d-4d+40$$
$$6d+40=5d+40$$
$$-5d \quad -5d$$
$$d+40= \quad 40$$
$$-40 \quad -40$$
$$d=0$$

45. Let x = one of the numbers.
3x – 7 = the other number.

$$x+(3x-7)=41$$
$$4x-7=41$$
$$+7 \quad +7$$
$$4x=48$$
$$\frac{\cancel{4}x}{\cancel{4}} = \frac{48}{4}$$
$$x=12$$

Then 3x – 7 = 3(12) – 7 = 29.
Thus, the numbers are 12 and 29.

CHECK: 29 is 7 less than 3 times 12.
The sum of 12 and 29 is 41.

47. Let x = number of danishes that Louise bought.
18 – x = number of pastries that Louise bought.

$$40x+55(18-x)=825$$
$$40x+990-55x=825$$
$$-15x+990=825$$
$$-990 \quad -990$$
$$-15x=-165$$
$$\frac{\cancel{-15}x}{\cancel{-15}} = \frac{-165}{-15}$$
$$x=11$$

Then 18 – x = 18 – 11 = 7.
Thus, Louise bought 11 danishes and 7 pastries.

CHECK: 11 danishes cost 11($.40) = $4.40.
7 pastries cost 7($.55) = $3.85. The total cost of the assortment is $4.40 + $3.85 = $8.25.

Cumulative Practice Test:

Chapters 1–3

1. (a) $-3^4 + 3(-2)^3$

$= -(3\cdot3\cdot3\cdot3) + 3((-2)(-2)(-2))$

$= -81 + 3(-8)$

$= -81 - 24$

$= -105$

(b) $(3-7-2)^2 = (-4-2)^2 = (-6)^2$

$= (-6)(-6) = 36$

3. (a) $5x^2 - 4x - 8 - 7x^2 - x + 11$

$= 5x^2 - 7x^2 - 4x - x - 8 + 11$

$= -2x^2 - 5x + 3$

(b) $2(x-3y) + 5(y-x)$

$= 2x - 6y + 5y - 5x$

$= 2x - 5x - 6y + 5y$

$= -3x - y$

(c) $3a(a^2 - 2b) - 5(a^3 - ab)$

$= 3a^3 - 6ab - 5a^3 + 5ab$

$= 3a^3 - 5a^3 - 6ab + 5ab$

$= -2a^3 - ab$

(d) $2(u-4v) - 3(v-2u) - (8u - 11v)$

$= 2u - 8v - 3v + 6u - 8u + 11v$

$= 2u + 6u - 8u - 8v - 3v + 11v$

$= 0u + 0v = 0$

(e) $4x^2y^3(x-5y) - 2xy(5y^2)(-2xy)$

$= 4x^3y^3 - 20x^2y^4 - 10xy^3(-2xy)$

$= 4x^3y^3 - 20x^2y^4 + 20x^2y^4$

$= 4x^3y^3$

(f) $6 - a[6 - a(6-a)]$

$= 6 - a[6 - 6a + a^2]$

$= 6 - 6a + 6a^2 - a^3$

5. (a) $7 - 3a \geq 13$

$\underline{-7 \qquad -7}$

$-3a \geq 6$

$\frac{-3a}{-3} \leq \frac{6}{-3}$

$a \leq -2$

-6 -5 -4 -3 -2 -1 0 1 2 3 4 5 6

(b) $1 \leq 4x - 3 < 17$

$\underline{+3 \qquad +3 \quad +3}$

$4 \leq 4x < 20$

$\frac{4}{4} \leq \frac{4x}{4} < \frac{20}{4}$

$1 \leq x < 5$

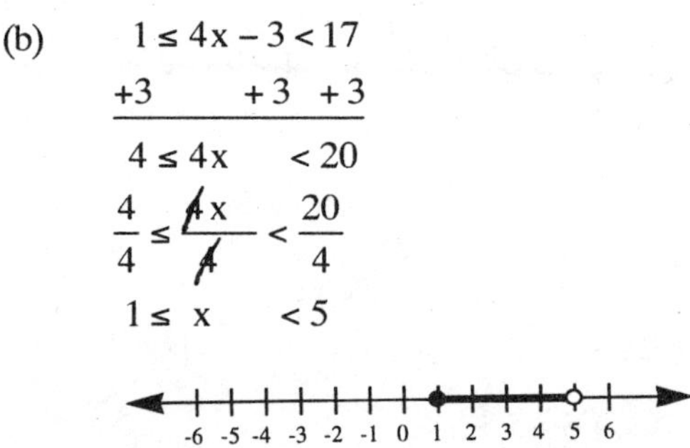

CHAPTER 4
RATIONAL EXPRESSIONS

Exercises 4.1

1. $\dfrac{18}{30} = \dfrac{3\cdot\cancel{6}}{5\cdot\cancel{6}} = \dfrac{3}{5}$

3. $\dfrac{-9}{21} = \dfrac{-3\cdot\cancel{3}}{7\cdot\cancel{3}} = \dfrac{-3}{7} = -\dfrac{3}{7}$

5. $\dfrac{-15}{-6} = \dfrac{5(\cancel{-3})}{2(\cancel{-3})} = \dfrac{5}{2}$

7. $\dfrac{-5+8}{10-4} = \dfrac{3}{6} = \dfrac{1\cdot\cancel{3}}{2\cdot\cancel{3}} = \dfrac{1}{2}$

9. $\dfrac{8-5(2)}{6-4(3)} = \dfrac{8-10}{6-12} = \dfrac{-2}{-6}$

$= \dfrac{1(\cancel{-2})}{3(\cancel{-2})} = \dfrac{1}{3}$

11. $\dfrac{x^3}{x} = \dfrac{x^2\cdot\cancel{x}}{1\cdot\cancel{x}} = \dfrac{x^2}{1} = x^2$

13. $\dfrac{x}{x^3} = \dfrac{1\cdot\cancel{x}}{x^2\cdot\cancel{x}} = \dfrac{1}{x^2}$

15. $\dfrac{10x}{4x^2} = \dfrac{5\cdot\cancel{2x}}{2x\cdot\cancel{2x}} = \dfrac{5}{2x}$

17. $\dfrac{-3z^6}{5z^2} = \dfrac{-3z^4\cdot\cancel{z^2}}{5\cancel{z^2}} = \dfrac{-3z^4}{5} = -\dfrac{3z^4}{5}$

19. $\dfrac{12t^5}{30t^{10}} = \dfrac{2\cdot\cancel{6t^5}}{5t^5\cdot\cancel{6t^5}} = \dfrac{2}{5t^5}$

21. $\dfrac{6ab^5}{-2a^3b^2} = \dfrac{3b^3\cdot\cancel{2ab^2}}{-a^2\cdot\cancel{2ab^2}} = \dfrac{3b^3}{-a^2} = -\dfrac{3b^3}{a^2}$

23. $\dfrac{(2x^3)(6x^2)}{(4x)(3x^4)} = \dfrac{1\cdot\cancel{12x^5}}{1\cdot\cancel{12x^5}} = \dfrac{1}{1} = 1$

25. $\dfrac{(r^3t^2)(-rt^3)}{2r^2t^7} = \dfrac{-r^4t^5}{2r^2t^7}$

$= \dfrac{-r^2 \cdot \cancel{r^2t^5}}{2t^2 \cdot \cancel{r^2t^5}} = \dfrac{-r^2}{2t^2} = -\dfrac{r^2}{2t^2}$

27. $\dfrac{3a(5b)(-4ab^3)}{6ab(2a^2b^2)} = \dfrac{-60a^2b^4}{12a^3b^3}$

$= \dfrac{-5b \cdot \cancel{12a^2b^3}}{a \cdot \cancel{12a^2b^3}} = \dfrac{-5b}{a} = -\dfrac{5b}{a}$

29. $\dfrac{(2x)^5}{(4x)^3} = \dfrac{32x^5}{64x^3} = \dfrac{x^2 \cdot \cancel{32x^3}}{2 \cdot \cancel{32x^3}} = \dfrac{x^2}{2}$

31. $\dfrac{(-4x^3)^2}{(-2x^4)^3} = \dfrac{16x^6}{-8x^{12}} = \dfrac{2 \cdot \cancel{8x^6}}{-x^6 \cdot \cancel{8x^6}}$

$= \dfrac{2}{-x^6} = -\dfrac{2}{x^6}$

33. $\dfrac{(2xy^2)^3}{(4x^2y^3)^3} = \dfrac{8x^3y^6}{64x^6y^9} = \dfrac{1 \cdot \cancel{8x^3y^6}}{8x^3y^3 \cdot \cancel{8x^3y^6}}$

$= \dfrac{1}{8x^3y^3}$

35. $\dfrac{5x - 2x}{10x - 4x} = \dfrac{3x}{6x} = \dfrac{1 \cdot \cancel{3x}}{2 \cdot \cancel{3x}} = \dfrac{1}{2}$

37. $\dfrac{5a(2x)}{15a(8x)} = \dfrac{10ax}{120ax} = \dfrac{1 \cdot \cancel{10ax}}{12 \cdot \cancel{10ax}}$

$= \dfrac{1}{12}$

39. $\dfrac{4s - 3t}{8s - 9t}$ cannot be reduced.

41. $\dfrac{7a^2 - 5a^2 - 6a^2}{4a - 8a} = \dfrac{-4a^2}{-4a}$

$= \dfrac{a\cancel{(-4a)}}{1\cancel{(-4a)}} = \dfrac{a}{1} = a$

43. $\dfrac{5x^2 - 3x - x^2 + 3x}{6x^2 - 5x - 2x^2 + 5x} = \dfrac{\cancel{4x^2}\,1}{\cancel{4x^2}\,1}$

$= \dfrac{1}{1} = 1$

45. The value of a fraction is unchanged when its numerator and denominator are both either multiplied or divided by the same nonzero quantity.

46. We would conclude that 2 = 4. It was incorrect to write

$$\frac{4+\cancel{2}}{1+\cancel{2}}=\frac{4}{1}.$$

The 2's cannot be crossed out, since 2 is a common term, not a common factor.

47. (a) The factor of x in the reduced fraction should be in its numerator.
(b) When both factors in the numerator are crossed out, a factor of 1 remains.
(c) As in part (b), a factor of 1 remains in the numerator, so that the reduced fraction is $\frac{1}{3x^2y}$
(d) Crossing out x's amounts to crossing out terms rather than factors.

48. The fractions in (a), (c), (d), and (e) are equivalent. So are the fractions in (b), (f), and (g).

49. Let w = width of the rectangle. 2w + 5 = length of the rectangle.

$$2w+2(2w+5)=34$$
$$2w+4w+10=34$$
$$6w+10=34$$
$$\underline{-10=-10}$$
$$6w=24$$
$$\frac{\cancel{6}w}{\cancel{6}}=\frac{24}{6}$$
$$w=4$$

Then 2w +5 = 2(4) + 5 = 13. Thus, the rectangle has a width of 4 cm and a length of 13 cm.

Exercises 4.2

1. $\frac{-4}{9}\cdot\frac{-2}{3}=\frac{(-4)(-2)}{9\cdot 3}=\frac{8}{27}$

3. $\frac{\cancel{-6}}{\cancel{10}}\cdot\frac{\cancel{15}}{\cancel{9}}=\frac{-1}{1}=-1$

5. $\frac{2}{3y}\cdot\frac{x}{5}=\frac{2x}{3y\cdot 5}=\frac{2x}{15y}$

7. $\frac{x^2}{4y}\cdot\frac{5x}{3y}=\frac{x^2\cdot 5x}{4y\cdot 3y}=\frac{5x^3}{12y^2}$

9. $\frac{4}{5}\div\frac{5}{4}=\frac{4}{5}\cdot\frac{4}{5}$

$=\frac{4\cdot 4}{5\cdot 5}=\frac{16}{25}$

11. $\dfrac{6x}{y} \div \dfrac{y^2}{2x^2} = \dfrac{6x}{y} \cdot \dfrac{2x^2}{y^2}$

$= \dfrac{6x \cdot 2x^2}{y \cdot y^2} = \dfrac{12x^3}{y^3}$

13. $\dfrac{3}{2x} \cdot \dfrac{xw}{6} = \dfrac{1 \cdot w}{2 \cdot 2} = \dfrac{w}{4}$

15. $4 \cdot \dfrac{x}{12} = \dfrac{4}{1} \cdot \dfrac{x}{12} = \dfrac{1 \cdot x}{1 \cdot 3} = \dfrac{x}{3}$

17. $4 \div \dfrac{x}{12} = \dfrac{4}{1} \div \dfrac{x}{12} = \dfrac{4}{1} \cdot \dfrac{12}{x}$

$= \dfrac{4 \cdot 12}{1 \cdot x} = \dfrac{48}{x}$

19. $\dfrac{x}{12} \div 4 = \dfrac{x}{12} \div \dfrac{4}{1} = \dfrac{x}{12} \cdot \dfrac{1}{4}$

$= \dfrac{x \cdot 1}{12 \cdot 4} = \dfrac{x}{48}$

21. $\dfrac{-2x}{3y^2} \cdot \dfrac{-9y}{4x} = \dfrac{(-1)(-3)}{2y} = \dfrac{3}{2y}$

23. $\dfrac{m^3n^2}{2m} \cdot \dfrac{6}{n^3} = \dfrac{m^2 \cdot 3}{1 \cdot n} = \dfrac{3m^2}{n}$

25. $\dfrac{3uv^2}{5w} \div \dfrac{6u^2v}{15w} = \dfrac{3uv^2}{5w} \cdot \dfrac{15w}{6u^2v}$

$= \dfrac{v \cdot 3}{2 \cdot u} = \dfrac{3v}{2u}$

27. $6xy \cdot \dfrac{2x}{3y} = \dfrac{6xy}{1} \cdot \dfrac{2x}{3y}$

$= \dfrac{2x \cdot 2x}{1 \cdot 1} = \dfrac{4x^2}{1} = 4x^2$

29. $6xy \div \dfrac{2x}{3y} = \dfrac{6xy}{1} \cdot \dfrac{3y}{2x}$

$= \dfrac{3y \cdot 3y}{1 \cdot 1} = \dfrac{9y^2}{1} = 9y^2$

31. $\dfrac{2x}{3y} \div (6xy) = \dfrac{2x}{3y} \div \dfrac{6xy}{1}$

$= \dfrac{2x}{3y} \cdot \dfrac{1}{6xy} = \dfrac{1 \cdot 1}{3y \cdot 3y} = \dfrac{1}{9y^2}$

33. $\frac{-4x}{9y} \cdot \frac{x^2}{y^2} \cdot \frac{3y}{2x} = \frac{-2 \cdot x^2}{3 \cdot y^2}$

$= -\frac{2x^2}{3y^2}$

35. $\frac{9}{a^2} \cdot \left(\frac{a}{3} \div \frac{3}{a}\right) = \frac{9}{a^2} \cdot \left(\frac{a}{3} \cdot \frac{a}{3}\right)$

$= \frac{9}{a^2} \cdot \frac{a^2}{9} = \frac{1 \cdot 1}{1 \cdot 1} = \frac{1}{1} = 1$

37. $\frac{9}{a^2} \div \left(\frac{a}{3} \cdot \frac{3}{a}\right) = \frac{9}{a^2} \div \frac{1}{1} = \frac{9}{a^2} \cdot \frac{1}{1}$

$= \frac{9 \cdot 1}{a^2 \cdot 1} = \frac{9}{a^2}$

39. $\frac{9}{a^2} \div \left(\frac{a}{3} \div \frac{3}{a}\right) = \frac{9}{a^2} \div \left(\frac{a}{3} \cdot \frac{a}{3}\right)$

$= \frac{9}{a^2} \div \frac{a^2}{9} = \frac{9}{a^2} \cdot \frac{9}{a^2} = \frac{9 \cdot 9}{a^2 \cdot a^2}$

$= \frac{81}{a^4}$

41. $\left(\frac{9}{a^2} \div \frac{a}{3}\right) \div \frac{a}{3} = \left(\frac{9}{a^2} \cdot \frac{3}{a}\right) \div \frac{a}{3}$

$= \frac{27}{a^3} \div \frac{a}{3} = \frac{27}{a^3} \cdot \frac{3}{a} = \frac{27 \cdot 3}{a^3 \cdot a}$

$= \frac{81}{a^4}$

43. $\frac{\frac{2x}{3}}{\frac{10x}{9}} = \frac{2x}{3} \div \frac{10x}{9} = \frac{2x}{3} \cdot \frac{9}{10x}$

$= \frac{1 \cdot 3}{1 \cdot 5} = \frac{3}{5}$

45. $\frac{\frac{x^2}{3}}{\frac{x}{6}} = \frac{x^2}{3} \div \frac{x}{6} = \frac{x^2}{3} \cdot \frac{6}{x}$

$= \frac{x \cdot 2}{1 \cdot 1} = \frac{2x}{1} = 2x$

47. $\frac{\frac{x}{y^2}}{\frac{y}{x^2}} = \frac{x}{y^2} \div \frac{y}{x^2} = \frac{x}{y^2} \cdot \frac{x^2}{y}$

$= \frac{x \cdot x^2}{y^2 \cdot y} = \frac{x^3}{y^3}$

49. $\dfrac{\frac{2u}{z^2}}{\frac{4z}{u}} = \dfrac{2u}{z^2} \div \dfrac{4z}{u} = \dfrac{2u}{z^2} \cdot \dfrac{u}{4z}$

$= \dfrac{u \cdot u}{z^2 \cdot 2z} = \dfrac{u^2}{2z^3}$

51. $\dfrac{3x^2 - x^2}{4y^2 - y^2} \cdot \dfrac{2y + y}{x^2 + x^2}$

$= \dfrac{2x^2}{3y^2} \cdot \dfrac{3y}{2x^2} = \dfrac{1 \cdot 1}{y \cdot 1} = \dfrac{1}{y}$

53. $\dfrac{3x^2 \cdot x^2}{4y^2 \cdot y^2} \cdot \dfrac{2y \cdot y}{x^2 \cdot x^2} = \dfrac{3x^4}{4y^4} \cdot \dfrac{2y^2}{x^4}$

$= \dfrac{3}{2y^2}$

55. To divide by a fraction, multiply by its reciprocal. This rule works because division is defined to be the inverse of multiplication.

56. (a) The fraction $\dfrac{2y}{3x}$ must be inverted, leading to $\dfrac{3x}{2y} \cdot \dfrac{3x}{2y} = \dfrac{9x^2}{4y^2}$.

(b) The factor of 5 is $\dfrac{5}{1}$, not $\dfrac{5}{5}$. Thus, $5 \cdot \dfrac{3x}{2} = \dfrac{5}{1} \cdot \dfrac{3x}{2} = \dfrac{5 \cdot 3x}{1 \cdot 2} = \dfrac{15x}{2}$.

57. A collection of dimes and quarters has a total value of \$2.75. If there are three more dimes than quarters, how many of each type of coin are in the collection?

Let x = number of quarters.

59. Let x = smallest angle
2x = largest angle
x + 20 = third angle

$$\begin{aligned} x + 2x + (x + 20) &= 180 \\ 4x + 20 &= 180 \\ -20 &= -20 \\ \hline 4x &= 160 \\ \frac{4x}{4} &= \frac{160}{4} \\ x &= 40 \end{aligned}$$

Then 2x = 2(40) = 80 and x + 20 = 40 + 20 = 60. So the three angles are 40°, 60°, and 80°.

Exercises 4.3

1. $\dfrac{5}{3} + \dfrac{4}{3} = \dfrac{5+4}{3} = \dfrac{9}{3} = \dfrac{3}{1} = 3$

3. $\frac{3}{5} - \frac{7}{5} = \frac{3-7}{5} = \frac{-4}{5} = -\frac{4}{5}$

5. $\frac{7}{9} - \frac{5}{9} - \frac{8}{9} = \frac{7-5-8}{9}$

$= \frac{-6}{9} = -\frac{2}{3}$

7. $\frac{2}{3} + \frac{4}{5} = \frac{2(5)}{3(5)} + \frac{4(3)}{5(3)}$

$= \frac{10}{15} + \frac{12}{15} = \frac{10+12}{15} = \frac{22}{15}$

9. $\frac{2}{3} \cdot \frac{4}{5} = \frac{2 \cdot 4}{3 \cdot 5} = \frac{8}{15}$

11. $\frac{2}{3} - \frac{5}{6} = \frac{2(2)}{3(2)} - \frac{5}{6}$

$= \frac{4}{6} - \frac{5}{6} = \frac{4-5}{6} = \frac{-1}{6} = -\frac{1}{6}$

13. $3 + \frac{3}{4} - \frac{3}{8} = \frac{3}{1} + \frac{3}{4} - \frac{3}{8}$

$= \frac{3(8)}{1(8)} + \frac{3(2)}{4(2)} - \frac{3}{8}$

$= \frac{24}{8} + \frac{6}{8} - \frac{3}{8}$

$= \frac{24+6-3}{8} = \frac{27}{8}$

15. $\frac{3}{2} - \frac{4}{5} + \frac{7}{10} = \frac{3(5)}{2(5)} - \frac{4(2)}{5(2)} + \frac{7}{10} = \frac{15}{10} - \frac{8}{10} + \frac{7}{10}$

$= \frac{15-8+7}{10} = \frac{14}{10} = \frac{7}{5}$

17. $\frac{5}{6} - \frac{3}{8} + \frac{1}{4} = \frac{5(4)}{6(4)} - \frac{3(3)}{8(3)} + \frac{1(6)}{4(6)}$

$= \frac{20}{24} - \frac{9}{24} + \frac{6}{24}$

$= \frac{20-9+6}{24} = \frac{17}{24}$

19. $2+\frac{1}{3}-\frac{1}{2}=\frac{2}{1}+\frac{1}{3}-\frac{1}{2}$

$$=\frac{2(6)}{1(6)}+\frac{1(2)}{3(2)}-\frac{1(3)}{2(3)}$$

$$=\frac{12}{6}+\frac{2}{6}-\frac{3}{6}$$

$$=\frac{12+2-3}{6}=\frac{11}{6}$$

21. $\frac{8}{3x}+\frac{4}{3x}=\frac{8+4}{3x}=\frac{12}{3x}=\frac{4}{x}$

23. $\frac{8}{3x}\cdot\frac{4}{3x}=\frac{8\cdot 4}{3x\cdot 3x}=\frac{32}{9x^2}$

25. $\frac{7}{6x}+\frac{13}{6x}-\frac{11}{6x}=\frac{7+13-11}{6x}$

$$=\frac{9}{6x}=\frac{3}{2x}$$

27. $\frac{3y}{7x}-\frac{5y}{7x}+\frac{4y}{7x}$

$$=\frac{3y-5y+4y}{7x}=\frac{2y}{7x}$$

29. $\frac{w}{9z}-\frac{5w}{9z}+\frac{4w}{9z}$

$$=\frac{w-5w+4w}{9z}=\frac{0}{9z}=0$$

31. $\frac{-5a}{7b}+\frac{3a}{7b}-\frac{12a}{7b}$

$$=\frac{-5a+3a-12a}{7b}=\frac{-14a}{7b}=\frac{-2a}{b}$$

$$=-\frac{2a}{b}$$

33. $\frac{x+3}{3x}+\frac{x-6}{3x}=\frac{x+3+x-6}{3x}$

$$=\frac{2x-3}{3x}$$

35. $\frac{3y^2-5}{4y}+\frac{5-4y^2}{4y}$

$$=\frac{3y^2-5+5-4y^2}{4y}=\frac{-y^2}{4y}$$

$$=\frac{-y}{4}=-\frac{y}{4}$$

$\frac{5x+2}{10x}-\frac{x+2}{10x}$
$=\frac{5x+2-(x+2)}{10x}$
$=\frac{5x+2-x-2}{10x}=\frac{4x}{10x}=\frac{2}{5}$

39. $\frac{2a-1}{3a}-\frac{5a-1}{3a}$
$=\frac{2a-1-(5a-1)}{3a}$
$=\frac{2a-1-5a+1}{3a}=\frac{-3a}{3a}$
$=\frac{-1}{1}=-1$

41. $\frac{w-4}{6w}-\frac{w-3}{6w}+\frac{5}{6w}$
$=\frac{w-4-(w-3)+5}{6w}$
$=\frac{w-4-w+3+5}{6w}=\frac{4}{6w}$
$=\frac{2}{3w}$

43. $\frac{t^2-3t+2}{5t^2}-\frac{7t+t^2}{5t^2}$
$=\frac{t^2-3t+2-(7t+t^2)}{5t^2}$
$=\frac{t^2-3t+2-7t-t^2}{5t^2}=\frac{-10t+2}{5t^2}$

45. $\frac{3}{x}+\frac{2}{y}=\frac{3(y)}{x(y)}+\frac{2(x)}{y(x)}$
$=\frac{3y}{xy}+\frac{2x}{xy}=\frac{3y+2x}{xy}$

47. $\frac{3}{x}\cdot\frac{2}{y}=\frac{3\cdot 2}{x\cdot y}=\frac{6}{xy}$

49. $\frac{5}{3x}-\frac{7}{2}=\frac{5(2)}{3x(2)}-\frac{7(3x)}{2(3x)}$
$=\frac{10}{6x}-\frac{21x}{6x}=\frac{10-21x}{6x}$

51. $\frac{5}{4x}+\frac{3}{2y}=\frac{5(y)}{4x(y)}+\frac{3(2x)}{2y(2x)}$

$=\frac{5y}{4xy}+\frac{6x}{4xy}=\frac{5y+6x}{4xy}$

53. $\frac{4}{x^2}-\frac{3}{2x}=\frac{4(2)}{x^2(2)}-\frac{3(x)}{2x(x)}$

$=\frac{8}{2x^2}-\frac{3x}{2x^2}=\frac{8-3x}{2x^2}$

55. $\frac{4}{x^2}\cdot\frac{3}{2x}=\frac{2\cdot 3}{x^2\cdot x}=\frac{6}{x^3}$

57. $\frac{2}{3x^2}+\frac{3}{2x^2}=\frac{2(2)}{3x^2(2)}+\frac{3(3)}{2x^2(3)}$

$=\frac{4}{6x^2}+\frac{9}{6x^2}=\frac{4+9}{6x^2}=\frac{13}{6x^2}$

59. $\frac{7}{4a^2}-\frac{9}{20a}=\frac{7(5)}{4a^2(5)}-\frac{9(a)}{20a(a)}$

$=\frac{35}{20a^2}-\frac{9a}{20a^2}=\frac{35-9a}{20a^2}$

61. $\frac{1}{x}+2=\frac{1}{x}+\frac{2}{1}=\frac{1}{x}+\frac{2(x)}{1(x)}$

$=\frac{1}{x}+\frac{2x}{x}=\frac{1+2x}{x}$

63. $\frac{5}{3xy}+\frac{1}{6y^2}=\frac{5(2y)}{3xy(2y)}$

$+\frac{1(x)}{6y^2(x)}=\frac{10y}{6xy^2}+\frac{x}{6xy^2}$

$=\frac{10y+x}{6xy^2}$

65. $\frac{7}{6a^2b}+\frac{3}{4ab^3}=\frac{7(2b^2)}{6a^2b(2b^2)}+\frac{3(3a)}{4ab^3(3a)}$

$=\frac{14b^2}{12a^2b^3}+\frac{9a}{12a^2b^3}$

$=\frac{14b^2+9a}{12a^2b^3}$

67. $\frac{7}{6a^2b}\cdot\frac{3}{4ab^3}=\frac{7\cdot 1}{2a^2b\cdot 4ab^3}$

$=\frac{7}{8a^3b^4}$

69. $\frac{4y}{3x^2} - \frac{3}{2x} + \frac{y}{x^2} = \frac{4y(2)}{3x^2(2)}$

$- \frac{3(3x)}{2x(3x)} + \frac{y(6)}{x^2(6)} = \frac{8y}{6x^2} - \frac{9x}{6x^2}$

$+ \frac{6y}{6x^2} = \frac{8y - 9x + 6y}{6x^2} = \frac{14y - 9x}{6x^2}$

71. $\frac{3}{4m^2n} - \frac{5}{6mn^3} + \frac{1}{8n^2}$

$= \frac{3(6n^2)}{4m^2n(6n^2)} - \frac{5(4m)}{6mn^3(4m)}$

$+ \frac{1(3m^2n)}{8n^2(3m^2n)} = \frac{18n^2}{24m^2n^3}$

$- \frac{20m}{24m^2n^3} + \frac{3m^2n}{24m^2n^3}$

$= \frac{18n^2 - 20m + 3m^2n}{24m^2n^3}$

73. $\frac{x}{y} + \frac{y}{x} + \frac{3x}{2y}$

$= \frac{x(2x)}{y(2x)} + \frac{y(2y)}{x(2y)} + \frac{3x(x)}{2y(x)}$

$= \frac{2x^2}{2xy} + \frac{2y^2}{2xy} + \frac{3x^2}{2xy}$

$= \frac{2x^2 + 2y^2 + 3x^2}{2xy} = \frac{5x^2 + 2y^2}{2xy}$

75. $t - \frac{3}{t} = \frac{t}{1} - \frac{3}{t} = \frac{t(t)}{1(t)} - \frac{3}{t}$

$= \frac{t^2}{t} - \frac{3}{t} = \frac{t^2 - 3}{t}$

77. $3x^2 + \frac{1}{x} - \frac{2}{x^2} = \frac{3x^2}{1} + \frac{1}{x} - \frac{2}{x^2}$

$= \frac{3x^2(x^2)}{1(x^2)} + \frac{1(x)}{x(x)} - \frac{2}{x^2}$

$= \frac{3x^4}{x^2} + \frac{x}{x^2} - \frac{2}{x^2} = \frac{3x^4 + x - 2}{x^2}$

79. $\frac{2x + 3}{x} + \frac{x}{2}$

$= \frac{(2x + 3)(2)}{x(2)} + \frac{x(x)}{2(x)}$

$= \frac{4x + 6}{2x} + \frac{x^2}{2x} = \frac{4x + 6 + x^2}{2x}$

81. $\frac{a-5}{2}+\frac{3}{a}$

$= \frac{(a-5)(a)}{2(a)}+\frac{3(2)}{a(2)}$

$= \frac{a^2-5a}{2a}+\frac{6}{2a}=\frac{a^2-5a+6}{2a}$

83. The least common denominator (LCD) is the "smallest" expression that is exactly divisible by each of the denominators in a problem. We need the LCD in order to add or subtract two or more fractions. (Actually, a common denominator is really all that is needed, but the LCD is the most efficient one to use.)

84. Each of the original denominators is a factor of the LCD, so this is a common denominator. Since each factor of the LCD was chosen the maximum number of times that it appears as a factor in any one of the denominators, there are no extra factors. Thus, this must be the smallest common denominator possible.

85. The LCD for $\frac{3}{10}$ and $\frac{7}{9}$ is 90. The LCD for $\frac{5}{6}$ and $\frac{3}{4}$ is 12. The LCD of two fractions will simply be the product of the two denominators when these denominators have no prime factor in common.

86. (a) We must subtract the entire expression $(5-x)$ from $x+3$ in the numerator. This gives

$\frac{x+3-(5-x)}{x}$

$= \frac{x+3-5+x}{x}=\frac{2x-2}{x}.$

(b) We cannot cancel when performing addition.

(c) When building fractions, we must multiply both numerator and denominator by the same quantity. So $\frac{5x}{2y}=\frac{5x(3x)}{2y(3x)}=\frac{15x^2}{6xy}$ and $\frac{7y}{6x}=\frac{7y(y)}{6x(y)}=\frac{7y^2}{6xy}$. Then $\frac{5x}{2y}+\frac{7y}{6x}$

$= \frac{15x^2}{6xy}+\frac{7y^2}{6xy}=\frac{15x^2+7y^2}{6xy}.$

(d) The cancellation step undoes the building step and brings the problem back to its original form. The final step is incorrect, since we cannot add two fractions with unlike denominators.

(e) The cancellation is not allowed, since there are no common factors to be crossed out.

87. Reducing these fractions would reverse the building process and bring us back to the original problem.

89. Let p = weight of package in pounds.

$$\begin{aligned} 10 + 4p &= 28 \\ -10 \quad &= -10 \\ 4p &= 18 \\ \frac{\cancel{4}p}{\cancel{4}} &= \frac{18}{4} \\ p &= 4\frac{1}{2} \end{aligned}$$

The package weighed $4\frac{1}{2}$ pounds.

Exercises 4.4

1. $$\frac{x}{3} = 9 \quad \text{LCD} = 3$$
$$3\left(\frac{x}{3}\right) = 3 \cdot 9$$
$$\frac{\overset{1}{\cancel{3}}}{1} \cdot \frac{x}{\underset{1}{\cancel{3}}} = 27$$
$$x = 27$$

3. $$\frac{a}{4} = -6 \quad \text{LCD} = 4$$
$$4\left(\frac{a}{4}\right) = 4 \cdot (-6)$$
$$\frac{\overset{1}{\cancel{4}}}{1} \cdot \frac{a}{\underset{1}{\cancel{4}}} = -24$$
$$a = -24$$

5. $$\frac{y}{6} = \frac{5}{4} \quad \text{LCD} = 12$$
$$12\left(\frac{y}{6}\right) = 12\left(\frac{5}{4}\right)$$
$$\frac{\overset{2}{\cancel{12}}}{1} \cdot \frac{y}{\underset{1}{\cancel{6}}} = \frac{\overset{3}{\cancel{12}}}{1} \cdot \frac{5}{\underset{1}{\cancel{4}}}$$
$$2y = 15$$
$$y = \frac{15}{2}$$

7. $$\frac{w}{8} = \frac{-7}{6} \quad \text{LCD} = 24$$
$$24\left(\frac{w}{8}\right) = 24\left(\frac{-7}{6}\right)$$
$$\frac{\overset{3}{\cancel{24}}}{1} \cdot \frac{w}{\underset{1}{\cancel{8}}} = \frac{\overset{4}{\cancel{24}}}{1} \cdot \frac{-7}{\underset{1}{\cancel{6}}}$$
$$3w = -28$$
$$w = \frac{-28}{3}$$

9. $$\frac{3x}{2} = -18 \quad \text{LCD} = 2$$
$$2\left(\frac{3x}{2}\right) = 2 \cdot (-18)$$
$$\frac{\overset{1}{\cancel{2}}}{1} \cdot \frac{3x}{\underset{1}{\cancel{2}}} = -36$$
$$3x = -36$$
$$x = -12$$

11. $16 = \frac{4}{5}x$ LCD = 5

$$5 \cdot 16 = 5\left(\frac{4}{5}x\right)$$

$$80 = \frac{5}{1} \cdot \frac{4}{5}x$$

$$80 = 4x$$

$$20 = x$$

13. $\frac{x}{3} - 2 = \frac{2}{3}$ LCD = 3

$$3\left(\frac{x}{3} - 2\right) = 3 \cdot \frac{2}{3}$$

$$\frac{3}{1} \cdot \frac{x}{3} - 3 \cdot 2 = \frac{3}{1} \cdot \frac{2}{3}$$

$$x - 6 = 2$$

$$x = 8$$

15. $\frac{3a}{4} + 2 = \frac{5}{4}$ LCD = 4

$$4\left(\frac{3a}{4} + 2\right) = 4 \cdot \frac{5}{4}$$

$$\frac{4}{1} \cdot \frac{3a}{4} + 4 \cdot 2 = \frac{4}{1} \cdot \frac{5}{4}$$

$$3a + 8 = 5$$

$$3a = -3$$

$$a = -1$$

17. $\frac{u}{2} - \frac{u}{4} = 2$ LCD = 4

$$4\left(\frac{u}{2} - \frac{u}{4}\right) = 4 \cdot 2$$

$$\frac{4}{1} \cdot \frac{u}{2} - \frac{4}{1} \cdot \frac{u}{4} = 4 \cdot 2$$

$$2u - u = 8$$

$$u = 8$$

19. $\frac{y}{3} + \frac{y}{5} < \frac{6}{5}$ LCD = 15

$$15\left(\frac{y}{3} + \frac{y}{5}\right) < 15 \cdot \frac{6}{5}$$

$$\frac{15}{1} \cdot \frac{y}{3} + \frac{15}{1} \cdot \frac{y}{5} < \frac{15}{1} \cdot \frac{6}{5}$$

$$5y + 3y < 18$$

$$8y < 18$$

$$y < \frac{9}{4}$$

21. $3x-\frac{2}{3}x=\frac{4}{3}$ LCD = 3

$$3\left(3x-\frac{2}{3}x\right)=3\cdot\frac{4}{3}$$

$$3\cdot 3x-\frac{3}{1}\cdot\frac{2x}{3}=\frac{3}{1}\cdot\frac{4}{3}$$

$$9x-2x=4$$

$$7x=4$$

$$x=\frac{4}{7}$$

23. $\frac{a}{4}-\frac{a}{3}\geq\frac{5}{2}$ LCD = 12

$$12\left(\frac{a}{4}-\frac{a}{3}\right)\geq 12\cdot\frac{5}{2}$$

$$\frac{12}{1}\cdot\frac{a}{4}-\frac{12}{1}\cdot\frac{a}{3}\geq\frac{12}{1}\cdot\frac{5}{2}$$

$$3a-4a\geq 30$$

$$-a\geq 30$$

$$a\leq -30$$

(Remember to reverse the inequality symbol when multiplying or dividing by a negative quantity.)

25. $.7x+.4x=5.5$ LCD = 10

$$10(.7x+.4x)=10(5.5)$$

$$10(.7x)+10(.4x)=10(5.5)$$

$$7x+4x=55$$

$$11x=55$$

$$x=5$$

27. $.3x-.25x=2$ LCD = 100

$$100(.3x-.25x)=100(2)$$

$$100(.3x)-100(.25x)=100(2)$$

$$30x-25x=200$$

$$5x=200$$

$$x=40$$

29. $.8m+.05m=.34$ LCD = 100

$$100(.8m+.05m)=100(.34)$$

$$100(.8m)+100(.05m)=100(.34)$$

$$80m+5m=34$$

$$85m=34$$

$$m=\frac{34}{85}=\frac{2\cdot 17}{5\cdot 17}=\frac{2}{5}$$

31. $\dfrac{w+3}{4} = \dfrac{w+4}{3}$ LCD = 12

$$\frac{\overset{3}{\cancel{12}}}{1}\cdot\frac{w+3}{\underset{1}{\cancel{4}}} = \frac{\overset{4}{\cancel{12}}}{1}\cdot\frac{w+4}{\underset{1}{\cancel{3}}}$$

$$3(w+3) = 4(w+4)$$

$$3w+9 = 4w+16$$

$$9 = w+16$$

$$-7 = w$$

33. $\dfrac{w+3}{4} + 1 = \dfrac{w+4}{3}$ LCD = 12

$$12\left(\frac{w+3}{4}+1\right) = 12\cdot\frac{w+4}{3}$$

$$\frac{\overset{3}{\cancel{12}}}{1}\cdot\frac{w+3}{\underset{1}{\cancel{4}}} + 12\cdot 1 = \frac{\overset{4}{\cancel{12}}}{1}\cdot\frac{w+4}{\underset{1}{\cancel{3}}}$$

$$3(w+3)+12 = 4(w+4)$$

$$3w+9+12 = 4w+16$$

$$3w+21 = 4w+16$$

$$21 = w+16$$

$$5 = w$$

35. $\dfrac{x+1}{2} + x = 6$ LCD = 2

$$2\left(\frac{x+1}{2}+x\right) = 2\cdot 6$$

$$\frac{\overset{1}{\cancel{2}}}{1}\cdot\frac{x+1}{\underset{1}{\cancel{2}}} + 2\cdot x = 2\cdot 6$$

$$x+1+2x = 12$$

$$3x+1 = 12$$

$$3x = 11$$

$$x = \frac{11}{3}$$

37. $\dfrac{y}{6} - \dfrac{y-2}{4} > 1$ LCD = 12

$$12\left(\frac{y}{6}-\frac{y-2}{4}\right) > 12\cdot 1$$

$$\frac{\overset{2}{\cancel{12}}}{1}\cdot\frac{y}{\underset{1}{\cancel{6}}} - \frac{\overset{3}{\cancel{12}}}{1}\cdot\frac{y-2}{\underset{1}{\cancel{4}}} > 12\cdot 1$$

$$2y-3(y-2) > 12$$

$$2y-3y+6 > 12$$

$$-y+6 > 12$$

$$-y > 6$$

$$y < -6$$

39. $3-\frac{a+4}{4}=\frac{a+4}{2}$ LCD = 4

$$4\left(3-\frac{a+1}{4}\right)=4\cdot\frac{a+4}{2}$$

$$4\cdot 3-\frac{4}{1}\cdot\frac{a+1}{4}=\frac{4}{1}\cdot\frac{a+4}{2}$$

$$12-(a+1)=2(a+4)$$

$$12-a-1=2a+8$$

$$11-a=2a+8$$

$$11=3a+8$$

$$3=3a$$

$$1=a$$

41. $\frac{2y-3}{2}-\frac{y-5}{3}=\frac{1}{6}$ LCD = 6

$$6\left(\frac{2y-3}{2}-\frac{y-5}{3}\right)=6\cdot\frac{1}{6}$$

$$\frac{6}{1}\cdot\frac{2y-3}{2}-\frac{6}{1}\cdot\frac{y-5}{3}=\frac{6}{1}\cdot\frac{1}{6}$$

$$3(2y-3)-2(y-5)=1$$

$$6y-9-2y+10=1$$

$$4y+1=1$$

$$4y=0$$

$$y=0$$

43. $\frac{x+2}{3}-\frac{2x+3}{4}=\frac{x+4}{8}$ LCD = 24

$$24\left(\frac{x+2}{3}-\frac{2x+3}{4}\right)=24\cdot\frac{x+4}{8}$$

$$\frac{24}{1}\cdot\frac{x+2}{3}-\frac{24}{1}\cdot\frac{2x+3}{4}=\frac{24}{1}\cdot\frac{x+4}{8}$$

$$8(x+2)-6(2x+3)=3(x+4)$$

$$8x+16-12x-18=3x+12$$

$$-4x-2=3x+12$$

$$-2=7x+12$$

$$-14=7x$$

$$-2=x$$

45. $\frac{t}{2}+\frac{t-1}{3}+\frac{t-6}{4}=t-2$ LCD = 12

$$12\left(\frac{t}{2}+\frac{t-1}{3}+\frac{t-6}{4}\right)=12(t-2)$$

$$\frac{12}{1}\cdot\frac{t}{2}-\frac{12}{1}\cdot\frac{t-1}{3}+\frac{12}{1}\cdot\frac{t-6}{4}=12(t-2)$$

$$6t + 4(t-1) + 3(t-6) = 12(t-2)$$
$$6t + 4t - 4 + 3t - 18 = 12t - 24$$
$$13t - 22 = 12t - 24$$
$$t - 22 = -24$$
$$t = -2$$

47. $.5(x+2) - .3(x-4) = 3 \quad \text{LCD} = 10$
$$10(.5(x+2) - .3(x-4)) = 10 \cdot 3$$
$$10(.5(x+2)) - 10(.3(x-4)) = 10 \cdot 3$$
$$5(x+2) - 3(x-4) = 30$$
$$5x + 10 - 3x + 12 = 30$$
$$2x + 22 = 30$$
$$2x = 8$$
$$x = 4$$

49. $3(y+2) + \frac{y+3}{5} = \frac{9y+8}{2} \quad \text{LCD} = 10$
$$10\left(3(y+2) + \frac{y+3}{5}\right) = 10 \cdot \frac{9y+8}{2}$$
$$10(3(y+2)) + \frac{\overset{2}{\cancel{10}}}{1} \cdot \frac{y+3}{\cancel{5}_1} = \frac{\overset{5}{\cancel{10}}}{1} \cdot \frac{9y+8}{\cancel{2}_1}$$
$$30(y+2) + 2(y+3) = 5(9y+8)$$
$$30y + 60 + 2y + 6 = 45y + 40$$
$$32y + 66 = 45y + 40$$
$$66 = 13y + 40$$
$$26 = 13y$$
$$2 = y$$

51. $z + \frac{z+5}{3} - \frac{z-2}{6} = \frac{z+4}{4} + 1 \quad \text{LCD} = 12$
$$12\left(z + \frac{z+5}{3} - \frac{z-2}{6}\right) = 12\left(\frac{z+4}{4} + 1\right)$$
$$12 \cdot z + \frac{\overset{4}{\cancel{12}}}{1} \cdot \frac{z+5}{\cancel{3}} - \frac{\overset{2}{\cancel{12}}}{1} \cdot \frac{z-2}{\cancel{6}_1} = \frac{\overset{3}{\cancel{12}}}{1} \cdot \frac{z+4}{\cancel{4}_1} + 12 \cdot 1$$
$$12z + 4(z+5) - 2(z-2) = 3(z+4) + 12$$
$$12z + 4z + 20 - 2z + 4 = 3z + 12 + 12$$
$$14z + 24 = 3z + 24$$
$$11z + 24 = 24$$
$$11z = 0$$
$$z = 0$$

$$3 \le \frac{x}{3} - \frac{x+1}{2} \le 6 \quad \text{LCD} = 6$$

$$6 \cdot 3 \le 6\left(\frac{x}{3} - \frac{x+1}{2}\right) \le 6 \cdot 6$$

$$18 \le \frac{6}{1} \cdot \frac{x}{3} - \frac{6}{1} \cdot \frac{x+1}{2} \le 36$$

$$18 \le 2x - 3(x+1) \le 36$$

$$18 \le 2x - 3x - 3 \le 36$$

$$18 \le -x - 3 \le 36$$

$$21 \le -x \le 39$$

$$-21 \ge x \ge -39$$

55. LCD of 3, 2, and 5 is 30. So

$$\frac{x}{3} + \frac{x}{2} + \frac{x}{5} = \frac{x(10)}{3(10)} + \frac{x(15)}{2(15)} + \frac{x(6)}{5(6)}$$

$$= \frac{10x}{30} + \frac{15x}{30} + \frac{6x}{30}$$

$$= \frac{10x + 15x + 6x}{30} = \frac{31x}{30}$$

57. $\frac{x}{3} + \frac{x}{2} + \frac{x}{5} = 62 \quad \text{LCD} = 30$

$$30\left(\frac{x}{3} + \frac{x}{2} + \frac{x}{5}\right) = 30 \cdot 62$$

$$\frac{30}{1} \cdot \frac{x}{3} + \frac{30}{1} \cdot \frac{x}{2} + \frac{30}{1} \cdot \frac{x}{5} = 1860$$

$$10x + 15x + 6x = 1860$$

$$31x = 1860$$

$$x = 60$$

59. $\frac{x+5}{2} - \frac{x-1}{4} = 2 \quad \text{LCD} = 4$

$$4\left(\frac{x+5}{2} - \frac{x-1}{4}\right) = 4 \cdot 2$$

$$\frac{4}{1} \cdot \frac{x+5}{2} - \frac{4}{1} \cdot \frac{x-1}{4} = 4 \cdot 2$$

$$2(x+5) - (x-1) = 8$$

$$2x + 10 - x + 1 = 8$$

$$x + 11 = 8$$

$$x = -3$$

61. LCD of 2 and 4 is 4. So

$$\frac{x+5}{2}-\frac{x-1}{4}$$
$$=\frac{(x+5)(2)}{2(2)}-\frac{x-1}{4}$$
$$=\frac{2(x+5)}{4}-\frac{x-1}{4}$$
$$=\frac{2(x+5)-(x-1)}{4}$$
$$=\frac{2x+10-x+1}{4}=\frac{x+11}{4}$$

63. A shipment of wooden boxes contains 80 pieces altogether. If oak boxes sell for \$30 apiece and mahogany boxes sell for \$50 apiece, how many of each type are in the shipment if its total value is \$3600?

Let x = number of oak boxes

65. Let x = weight of package

$$7+4(x-1)=43$$
$$7+4x-4=43$$
$$4x+3=43$$
$$4x=40$$
$$x=10$$

The package weighs 10 lbs.

Exercises 4.5

1. $\frac{\#\text{ red}}{\#\text{ black}}=\frac{7}{5}$

3. $\frac{\#\text{ black}}{\#\text{ red}}=\frac{5}{7}$

5. $\frac{\#\text{ with}}{\#\text{ without}}=\frac{11}{5}$

7. $\frac{\overset{1}{\cancel{x}}}{3\cancel{x}}=\frac{1}{3}$

9. $\frac{a}{b+c}$

11.
$$\frac{x}{5}=\frac{12}{3}$$
$$15\cdot\frac{x}{5}=15\cdot\frac{12}{3}$$
$$\frac{15}{1}\cdot\frac{x}{5}=\frac{15}{1}\cdot\frac{12}{3}$$
$$3x=60$$
$$x=20$$

13.
$$\frac{a}{6}=\frac{5}{3}$$
$$6\cdot\frac{a}{6}=6\cdot\frac{5}{3}$$
$$\frac{6}{1}\cdot\frac{a}{6}=\frac{6}{1}\cdot\frac{5}{3}$$
$$a=10$$

15.
$$\frac{y}{15}=\frac{20}{6}$$
$$30\cdot\frac{y}{15}=30\cdot\frac{20}{6}$$
$$\frac{30}{1}\cdot\frac{y}{15}=\frac{30}{1}\cdot\frac{20}{6}$$
$$2y=100$$
$$y=50$$

17.
$$\frac{y}{9}=\frac{4}{3}$$
$$9\cdot\frac{y}{9}=9\cdot\frac{4}{3}$$
$$\frac{9}{1}\cdot\frac{y}{9}=\frac{9}{1}\cdot\frac{4}{3}$$
$$y=12$$

19. Let x = number of red marbles in the jar.
$$\frac{x}{210}=\frac{7}{5}$$
$$210\cdot\frac{x}{210}=210\cdot\frac{7}{5}$$
$$\frac{210}{1}\cdot\frac{x}{210}=\frac{210}{1}\cdot\frac{7}{5}$$
$$x=294$$
There are 294 red marbles in the jar.

21. Let x = number of students who got 90 or above.

$$\frac{x}{24} = \frac{3}{8}$$

$$24 \cdot \frac{x}{24} = 24 \cdot \frac{3}{8}$$

$$\frac{24}{1} \cdot \frac{x}{24} = \frac{24}{1} \cdot \frac{3}{8}$$

$$x = 9$$

There were 9 students who got 90 or above.

23. Let W = width of the rectangle.

$$\frac{18}{W} = \frac{9}{4}$$

$$4W \cdot \frac{18}{W} = 4W \cdot \frac{9}{4}$$

$$\frac{4W}{1} \cdot \frac{18}{W} = \frac{4W}{1} \cdot \frac{9}{4}$$

$$72 = 9W$$

$$8 = W$$

The width of the rectangle is 8 cm.

25. $\frac{\text{shorter side}}{\text{longer side}} = \frac{6x}{15x} = \frac{2}{5}$

27. Let x = length of scale drawing of the 20-meter wall (in cm).

$$\frac{x}{20} = \frac{5}{12}$$

$$60 \cdot \frac{x}{20} = 60 \cdot \frac{5}{12}$$

$$\frac{60}{1} \cdot \frac{x}{20} = \frac{60}{1} \cdot \frac{5}{12}$$

$$3x = 25$$

$$x = \frac{25}{3} = 8.33$$

The scale drawing of the 20-meter wall is 8.33 cm.

9. Let x = number of kilograms in 10 lbs.

$$\frac{10}{x} = \frac{2.2}{1}$$

$$x \cdot \frac{10}{x} = x \cdot \frac{2.2}{1}$$

$$\frac{\cancel{x}}{1} \cdot \frac{10}{\cancel{x}} = 2.2x$$

$$10 = 2.2x$$

$$\frac{10}{2.2} = x$$

$$4.55 = x$$

There are 4.55 kilograms in 10 lbs.

31. Let x = number of yards in 100 meters.

$$\frac{100}{x} = \frac{.92}{1}$$

$$x \cdot \frac{100}{x} = x \cdot \frac{.92}{1}$$

$$\frac{\cancel{x}}{1} \cdot \frac{100}{\cancel{x}} = .92x$$

$$100 = .92x$$

$$\frac{100}{.92} = x$$

$$108.70 = x$$

There are 108.70 yards in 100 meters.

33. Let x = distance between the two cities (in miles).

$$\frac{12.6}{x} = \frac{5}{10}$$

$$10\cancel{x} \cdot \frac{12.6}{\cancel{x}} = \cancel{10}x \cdot \frac{5}{\cancel{10}}$$

$$126 = 5x$$

$$25.2 = x$$

The distance between the two cities is 25.2 miles.

35. Let x = Rolfe's share of the profits (in dollars).

$$\frac{18,500}{2,800} = \frac{12,800}{x}$$

$$\frac{185}{28} = \frac{12,800}{x}$$

$$\cancel{28}x \cdot \frac{185}{\cancel{28}} = 28\cancel{x} \cdot \frac{12,800}{\cancel{x}}$$

$$185x = 358,400$$

$$x = 1,937.30$$

Rolfe's share of the profits was $1,937.30.

37. Let x = number of untagged deer in the entire area.

$$\frac{48}{x} = \frac{18}{42}$$

(42 = 60 – 18 untagged deer in the second group.)

$$42x \cdot \frac{48}{x} = 42x \cdot \frac{18}{42}$$

$$\frac{42x}{1} \cdot \frac{48}{x} = \frac{42x}{1} \cdot \frac{18}{42}$$

$$2016 = 18x$$

$$112 = x$$

Then the total number of deer estimated in the entire area is 48 + 112 = 160.

39.

$$\frac{x}{6} = \frac{18}{9}$$

$$18 \cdot \frac{x}{6} = 18 \cdot \frac{18}{9}$$

$$3x = 36$$

$$x = 12$$

$$\frac{y}{8} = \frac{9}{6}$$

$$24 \cdot \frac{y}{8} = 24 \cdot \frac{9}{6}$$

$$3y = 36$$

$$y = 12$$

41.

$$\frac{x}{12.5} = \frac{5.2}{8.4}$$

$$(8.4)(12.5) \cdot \frac{x}{12.5} = (8.4)(12.5) \cdot \frac{5.2}{8.4}$$

$$8.4x = 65$$

$$x = 7.74$$

$$\frac{y}{8.4} = \frac{9.6}{12.5}$$

$$(8.4)(12.5) \cdot \frac{y}{8.4} = (8.4)(12.5) \cdot \frac{9.6}{12.5}$$

$$12.5y = 80.64$$

$$y = 6.45$$

43. Let x = height of the street light (in feet).

$$\frac{6}{8} = \frac{x}{23}$$

$$184 \cdot \frac{6}{8} = 184 \cdot \frac{x}{23}$$

$$138 = 8x$$

$$17.25 = x$$

The height of the street light is 17.25 feet.

45. $x = .12$

47. $t = 11.59$

49. $y = -2.05$

Exercises 4.6

1. Let x = the number.

$$\frac{2}{3}x + 5 = 9$$

$$3\left(\frac{2}{3}x + 5\right) = 3\cdot 9$$

$$\frac{3}{1}\cdot\frac{2}{3}x + 3\cdot 5 = 3\cdot 9$$

$$2x + 15 = 27$$

$$2x = 12$$

$$x = 6$$

Thus, the number is 6.

CHECK: $\frac{2}{3}$ of 6 is 4, and 5 more than 4 is 9.

3. Let x = the number.

$$\frac{3}{4}x - 2 = \frac{1}{8}x - 7$$

$$8\left(\frac{3}{4}x - 2\right) = 8\left(\frac{1}{8}x - 7\right)$$

$$\frac{8}{1}\cdot\frac{3}{4}x - 8\cdot 2 = \frac{8}{1}\cdot\frac{1}{8}x - 8\cdot 7$$

$$6x - 16 = x - 56$$

$$5x - 16 = -56$$

$$5x = -40$$

$$x = -8$$

Thus, the number is –8.

CHECK: 2 less than $\frac{3}{4}$ of –8 is 2 less than –6 or –8. $\frac{1}{8}$ of –8 is –1, and –8 is 7 less than –1.

5. Let L = length of the rectangle (in meters).

$\frac{1}{2}L$ = width of the rectangle (in meters).

$$2(L) + 2\left(\frac{1}{2}L\right) = 36$$

$$2L + \frac{2}{1}\cdot\frac{1}{2}L = 36$$

$$2L + L = 36$$

$$3L = 36$$

$$L = 12$$

Then $\frac{1}{2}L = \frac{1}{2}(12) = 6$.

Thus, the dimensions of the rectangle are 12 meters and 6 meters.

CHECK: The width of the $6\text{ m} \times 12\text{ m}$ rectangle is half its length. The perimeter of this rectangle is 2(6) + 2(12) = 12 + 24 = 36 meters.

7. Let x = length of the longest side of the triangle (in inches).

$\frac{3}{4}x$ = length of the medium side of the triangle (in inches).

$\frac{1}{2}\left(\frac{3}{4}x\right)$ = length of the shortest side of the triangle (in inches).

$$x + \frac{3}{4}x + \frac{1}{2}\left(\frac{3}{4}x\right) = 17$$

$$x + \frac{3}{4}x + \frac{3}{8}x = 17$$

$$8\left(x + \frac{3}{4}x + \frac{3}{8}x\right) = 8\cdot 17$$

$$8\cdot x + \frac{8}{1}\cdot\frac{3}{4}x + \frac{8}{1}\cdot\frac{3}{8}x = 8\cdot 17$$

$$8x + 6x + 3x = 136$$

$$17x = 136$$

$$x = 8$$

Then $\frac{3}{4}x = \frac{3}{4}(8) = 6$ and $\frac{1}{2}\left(\frac{3}{4}x\right) = \frac{1}{2}\left(\frac{3}{4}(8)\right) = 3$. Thus, the sides of the triangle are 8 inches, 6 inches, and 3 inches.

CHECK: 6 is $\frac{3}{4}$ of 8, and 3 is $\frac{1}{2}$ of 6. The sum of 8, 6, and 3 is 17.

9. Let x = number of regular tickets sold. 350 – x = number of combination tickets sold.

$$15x + 22(350 - x) = 6895$$

$$15x + 7700 - 22x = 6895$$

$$-7x + 7700 = 6895$$

$$-7x = -805$$

$$x = 115$$

Then 350 – x = 350 – 115 = 235. Thus, there were 115 regular tickets and 235 combination tickets sold.

CHECK: 115 regular tickets at \$15 each gives \$1725. 235 combination tickets at \$22 each gives \$5170. The total amount collected is \$1725 + \$5170 = \$6895.

11. Let x = number of quarters in the collection. 2x + 3 = number of dimes in the collection.

$$\begin{aligned} 25x + 10(2x + 3) &= 255 \\ 25x + 20x + 30 &= 255 \\ 45x + 30 &= 255 \\ 45x &= 225 \\ x &= 5 \end{aligned}$$

Then 2x + 3 = 2(5) + 3 = 13. Thus, there are 5 quarters and 13 dimes in the collection.

CHECK: 13 is 3 more than twice 5. 5 quarters are worth 5(\$.25) = \$1.25, while 13 dimes are worth 13(\$.10) = \$1.30. Total value = \$1.25 + \$1.30 = \$2.55.

13. Let t = number of minutes that the older machine works. t – 15 = number of minutes that the newer machine works.

$$\begin{aligned} 175t + 250(t - 15) &= 13675 \\ 175t + 250t - 3750 &= 13675 \\ 425t - 3750 &= 13675 \\ 425t &= 17425 \\ t &= 41 \end{aligned}$$

Since the older machine began the sorting process at 10:00 a.m. and worked for 41 minutes, the sorting is completed at 10:41 a.m.

CHECK: The older machine sorts 175(41) = 7175 screws, while the newer one sorts 250(41 – 15) = 250(26) = 6500 screws. In all, 7175 + 6500 = 13,675 screws are sorted.

15. Let x = amount of money invested at 8%. x + 4000 = amount of money invested at 11%.

$$\begin{aligned} .08x + .11(x + 4000) &= 1390 \\ 100(.08x + .11(x + 4000)) &= 100(1390) \\ 100(.08x) + 100(.11(x + 4000)) &= 100(1390) \\ 8x + 11(x + 4000) &= 139000 \\ 8x + 11x + 44000 &= 139000 \\ 19x + 44000 &= 139000 \\ 19x &= 95000 \\ x &= 5000 \end{aligned}$$

Then x + 4000 = 5000 + 4000 = 9000. Thus, \$9000 was invested at 11%.

CHECK: \$5000 at 8% yields \$400 interest. \$9000 at 11% yields \$990 interest. The total interest earned is \$400 + \$990 = \$1390.

17. Let x = amount of money invested at 9%. 800 – x = amount of money invested at 6%.

$$\begin{aligned}
.09x+.06(800-x) &= 67.50\\
100(.09x+.06(800-x)) &= 100(67.50)\\
100(.09x)+100(.06(800-x)) &= 100(67.50)\\
9x+6(800-x) &= 6750\\
9x+4800-6x &= 6750\\
3x+4800 &= 6750\\
3x &= 1950\\
x &= 650
\end{aligned}$$

Then 800 – x = 800 – 650 = 150. Thus, $650 was invested at 9% and $150 was invested at 6%.

CHECK: The interest on the 9% investment is .09($650) = $58.50. The interest on the 6% investment is .06($150) = $9.00. Total interest = $58.50 + $9.00 = $67.50.

19. Let x = amount of money invested at 8%. 6000 – x = amount of money invested at 12%.

$$\begin{aligned}
.08x+.12(6000-x) &= .09(6000)\\
100(.08x+.12(6000-x)) &= 100(.09(6000))\\
100(.08x)+100(.12(6000-x)) &= 100(.09(6000))\\
8x+12(6000-x) &= 9(6000)\\
8x+72000-12x &= 54000\\
-4x+72000 &= 54000\\
-4x &= -18000\\
x &= 4500
\end{aligned}$$

Then 6000 – x = 6000 – 4500 = 1500. Thus, $4500 should be invested at 12%.

CHECK: The interest on the 8% investment is .08($4500) = $360. The interest on the 12% investment is .12($1500) = $180. The total interest $360 + $180 = $540 is 9% of $6000.

21. Let x = number of ml of 30% hydrochloric acid solution.

$$\begin{aligned}
.30x+.50(30) &= .45(x+30)\\
100(.30x+.50(30)) &= 100(.45(x+30))\\
100(.30x)+100(.50(30)) &= 100(.45(x+30))\\
30x+50(30) &= 45(x+30)\\
30x+1500 &= 45x+1350\\
1500 &= 15x+1350\\
150 &= 15x\\
10 &= x
\end{aligned}$$

Thus, 10 ml of 30% hydrochloric acid solution should be used in the mixture.

CHECK: In 10 ml of a 30% solution, there are 3 ml of acid. In 30 ml of a 50% solution, there are 15 ml of acid. So there are 3 + 15 = 18 ml of acid in the mixture, which is 45% of 40 ml.

23. Let x = number of liters of 25% salt solution. 90 – x = number of liters of 55% salt solution.

$$
\begin{aligned}
.25x + .55(90 - x) &= .50(90) \\
100(.25x + .55(90 - x)) &= 100(.50(90)) \\
100(.25x) + 100(.55(90 - x)) &= 100(.50(90)) \\
25x + 55(90 - x) &= 50(90) \\
25x + 4950 - 55x &= 4500 \\
-30x + 4950 &= 4500 \\
-30x &= -450 \\
x &= 15
\end{aligned}
$$

Then 90 – x = 90 – 15 = 75. Thus, 15 liters of the 25% solution should be mixed with 75 liters of the 55% solution.

CHECK: In 15 liters of a 25% salt solution, there are .25(15) = 3.75 liters of salt. In 75 liters of a 55% salt solution, there are .55(75) = 41.25 liters of salt. In the mixture, there are 3.75 + 41.25 = 45 liters of salt, which is 50% of 90 liters.

25. Let x = number of gallons of pure anti-freeze to be used.

$$
\begin{aligned}
1.00x + .30(10) &= .50(x + 10) \\
100(1.00x + .30(10)) &= 100(.50(x + 10)) \\
100(1.00x) + 100(.30(10)) &= 100(.50(x + 10)) \\
100x + 30(10) &= 50(x + 10) \\
100x + 300 &= 50x + 500 \\
50x + 300 &= 500 \\
50x &= 200 \\
x &= 4
\end{aligned}
$$

Thus, 4 gallons of pure anti-freeze should be added to the radiator.

CHECK: 10 gallons of a 30% anti-freeze solution contains 3 gallons of anti-freeze. If we add 4 gallons of anti-freeze we get 14 gallons in the solution, 7 of which are anti-freeze. This is 50%.

27. Let x = number of pounds of candy that sells for \$3.75/lb.

$$
\begin{aligned}
3.75x + 5.00(35) &= 4.25(x + 35) \\
100(3.75x + 5.00(35)) &= 100(4.25(x + 35)) \\
100(3.75x) + 100(5.00(35)) &= 100(4.25(x + 35)) \\
375x + 500(35) &= 425(x + 35) \\
375x + 17500 &= 425x + 14875 \\
17500 &= 50x + 14875 \\
2625 &= 50x \\
52.5 &= x
\end{aligned}
$$

Thus, 52.5 pounds of candy that sells for \$3.75/lb should be added to the mixture.

CHECK: 52.5 lbs. at \$3.75/lb = \$196.87$\frac{1}{2}$. 35 lbs. at \$5/lb = \$175. Total = \$371.87$\frac{1}{2}$. 87.5 lbs at \$4.25/lb = \$371.87$\frac{1}{2}$.

29. Let t = number of hours that John and Susan must travel before they meet.

$4t + 8t = 9$

$12t = 9$

$t = \frac{9}{12} = \frac{3}{4}$

Since each must travel for $\frac{3}{4}$ of an hour or for 45 minutes, they will meet at 8:45 a.m.

CHECK: John walks $4\left(\frac{3}{4}\right) = 3$ miles and Susan jogs $8\left(\frac{3}{4}\right) = 6$ miles. $3 + 6 = 9$.

31. Let t = number of hours that John must travel before they meet. $t - \frac{1}{4}$ = number of hours that Susan must travel before they meet.

$4t + 8\left(t - \frac{1}{4}\right) = 9$

$4t + 8t - 8 \cdot \frac{1}{4} = 9$

$12t - 2 = 9$

$12t = 11$

$t = \frac{11}{12}$

Since John must travel $\frac{11}{12}$ hours or 55 minutes before he meets Susan, they will meet at 8:40 a.m.

CHECK: John walks $4\left(\frac{11}{12}\right) = \frac{11}{3} = 3\frac{2}{3}$ miles and Susan jogs $8\left(\frac{11}{12} - \frac{1}{4}\right) = 8\left(\frac{8}{12}\right) = \frac{16}{3} = 5\frac{1}{3}$ miles. $3\frac{2}{3} + 5\frac{1}{3} = 9$.

33. Let t = number of hours that David spends jogging. $2 - t$ = number of hours that David spends walking.

$5(2 - t) + 9t = 16$

$10 - 5t + 9t = 16$

$10 + 4t = 16$

$4t = 6$

$t = \frac{6}{4} = \frac{3}{2}$

Thus, David jogs for $1\frac{1}{2}$ hours.

CHECK: David jogs $9\left(\frac{3}{2}\right) = \frac{27}{2} = 13\frac{1}{2}$ miles and walks $5\left(\frac{1}{2}\right) = \frac{5}{2} = 2\frac{1}{2}$ miles. $13\frac{1}{2} + 2\frac{1}{2} = 16$.

5. Let r = rate at which additional money is invested.

$$.07(3200) + r(2800) = 476$$
$$224 + 2800r = 476$$
$$2800r = 252$$
$$r = \frac{252}{2800}$$
$$r = .09$$

Sal must invest the additional $2,800 at a 9% rate.

CHECK: An investment of $3,200 at 7% pays $224 interest. An investment of $2,800 at 9% pays $252 interest. $224 + $252 = $476.

37. Let p = percent solution for additional hydrochloric acid.

$$.20(30) + p(60) = .40(90)$$
$$6 + 60p = 36$$
$$60p = 30$$
$$p = \frac{30}{60}$$
$$p = .50$$

The chemist must use a 50% hydrochloric acid solution.

CHECK: 30 ml of a 20% solution contains 6 ml of acid. 60 ml of a 50% solution contains 30 ml of acid. 90 ml of a 40% solution contains 36 ml of acid, and

6 + 30 = 36.

39. Let w = width of the rectangle. 2w – 1 = length of the rectangle.

$$w = \frac{1}{7}(2(w) + 2(2w - 1)) + 1$$
$$w = \frac{1}{7}(2w + 4w - 2) + 1$$
$$7w = 7\left(\frac{1}{7}(6w - 2) + 1\right)$$
$$7w = 6w - 2 + 7$$
$$7w = 6w + 5$$
$$w = 5$$

Then 2w – 1 = 2(5) – 1 = 9. So the rectangle has a width of 5 and a length of 9.

CHECK: 9 is 1 less than twice 5. The perimeter of a 5×9 rectangle is 2(5) + 2(9) = 28, and 5 is 1 more than $\frac{1}{7}(28) = 4$.

41. $C = \frac{5}{9}(F - 32)$, where C is the temperature in degrees Celsius and F is the temperature in degrees Fahrenheit.

$$25 < C < 40$$
$$25 < \frac{5}{9}(F - 32) < 40$$
$$9(25) < 9\left(\frac{5}{9}(F - 32)\right) < 9(40)$$
$$225 < 5(F - 32) < 360$$
$$45 < F - 32 < 72$$
$$77 < F < 104$$

Thus, the corresponding temperature range in degrees Fahrenheit is between 77°F and 104°F.

43. Let x = original price of a skirt. .80x = sale price of the skirt.

$$12.60 < .80x < 20.76$$
$$100(12.60) < 100(.80x) < 100(20.76)$$
$$1260 < 80x < 2076$$
$$15.75 < x < 25.95$$

Thus, the original range of prices on the skirts was between \$15.75 and \$25.95.

Chapter 4 Review Exercises

1. $\frac{-18}{42} = \frac{-3 \cdot \cancel{6}}{7 \cdot \cancel{6}} = \frac{-3}{7} = -\frac{3}{7}$

3. $\frac{15x^6}{6x^2} = \frac{5x^4 \cdot \cancel{3x^2}}{2 \cdot \cancel{3x^2}} = \frac{5x^4}{2}$

5. $\frac{-10x^3y^5}{4xy^{10}} = \frac{-5x^2 \cdot \cancel{2xy^5}}{2y^5 \cdot \cancel{2xy^5}} = \frac{-5x^2}{2y^5}$

$$= -\frac{5x^2}{2y^5}$$

7. $\frac{3t - 7t - t}{-2t^2 - 3t^2} = \frac{-5t}{-5t^2} = \frac{1\cancel{(-5t)}}{t\cancel{(-5t)}}$

$$= \frac{1}{t}$$

9. $\frac{a}{4} \cdot \frac{a}{4} = \frac{a^2}{16}$

11. $\frac{7a}{6} - \frac{5a}{6} = \frac{7a - 5a}{6} = \frac{\overset{1}{\cancel{2}}a}{\cancel{6}\,3}$

$$= \frac{a}{3}$$

13. $\frac{4x-3}{6x}-\frac{x-1}{6x}$

$=\frac{4x-3-(x-1)}{6x}$

$=\frac{4x-3-x+1}{6x}=\frac{3x-2}{6x}$

15. $\frac{2y^2-3y}{4}-\frac{y^2-3y}{4}+\frac{y^2}{4}$

$=\frac{2y^2-3y-(y^2-3y)+y^2}{4}$

$=\frac{2y^2-3y-y^2+3y+y^2}{4}$

$=\frac{2y^2}{4}=\frac{y^2}{2}$

17. $\left(\frac{x^2}{4}\cdot\frac{6}{xy^2}\right)\div(2xy)=\frac{3x}{2y^2}\div\frac{2xy}{1}$

$=\frac{3x}{2y^2}\cdot\frac{1}{2xy}=\frac{3}{4y^3}$

19. $\frac{a}{2}\cdot\frac{a}{4}=\frac{a^2}{8}$

21. $\frac{x^2}{2}-\frac{x^2}{6}+\frac{x^2}{3}$

$=\frac{x^2(3)}{2(3)}-\frac{x^2}{6}+\frac{x^2(2)}{3(2)}$

$=\frac{3x^2}{6}-\frac{x^2}{6}+\frac{2x^2}{6}$

$=\frac{3x^2-x^2+2x^2}{6}=\frac{4x^2}{6}=\frac{2x^2}{3}$

23. $\frac{4}{x^2}+\frac{3}{2x}=\frac{4(2)}{x^2(2)}+\frac{3(x)}{2x(x)}$

$=\frac{8}{x^2}+\frac{3x}{2x^2}=\frac{8+3x}{2x^2}$

25. $\frac{3}{4a^2b}-\frac{5}{6ab}+\frac{7}{8b^3}=\frac{3(6b^2)}{4a^2b(6b^2)}-\frac{5(4ab^2)}{6ab(4ab^2)}+\frac{7(3a^2)}{8b^3(3a^2)}$

$=\frac{18b^2}{24a^2b^3}-\frac{20ab^2}{24a^2b^3}+\frac{21a^2}{24a^2b^3}$

$=\frac{18b^2-20ab^2+21a^2}{24a^2b^3}$

27.
$$\frac{x}{6}-\frac{1}{4}=\frac{7}{12}$$
$$12\left(\frac{x}{6}-\frac{1}{4}\right)=12\left(\frac{7}{12}\right)$$
$$\frac{\overset{2}{\cancel{12}}}{1}\cdot\frac{x}{\underset{1}{\cancel{6}}}-\frac{\overset{3}{\cancel{12}}}{1}\cdot\frac{1}{\underset{1}{\cancel{4}}}=\frac{\overset{1}{\cancel{12}}}{1}\cdot\frac{7}{\underset{1}{\cancel{12}}}$$
$$2x-3=7$$
$$2x=10$$
$$x=5$$

29.
$$\frac{t+1}{2}+\frac{t+2}{3}<\frac{t+7}{6}$$
$$6\left(\frac{t+1}{2}+\frac{t+2}{3}\right)<6\left(\frac{t+7}{6}\right)$$
$$\frac{\overset{3}{\cancel{6}}}{1}\cdot\frac{t+1}{\underset{1}{\cancel{2}}}+\frac{\overset{2}{\cancel{6}}}{1}\cdot\frac{t+2}{\underset{1}{\cancel{3}}}<\frac{\overset{1}{\cancel{6}}}{1}\cdot\frac{t+7}{\underset{1}{\cancel{6}}}$$
$$3(t+1)+2(t+2)<t+7$$
$$3t+3+2t+4<t+7$$
$$5t+7<t+7$$
$$4t+7<7$$
$$4t<0$$
$$t<0$$

31.
$$\frac{y+3}{5}-\frac{y-2}{3}=1$$
$$15\left(\frac{y+3}{5}-\frac{y-2}{3}\right)=15(1)$$
$$\frac{\overset{3}{\cancel{15}}}{1}\cdot\frac{y+3}{\underset{1}{\cancel{5}}}-\frac{\overset{5}{\cancel{15}}}{1}\cdot\frac{y-2}{\underset{1}{\cancel{3}}}=15$$
$$3(y+3)-5(y-2)=15$$
$$3y+9-5y+10=15$$
$$-2y+19=15$$
$$-2y=-4$$
$$y=2$$

33.
$$\frac{x}{3}=\frac{x+1}{6}$$
$$6\cdot\frac{x}{3}=6\cdot\frac{x+1}{6}$$
$$\frac{\overset{2}{\cancel{6}}}{1}\cdot\frac{x}{\underset{1}{\cancel{3}}}=\frac{\overset{1}{\cancel{6}}}{1}\cdot\frac{x+1}{\underset{1}{\cancel{6}}}$$
$$2x=x+1$$
$$x=1$$

35. $$\begin{aligned} 2x+.2(x+6) &= 10 \\ 10(2x+.2(x+6)) &= 10\cdot 10 \\ 10(2x)+10(.2(x+6)) &= 100 \\ 20x+2(x+6) &= 100 \\ 20x+2x+12 &= 100 \\ 22x+12 &= 100 \\ 22x &= 88 \\ x &= 4 \end{aligned}$$

37. Let x = number of ounces in 1 kilogram (1000 grams).

$$\frac{1000}{x} = \frac{28.4}{1}$$

$$x\cdot\frac{1000}{x} = x\cdot\frac{28.4}{1}$$

$$1000 = 28.4x$$

$$\frac{1000}{28.4} = x$$

35.21 = x (to the neareast hundredth). Thus, there are 35.21 ounces in 1 kilogram.

39. Let x = amount invested at 6%

2x = amount invested at 7%

7000 – 3x = amount invested at 8%

$$\begin{aligned} .06x+.07(2x)+.08(7000-3x) &\geq 500 \\ 100(.06x+.07(2x)+.08(7000-3x)) &\geq 100(500) \\ 100(.06x)+100(.07(2x))+100(.08(7000-3x)) &\geq 50000 \\ 6x+7(2x)+8(7000-3x) &\geq 50000 \\ 6x+14x+56000-24x &\geq 50000 \\ -4x+56000 &\geq 50000 \\ -4x &\geq -6000 \\ x &\leq 1500 \end{aligned}$$

Thus, the most that can be invested at 6% is $1500.

41. Let x = Bill's present speed

5x = 3(x + 20)

5x = 3x + 60

2x = 60

x = 30

Thus, Bill's present speed is 30 mph.

CHECK: After 5 hours, Bill covered (30)(5) = 150 miles. In 3 hours, he would cover (50)(3) = 150 miles at the faster rate as well.

Chapter 4 Practice Test

1. (a) $\dfrac{-10}{24} = \dfrac{2(-5)}{2(12)} = \dfrac{-5}{12} = -\dfrac{5}{12}$

(b) $\dfrac{x^{10}}{x^2} = \dfrac{x^8}{1} = x^8$

(c) $\dfrac{6a^6}{3a^3} = \dfrac{2a^3}{1} = 2a^3$

(d) $\dfrac{25r^2t^3}{-15r^4t} = \dfrac{5t^2}{-3r^2} = -\dfrac{5t^2}{3r^2}$

3. (a)
$$\frac{x}{3} + \frac{x}{5} = 8$$
$$15\left(\frac{x}{3} + \frac{x}{5}\right) = 15 \cdot 8$$
$$\frac{15}{1} \cdot \frac{x}{3} + \frac{15}{1} \cdot \frac{x}{5} = 120$$
$$5x + 3x = 120$$
$$8x = 120$$
$$x = 15$$

(b)
$$\frac{x-5}{2} + \frac{x}{5} \geq 8$$
$$10\left(\frac{x-5}{2} + \frac{x}{5}\right) \geq 10 \cdot 8$$
$$\frac{10}{1} \cdot \frac{x-5}{2} + \frac{10}{1} \cdot \frac{x}{5} \geq 80$$
$$5(x-5) + 2x \geq 80$$
$$5x - 25 + 2x \geq 80$$
$$7x - 25 \geq 80$$
$$+25 \quad +25$$
$$7x \geq 105$$
$$\frac{7x}{7} \geq \frac{105}{7}$$
$$x \geq 15$$

(c)
$$\frac{a+3}{5}-\frac{a-2}{4}=1$$
$$20\left(\frac{a+3}{5}-\frac{a-2}{4}\right)=20\cdot 1$$
$$\frac{\overset{4}{\cancel{20}}}{1}\cdot\frac{a+3}{\cancel{5}_1}-\frac{\overset{5}{\cancel{20}}}{1}\cdot\frac{a-2}{\cancel{4}_1}=20$$
$$4(a+3)-5(a-2)=20$$
$$4a+12-5a+10=20$$
$$-a+22=20$$
$$\underline{\quad -22 \quad -22}$$
$$-a=-2$$
$$\frac{-a}{-1}=\frac{-2}{-1}$$
$$a=2$$

(d)
$$.03t+.5t=10.6$$
$$100(.03t+.5t)=100(10.6)$$
$$100(.03t)+100(.5t)=100(10.6)$$
$$3t+50t=1060$$
$$53t=1060$$
$$\frac{\cancel{53}t}{\cancel{53}}=\frac{1060}{53}$$
$$t=20$$

5. Let x = the number
$$x+\frac{2}{3}x=2x-5$$
$$3\left(x+\frac{2}{3}x\right)=3(2x-5)$$
$$3x+\frac{\cancel{3}}{1}\cdot\frac{2}{\cancel{3}_1}x=3(2x-5)$$
$$3x+2x=6x-15$$
$$5x=6x-15$$
$$\underline{\quad -6x \quad -6x}$$
$$-x=-15$$
$$\frac{-x}{-1}=\frac{-15}{-1}$$
$$x=15$$

CHECK: $\frac{2}{3}$ of 15 $=\frac{2}{\cancel{3}}\cdot\frac{\overset{5}{\cancel{15}}}{1}=\frac{10}{1}$ = 10. Then 15 + 10 = 25, which is 5 less than twice 15.

7. Let x = amount invested at 8%. 7000 – x = amount invested at 13%.

$$.08x+.13(7000-x)=750$$
$$100(.08x+.13(7000-x))=100(750)$$
$$100(.08x)+100(.13(7000-x))=100(750)$$
$$8x+13(7000-x)=75000$$
$$8x+91000-13x=75000$$
$$-5x+91000=75000$$
$$\underline{-91000\quad -91000}$$
$$-5x=-16000$$
$$\frac{\cancel{-5}x}{\cancel{-5}}=\frac{-16000}{-5}$$
$$x=3200$$

Then 7000 – x = 7000 – 3200 = 3800. Thus, $3200 is invested at 8% and $3800 is invested at 13%.

CHECK: 3200 + 3800 = 7000. The interest on $3200 at 8% is $256 (= .08(3200)). The interest on $3800 at 13% is $494 (= .13(3800)). Then $256 + $494 = $750, as required.

9. Let x = # of hours that the first person drives. x – 4 = # of hours that the second person drives.

$$48x+55(x-4)=604$$
$$48x+55x-220=604$$
$$103x-220=604$$
$$\underline{+220\quad +220}$$
$$103x=824$$
$$\frac{\cancel{103}x}{\cancel{103}}=\frac{824}{103}$$
$$x=8$$

Thus, they will be 604 kilometers apart after the first person drives for 8 hours, or at 7:00 p.m.

CHECK: In 8 hours, the first person covers 48(8) = 384 km. In 8 – 4 = 4 hours, the second person covers

55(4) = 220 km. 384 + 220 = 604.

CHAPTER 5
GRAPHING STRAIGHT LINES

Exercises 5.1

1–19. (odd)

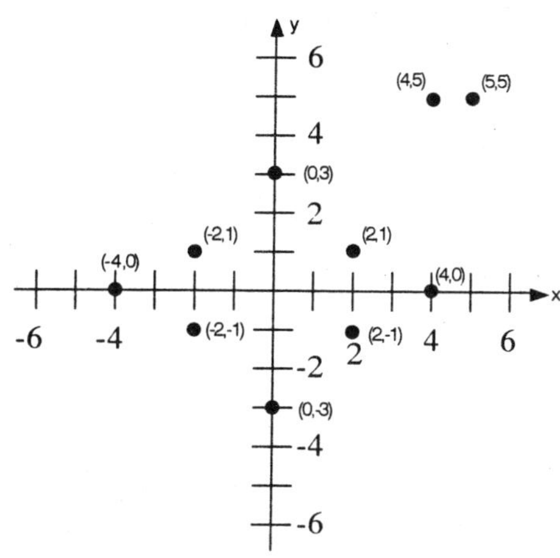

21. Quadrant I

23. Quadrant III

25. on y-axis

27. Quadrant II

29. on both x-axis and y-axis

31. Quadrant IV

33. (1, 7) satisfies the equation $2y - 4x = 10$, since $2(7) - 4(1) = 14 - 4 = 10$.

35. (–6, –10) satisfies the equation $\frac{2}{3}x - \frac{1}{2}y = 1$, since $\frac{2}{3}(-6) - \frac{1}{2}(-10) = -4 + 5 = 1$.

37. (1, –6) satisfies the equation $y = x^2 - 3x - 4$, since $1^2 - 3(1) - 4 = 1 - 3 - 4 = -6$

39. The last two ordered pairs satisfy the equation $y = \frac{1}{x+2}$, since $\frac{1}{1+2} = \frac{1}{3}$ and $\frac{1}{-1+2} = 1$.

41. {(x, y) | y = x + 2 and x = –3, 0, 4}

x	y	(x, y)
–3	–1	(-3, -1)
0	2	(0, 2)
4	6	(4, 6)

42. {(x, y) | x = y + 2 and x = –3, 0, 4}

x	y	(x, y)
–3	–5	(-3, -5)
0	–2	(0, -2)
4	2	(4, 2)

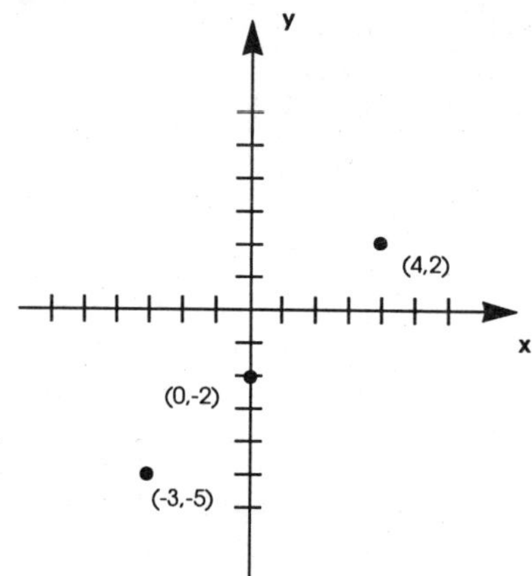

43. (2, 3), (0, 6), (6, –3), (4, 0)

44. (-2, –16), (0, –8), (3, 4), (2, 0)

45. An ordered pair (x, y) is a pair of numbers in which the order of appearance matters. For instance, (3, 4) and (4, 3) would be different as ordered pairs, even though both involve the same two numbers.

46. The notation {x, y} stands for the set containing the two elements x and y. Here, order does not matter. That is, {3, 4} and {4, 3} are equal sets. When we write (x, y), we mean that x comes first and y second.

x	y	(x, y)
−1	1	(−1, 1)
0	0	(0, 0)
1	1	(1, 1)
2	4	(2,4)

The graph of $y = x^2$ is not a straight line.

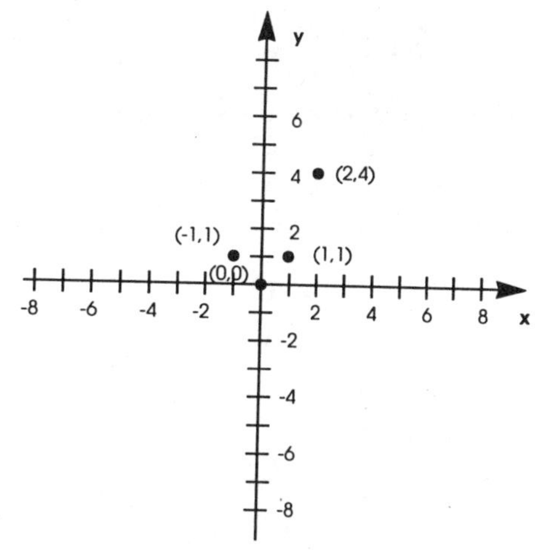

48.

x	y	(x, y)
−1	−1	(−1, −1)
0	0	(0, 0)
1	1	(1, 1)
2	8	(2, 8)

The graph of $y = x^3$ is not a straight line.

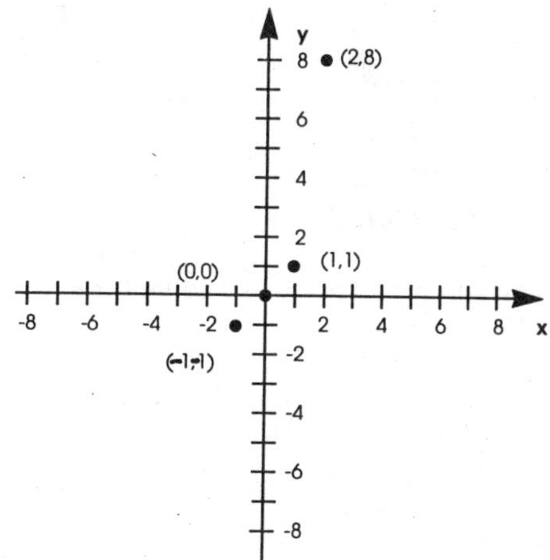

49. Let W = width of rectangle (in cm). 2W + 5 = length of rectangle (in cm).

$$2W + 2(2W + 5) = 34$$
$$2W + 4W + 10 = 34$$
$$6W + 10 = 34$$
$$6W = 24$$
$$W = 4$$

Then 2W + 5 = 2(4) + 5 = 13

Thus, the dimensions of the rectangle are 13 cm × 4 cm . .

Exercises 5.2

1. To find the x-intercept, we set y = 0 and solve for x.

$$x - y = 7$$
$$x - (0) = 7$$
$$x - 0 = 7$$
$$x = 7$$

Therefore, the x-intercept is 7.

To find the y-intercept, we set x = 0 and solve for y.

$$\begin{aligned} x - y &= 7 \\ (0) - y &= 7 \\ 0 - y &= 7 \\ -y &= 7 \\ y &= -7 \end{aligned}$$

Therefore, the y-intercept is –7.

5. To find the x-intercept, we set y = 0 and solve for x.

$$\begin{aligned} 2x + 3y &= 12 \\ 2x + 3(0) &= 12 \\ 2x + 0 &= 12 \\ 2x &= 12 \\ x &= 6 \end{aligned}$$

Therefore, the x-intercept is 6.

To find the y-intercept, we set x = 0 and solve for y.

$$\begin{aligned} 2x + 3y &= 12 \\ 2(0) + 3y &= 12 \\ 0 + 3y &= 12 \\ 3y &= 12 \\ y &= 4 \end{aligned}$$

Therefore, the y-intercept is 4.

9. To find the x-intercept, we set y = 0 and solve for x.

$$\begin{aligned} 2x &= 5y \\ 2x &= 5(0) \\ 2x &= 0 \\ x &= 0 \end{aligned}$$

Therefore, the x-intercept is 0.

To find the y-intercept, we set x = 0 and solve for y.

$$\begin{aligned} 2x &= 5y \\ 2(0) &= 5y \\ 0 &= 5y \\ 0 &= y \end{aligned}$$

Therefore, the y-intercept is 0.

11. To find the x-intercept, we set y = 0 and solve for x. Since the equation x = 5 does not contain y, its solution is just x = 5. So the x-intercept is 5. To find the y-intercept, we set x = 0 and solve for y. But this gives 0 = 5, which is a contradiction. So the equation x = 5 has no y-intercept.

5. $x + y = -5$

x - intercept:

$x + (0) = -5$

$x + 0 = -5$

$x = -5$

Plot $(-5,\ 0)$

y - intercept:

$(0) + y = -5$

$0 + y = -5$

$y = -5$

Plot $(0,\ -5)$

check point: choose $x = 1$

$(1) + y = -5$

$1 + y = -5$

$y = -6$

Plot $(1,\ -6)$

The graph of $x + y = -5$

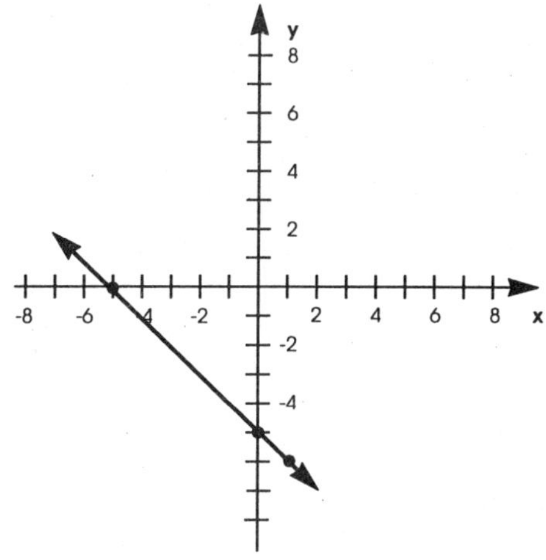

19. $3x - 4y = 12$

x-intercept:

$3x - 4(0) = 12$

$3x - 0 = 12$

$3x = 12$

$x = 4$

Plot $(4, 0)$

y-intercept:

$3(0) - 4y = 12$
$0 - 4y = 12$
$-4y = 12$
$y = -3$

Plot (0, –3)

check point: choose x = 8

$3(8) - 4y = 12$
$24 - 4y = 12$
$-4y = -12$
$y = 3$

Plot (8, 3)

The graph of $3x - 4y = 12$

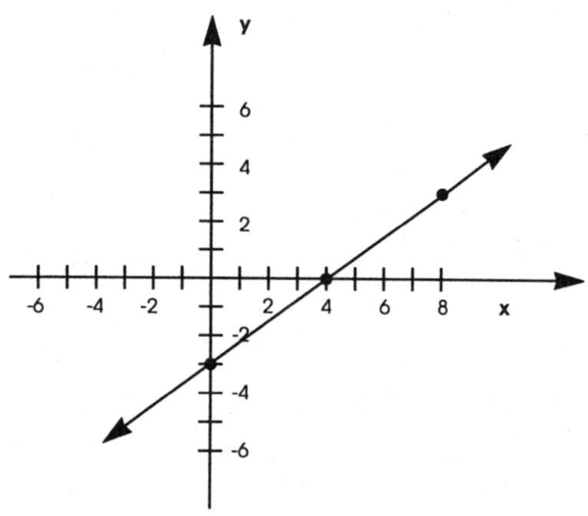

23. $y = -4x$

x-intercept:

$(0) = -4x$
$0 = -4x$
$0 = x$

Plot (0, 0)

second point: choose x = 1

$y = -4(1)$
$y = -4$

Plot (1, –4)

check point: choose x = –1

$y = -4(-1)$
$y = 4$

Plot $(-1, 4)$

(As noted in the text, the y-intercept of this equation will also be equal to 0.)

The graph of $y = -4x$

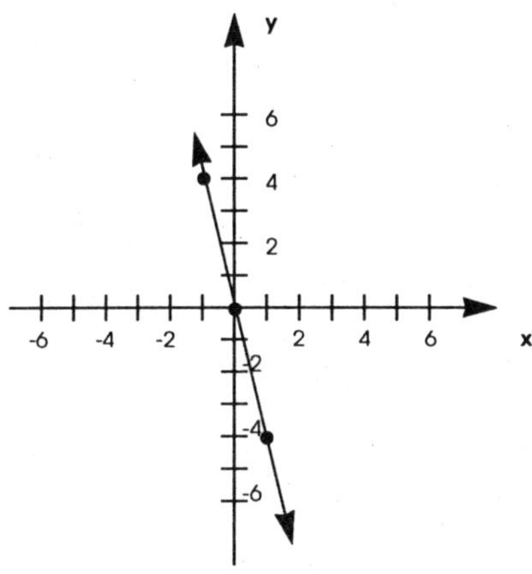

27. $y = 5$

Its graph is a line parallel to the x-axis and 5 units above it.

The graph of $y = 5$

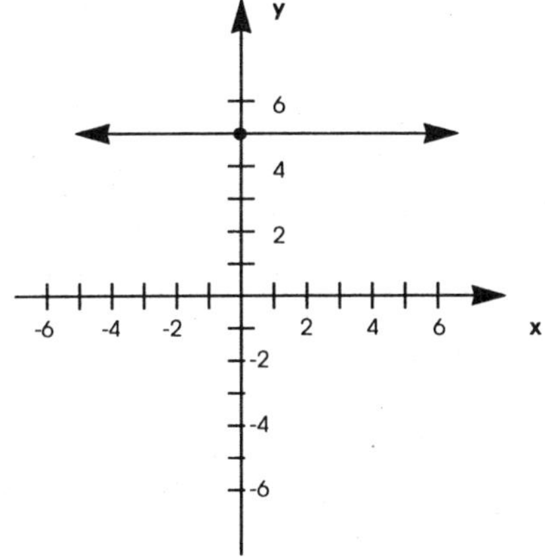

29. $x = -4$

Its graph is a line parallel to the y-axis and 4 units to the left of it.

The graph of $x = -4$

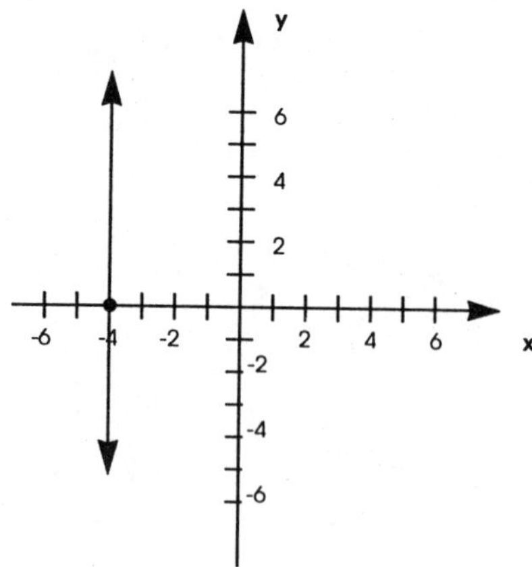

33. $2(x - 3) = 4(y + 2)$

x-intercept:

$$2(x-3) = 4((0)+2)$$
$$2(x-3) = 4(0+2)$$
$$2x - 6 = 8$$
$$2x = 14$$
$$x = 7$$

Plot (7, 0)

y-intercept:

$$2((0)-3) = 4(y+2)$$
$$2(0-3) = 4(y+2)$$
$$-6 = 4y + 8$$
$$-14 = 4y$$
$$-\frac{7}{2} = y$$

Plot $\left(0,\ -\frac{7}{2}\right)$

check point: choose $x = 1$

$$2((1) - 3) = 4(y + 2)$$
$$2(1 - 3) = 4(y + 2)$$
$$2(-2) = 4(y + 2)$$
$$-4 = 4y + 8$$
$$-12 = 4y$$
$$-3 = y$$

Plot (1, –3)

The graph of $2(x - 3) = 4(y + 2)$

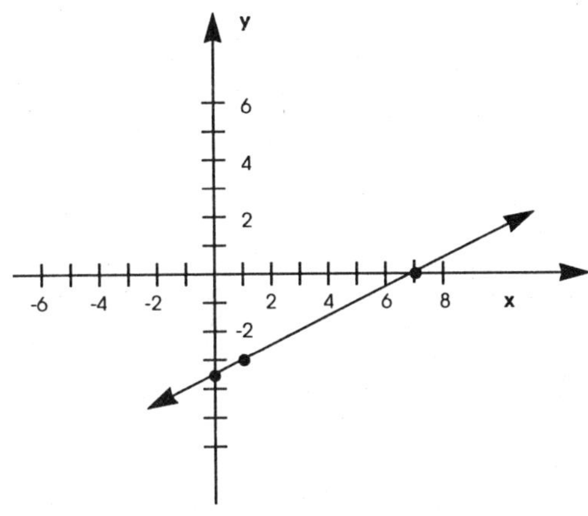

37. $d = 3t + 4$

d-intercept:
$$d = 3(0) + 4$$
$$d = 0 + 4$$
$$d = 4$$

Plot (0, 4)

t-intercept:
$$0 = 3t + 4$$
$$-4 = 3t$$
$$-\frac{4}{3} = t$$

Plot $(-\frac{4}{3}, 0)$

check point: choose $t = -1$
$$d = 3(-1) + 4$$
$$d = -3 + 4$$
$$d = 1$$

Plot (–1, 1)

The graph of $d = 3t + 4$

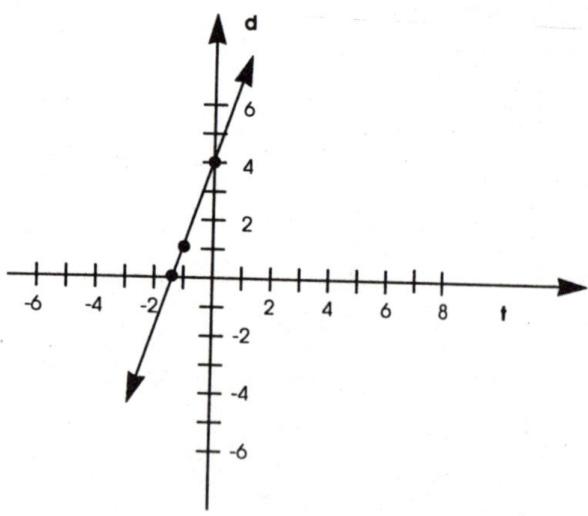

41. $u - 4v = 8$

u-intercept:

$u - 4(0) = 8$

$u - 0 = 8$

$u = 8$

Plot (8, 0)

v-intercept:

$0 - 4v = 8$

$-4v = 8$

$v = -2$

Plot (0, –2)

check point: choose $v = -1$

$u - 4(-1) = 8$

$u + 4 = 8$

$u = 4$

Plot (4, –1)

The graph of u – 4v = 8

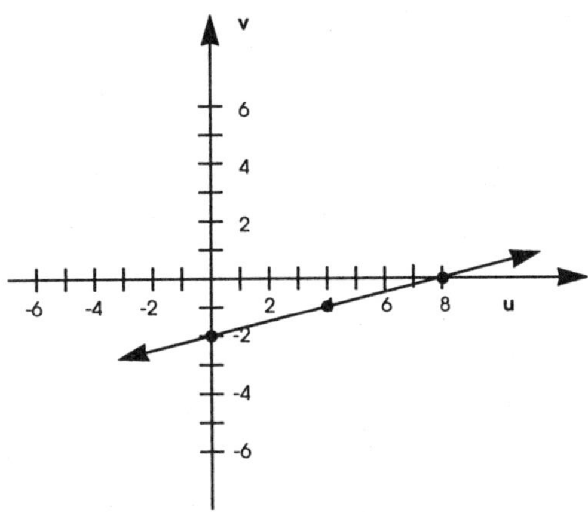

43.

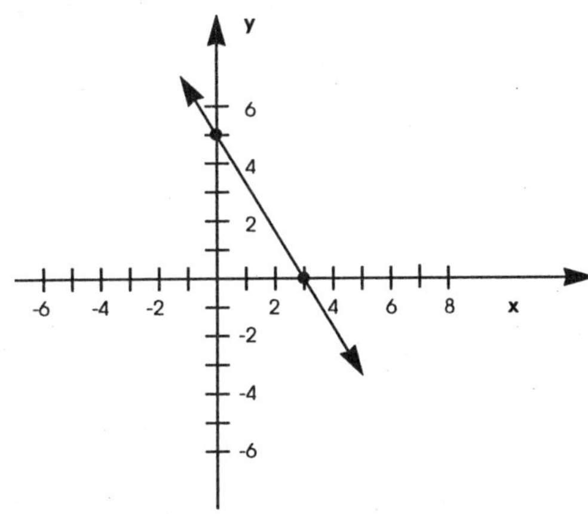

44. Some points on this line are (–4, –4), (–3, –3), (0, 0), (2, 2), and (5, 5). Since the y-coordinate of any such point is equal to its x-coordinate, an equation of this line would be y = x.

The graph of $y = x$

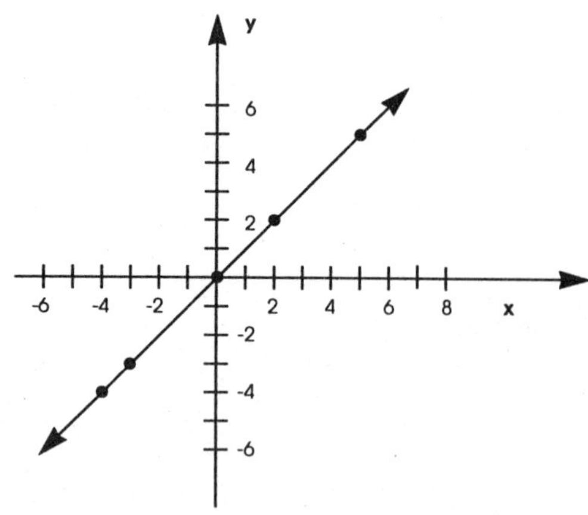

45. Some points on this line are (–4, 4), (–3, 3), (0, 0), (2, –2), and (5, –5). Since the y-coordinate of any such point is equal to the negative of its x-coordinate, an equation of this line would be $y = -x$.

The graph of $y = -x$

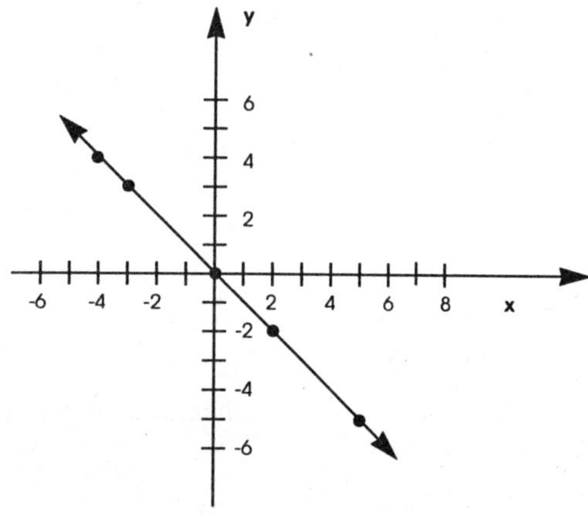

46. The x-coordinate of the point where a graph intersects the x-axis is called an x-intercept of the graph. To find an x-intercept of an equation, set $y = 0$ and solve for x. The y-coordinate of the point where a graph intersects the y-axis is called a y-intercept of this graph. To find a y-intercept of an equation, set $x = 0$ and solve for y.

47. Let x = number of miles driven

$$2(15)+.20x = 65$$
$$30+.20x = 65$$
$$.20x = 35$$
$$x = 175$$

Thus, 175 miles were driven.

Exercises 5.3

1. $m = \dfrac{y_2 - y_1}{x_2 - x_1} = \dfrac{9-5}{6-3} = \dfrac{4}{3}$

7. $m = \dfrac{y_2 - y_1}{x_2 - x_1} = \dfrac{-4-(-2)}{-3-(-1)}$

$= \dfrac{-4+2}{-3+1} = \dfrac{-2}{-2} = 1$

11. $m = \dfrac{y_2 - y_1}{x_2 - x_1} = \dfrac{7-7}{-3-4} = \dfrac{0}{-7} = 0$

13. m is undefined. (If we tried to use the formula, we would have obtained $m = \dfrac{9-6}{2-2} = \dfrac{3}{0}$, and division by zero is undefined.)

17. $m = \dfrac{y_2 - y_1}{x_2 - x_1} = \dfrac{\frac{1}{4}-\frac{1}{5}}{\frac{3}{2}-(-\frac{1}{3})}$

$= \dfrac{\frac{1}{4}-\frac{1}{5}}{\frac{3}{2}+\frac{1}{3}} = \dfrac{\frac{5}{20}-\frac{4}{20}}{\frac{9}{6}+\frac{2}{6}}$

$= \dfrac{\frac{1}{20}}{\frac{11}{6}} = \dfrac{1}{20}\cdot\dfrac{6}{11} = \dfrac{3}{110}$

23. $m = \dfrac{y_2 - y_1}{x_2 - x_1} = \dfrac{0-a}{a-0} = \dfrac{-a}{a} = -1$

31.

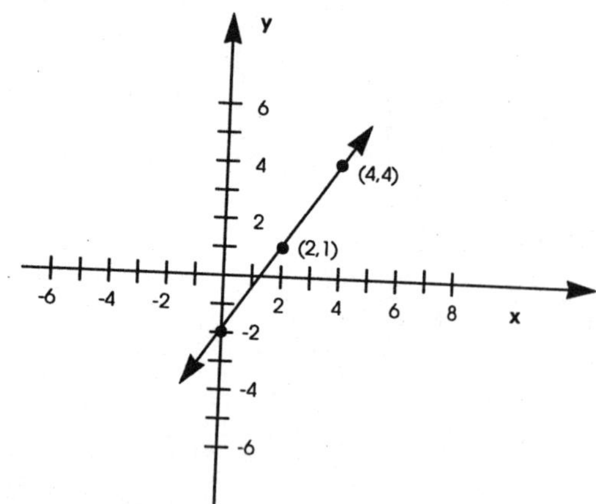

35.

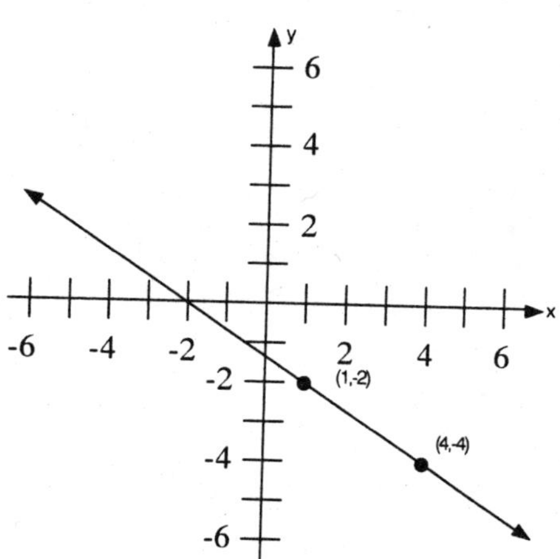

[illegible]. A slope of 0 means that the line is horizontal.

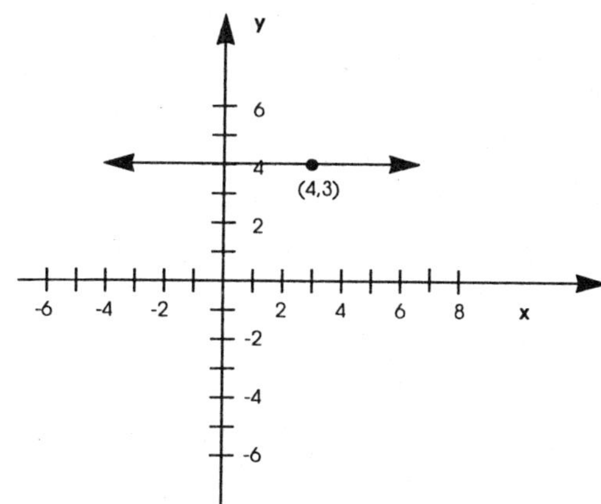

39. No slope means that the line is vertical.

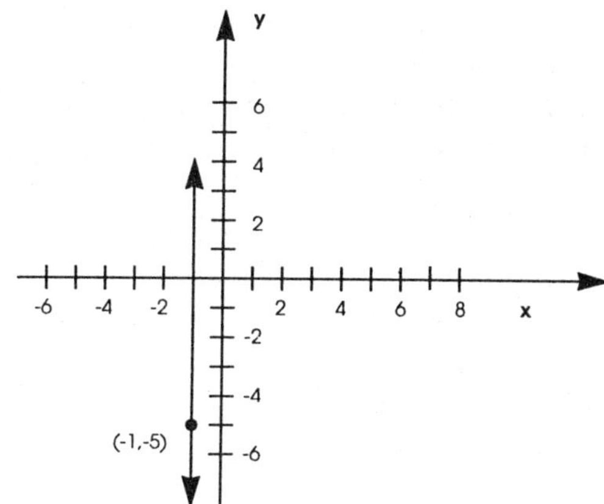

41. $\frac{-5}{4}$	43. 5	45. $\frac{7}{3}$	47. 0
49. positive	51. zero	53. M_3, M_2, M_1	55. 4.14
57. −1.02	59. 0		

61. Let n = number of nickels

$$28 - n = \text{number of dimes}$$
$$5n + 10(28 - n) = 200$$
$$5n + 280 - 10n = 200$$
$$-5n + 280 = 200$$
$$-5n = -80$$
$$n = 16$$
$$\text{Then } 28 - n = 12$$

Thus, there are 16 nickels and 12 dimes.

63. First, compute the slope of the line through (–1, –2) and (2, 0):

$$m = \frac{0-(-2)}{2-(-1)} = \frac{2}{3}.$$

Next, compute the slope of the line through (2, 0) and (5, 2):

$$m = \frac{2-0}{5-2} = \frac{2}{3}.$$

Since these two lines have equal slopes, they are either parallel or actually one line. But the two lines in question cannot be parallel, since (2, 0) is a point on each and parallel lines have no point in common. We can then conclude that the three points (–1, –2), (2, 0), and (5, 2) all lie on a straight line. (Points with this property are called collinear.)

Exercises 5.4

1.
$$y - y_1 = m(x - x_1)$$
$$(x_1, y_1) = (1, 5),\ m = 3$$
$$y - 5 = 3(x - 1) \text{ or}$$
$$y = 3x + 2$$

5.
$$y - y_1 = m(x - x_1)$$
$$(x_1,\ y_1) = (-3,\ -5),\ m = -\frac{2}{3}$$
$$y - (-5) = -\frac{2}{3}(x - (-3))$$
$$y + 5 = -\frac{2}{3}(x + 3) \text{ or}$$
$$y = -\frac{2}{3}x - 7$$

$y = mx + b$

$m = \frac{1}{4}, \ b = -2$

$y = \frac{1}{4}x + (-2)$

$y = \frac{1}{4}x - 2$

11. $y - y_1 = m(x - x_1)$

$(x_1, y_1) = (-2, 0), \ m = -\frac{3}{4}$

$y - 0 = -\frac{3}{4}(x - (-2))$

$y = -\frac{3}{4}(x + 2)$ or

$y = -\frac{3}{4}x - \frac{3}{2}$

15. $y - y_1 = m(x - x_1)$

$(x_1, y_1) = (5, 6), \ m = 0$

$y - 6 = 0(x - 5)$

$y - 6 = 0$

$y = 6$

19. $m = \frac{y_2 - y_1}{x_2 - x_1} = \frac{-2 - 4}{2 - (-1)} = \frac{-6}{3} = -2$

$(x_1, y_1) = (-1, 4)$

$y - y_1 = m(x - x_1)$

$y - 4 = -2(x - (-1))$

$y - 4 = -2(x + 1)$ or $y = -2x + 2$

23. $m = \frac{y_2 - y_1}{x_2 - x_1} = \frac{-1 - 0}{0 - (-1)} = \frac{-1}{1} = -1$

$b = -1$

$y = mx + b$

$y = (-1)x + (-1)$

$y = -x - 1$

27. Since the line is vertical, it has no slope. Therefore, we cannot use either the point-slope form or the slope-intercept form of an equation of a line. We know that all points on a vertical line have the same x-coordinate. Since our line passes through (4, –3), its equation is x = 4.

31. An x-intercept of –3 means (–3, 0) is on the line. A y-intercept of 4 means (0, 4) is on the line.

Then

$$m = \frac{y_2 - y_1}{x_2 - x_1} = \frac{4 - 0}{0 - (-3)} = \frac{4}{3}$$

and b = 4, so that an equation (in slope-intercept form) is

$$y = \frac{4}{3}x + 4.$$

33. $y = 5x + 7$

$y = mx + b$. So $m = 5$.

37. $x + y = 7$

$\underline{-x \qquad -x}$

$y = -x + 7$

$y = mx + b$. So $m = -1$.

41. $2x - 5y + 7 = 0$

$\underline{-2x \qquad\qquad -2x}$

$-5y + 7 = -2x$

$\underline{\qquad -7 \qquad -7}$

$$-5y = -2x - 7$$
$$\frac{-5y}{-5} = \frac{-2x - 7}{-5}$$
$$y = \frac{-2x}{-5} + \frac{-7}{-5}$$
$$y = \frac{2}{5}x + \frac{7}{5}$$

$y = mx + b$. So $m = \frac{2}{5}$.

43. $y = mx + b$

$$m = \frac{2 - 0}{3 - 0} = \frac{2}{3}$$

$b = 0$

So $y = \frac{2}{3}x$

7. $y - y_1 = m(x - x_1)$

$$m = \frac{3-(-1)}{6-(-6)} = \frac{4}{12} = \frac{1}{3}$$

$$y - 3 = \frac{1}{3}(x - 6)$$

$$y - 3 = \frac{1}{3}x - 2$$

$$y = \frac{1}{3}x + 1$$

51. By comparison with y = mx + b, the line whose equation is y = 3x – 8 has slope = 3. Since parallel lines have equal slopes, the line in question also has a slope of 3.

53. The line whose equation is y = –x + 4 has slope = –1. So the line in question also has a slope of –1, and it passes through (–3, 0).

$$y - y_1 = m(x - x_1)$$

$$y - 0 = -1(x - (-3))$$

$$y = -1(x + 3)$$

$$y = -x - 3$$

55. (a)

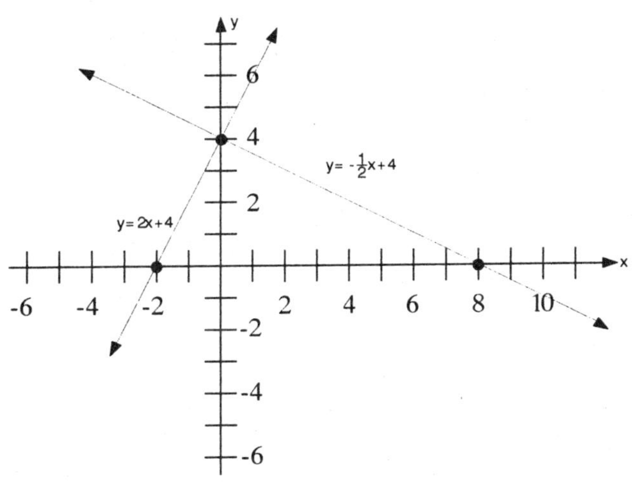

(b) From the graph, it appears that these lines are perpendicular.

(c) The line whose equation is y = 2x + 4 has slope = 2. The line whose equation is $y = -\frac{1}{2}x + 4$ has slope = $-\frac{1}{2}$. The product of these slopes is equal to –1. Put another way, the slopes are negative reciprocals of one another.

(d) The line whose equation is $y = \frac{2}{5}x + 7$ has slope $= \frac{2}{5}$. Therefore, any line that is perpendicular to this line must have slope $= -\frac{5}{2}$.

57. Let x = weight of the package (in lbs.)

$$
\begin{aligned}
7 + 4(x - 1) &= 43 \\
7 + 4x - 4 &= 43 \\
3 + 4x &= 43 \\
4x &= 40 \\
x &= 10
\end{aligned}
$$

Thus, the package weighs 10 lbs.

Chapter 5 Review Exercises

1, 3, 5.

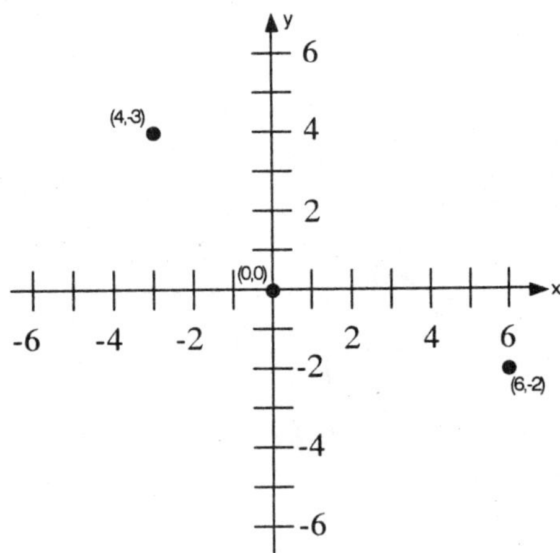

7.
$$
\begin{aligned}
2x + 4y &= 14 \\
2(3) + 4y &= 14 \\
6 + 4y &= 14 \\
4y &= 8 \\
y &= 2
\end{aligned}
$$

Therefore, the ordered pair is (3, 2).

9.
$$
\begin{aligned}
x + 2y &= 4 \\
x + 2(-3) &= 4 \\
x - 6 &= 4 \\
x &= 10
\end{aligned}
$$

Therefore, the ordered pair is (10, –3).

11.
$$x - y = 0$$
$$(0) - y = 0$$
$$0 - y = 0$$
$$-y = 0$$
$$y = 0$$

Therefore, the ordered pair is (0, 0).

13. $x - y = 8$

x intercept:
$$x - (0) = 8$$
$$x - 0 = 8$$
$$x = 8$$

Plot (8, 0)

y-intercept:
$$(0) - y = 8$$
$$0 - y = 8$$
$$-y = 8$$
$$y = -8$$

Plot (0, –8)

check point: choose $x = 4$
$$(4) - y = 8$$
$$4 - y = 8$$
$$-y = 4$$
$$y = -4$$

Plot (4, –4)

The graph of $x - y = 8$

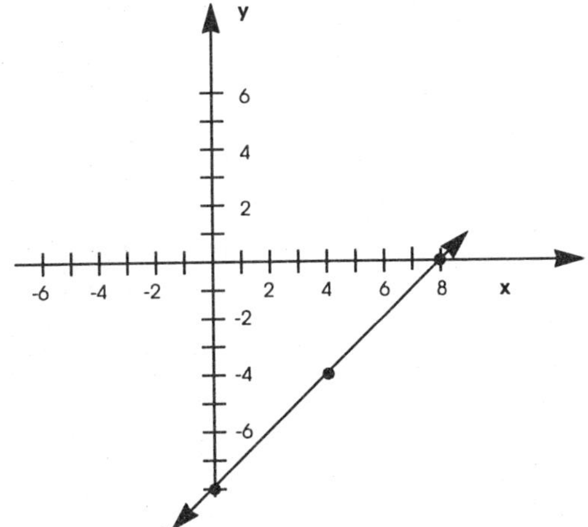

15. $3x + 7y = 21$

x-intercept:
$3x + 7(0) = 21$

$3x + 0 = 21$

$3x = 21$

$x = 7$

Plot (7, 0)

y-intercept:
$3(0) + 7y = 21$

$0 + 7y = 21$

$7y = 21$

$y = 3$

Plot (0, 3)

check point: choose $y = 6$
$3x + 7(6) = 21$

$3x + 42 = 21$

$3x = -21$

$x = -7$

Plot (–7, 6)

The graph of $3x + 7y = 21$

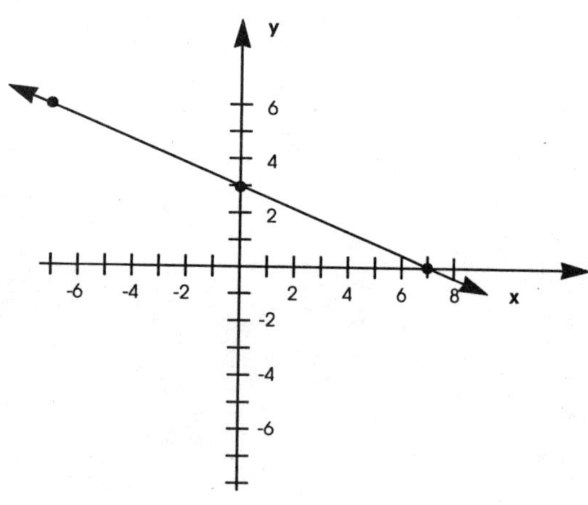

7. $y = 3x + 2$

x-intercept:

$$0 = 3x + 2$$

$$-2 = 3x$$

$$-\frac{2}{3} = x$$

Plot $(-\frac{2}{3}, 0)$

y-intercept:

$$y = 3(0) + 2$$

$$y = 0 + 2$$

$$y = 2$$

Plot (0, 2)

check point: choose $x = 1$

$$y = 3(1) + 2$$

$$y = 3 + 2$$

$$y = 5$$

Plot (1, 5)

The graph of $y = 3x + 2$

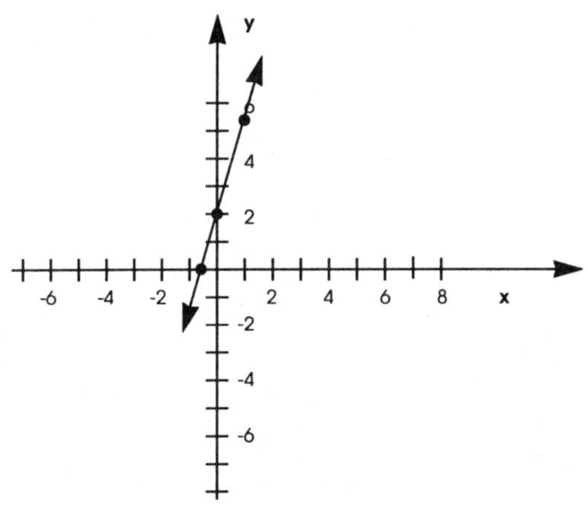

19. $x - 2y = 4$

x-intercept:

$$x - 2(0) = 4$$
$$x - 0 = 4$$
$$x = 4$$

Plot (4, 0)

y-intercept:

$$(0) - 2y = 4$$
$$0 - 2y = 4$$
$$-2y = 4$$
$$y = -2$$

Plot (0, –2)

check point: choose $y = -1$

$$x - 2(-1) = 4$$
$$x + 2 = 4$$
$$x = 2$$

Plot (2, –1)

The graph of $x - 2y = 4$

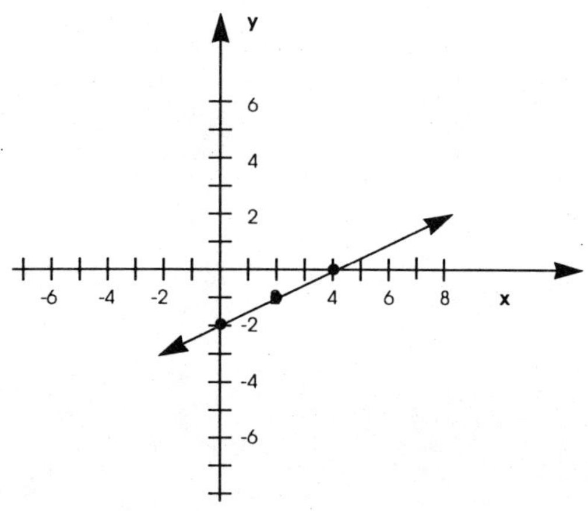

21. $x = 4$

Its graph is a line parallel to the y-axis and 4 units to the right of it.

The graph of $x = 4$

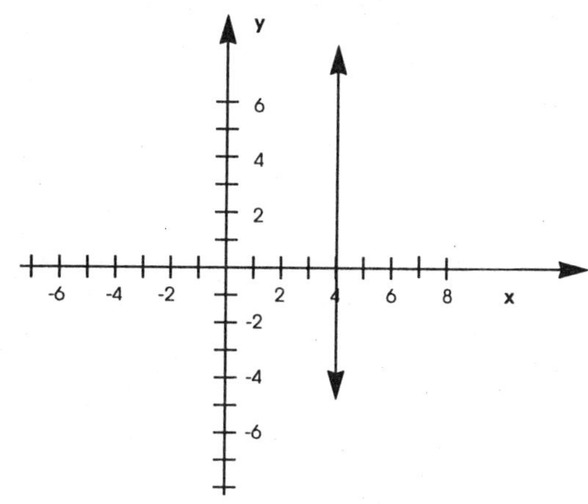

23. $3x + 2y = 12$

x-intercept:

$$3x + 2(0) = 12$$
$$3x + 0 = 12$$
$$3x = 12$$
$$x = 4$$

Plot (4, 0)

y-intercept:

$$3(0) + 2y = 12$$
$$0 + 2y = 12$$
$$2y = 12$$
$$y = 6$$

Plot (0, 6)

check point: choose $x = 2$

$$3(2) + 2y = 12$$
$$6 + 2y = 12$$
$$2y = 6$$
$$y = 3$$

Plot (2, 3)

The graph of $3x + 2y = 12$

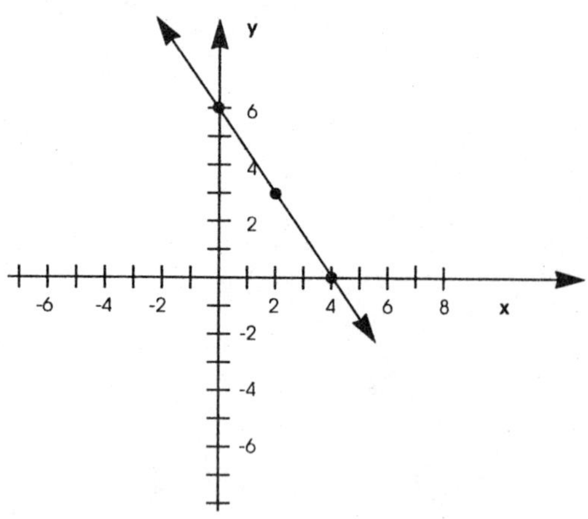

25. $3y = 6$

This equation is equivalent to $y = 2$ (divide both sides by 3), whose graph is a line parallel to the x-axis and 2 units above it.

The graph of $3y = 6$

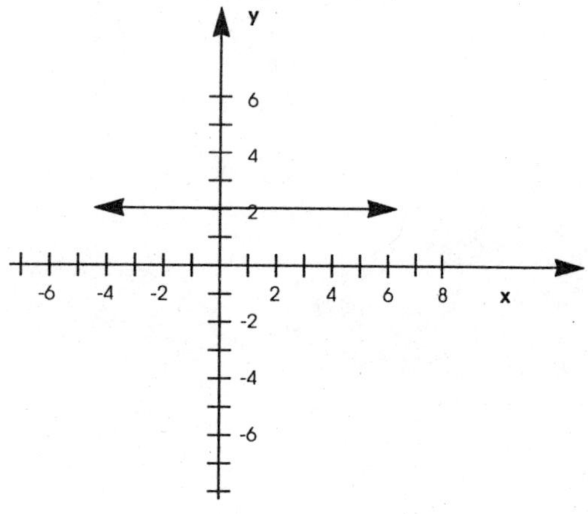

27. $m = \dfrac{y_2 - y_1}{x_2 - x_1} = \dfrac{7-5}{1-2} = \dfrac{2}{-1} = -2$

29. $m = \frac{y_2 - y_1}{x_2 - x_1} = \frac{2-(-2)}{3-(-3)}$

$= \frac{2+2}{3+3} = \frac{4}{6} = \frac{2}{3}$

31. $m = \frac{y_2 - y_1}{x_2 - x_1} = \frac{8-8}{-1-1} = \frac{0}{-2} = 0$

33. $(x_1, y_1) = (3, 2)$, $m = 4$

$y - y_1 = m(x - x_1)$

$y - 2 = 4(x-3)$ or $y = 4x - 10$

35. $m = \frac{y_2 - y_1}{x_2 - x_1} = \frac{0-(-6)}{1-2} = \frac{0+6}{1-2}$

$= \frac{6}{-1} = -6$

$(x_1, y_1) = (2, -6)$

$y - y_1 = m(x - x_1)$

$y - (-6) = -6(x-2)$

$y + 6 = -6(x-2)$ or $y = -6x + 6$

37. All points on a horizontal line have the same y-coordinate. Since the point (3, 8) is on our line, an equation for this horizontal line must be $y = 8$.

39. $-\frac{3}{4}$

41. The line whose equation is $y = 5x - 1$ has slope = 5. So the line in question has slope = 5, and it passes through (–3, 5).

$y - y_1 = m(x - x_1)$

$y - 5 = 5(x-(-3))$

$y - 5 = 5(x+3)$

$y - 5 = 5x + 15$

$y = 5x + 20$

43. $y = mx + b$

$m = \frac{4-0}{3-0} = \frac{4}{3}$

$b = 0$

So $y = \frac{4}{3}x$

Chapter 5 Practice Test

1. $$\frac{2y - x}{5} = x - y$$

$$\frac{2(-4) - (-3)}{5} \stackrel{?}{=} (-3) - (-4)$$

$$\frac{-8 + 3}{5} \stackrel{?}{=} -3 + 4$$

$$\frac{-5}{5} \stackrel{?}{=} 1$$

$$-1 \neq 1$$

So the point (–3, –4) does not satisfy the given equation.

3. (a) $3x - 5y = 15$

x-intercept:

$$3x - 5(0) = 15$$
$$3x - 0 = 15$$
$$3x = 15$$
$$x = 5$$

Plot (5, 0)

y-intercept:

$$3(0) - 5y = 15$$
$$0 - 5y = 15$$
$$-5y = 15$$
$$y = -3$$

Plot (0, –3)

check point: choose x = –5

$$3(-5) - 5y = 15$$
$$-15 - 5y = 15$$
$$-5y = 30$$
$$y = -6$$

Plot (–5, –6)

The graph of $3x - 5y = 15$

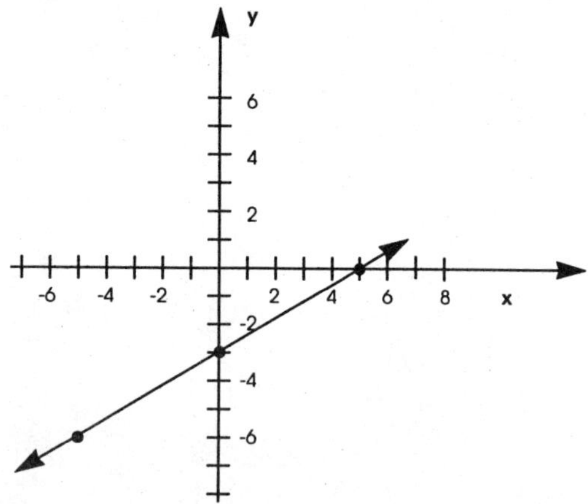

(b) $y = 3x - 12$

x-intercept:
$0 = 3x - 12$

$12 = 3x$

$4 = x$

Plot (4, 0)

y-intercept:
$y = 3(0) - 12$

$y = 0 - 12$

$y = -12$

Plot (0, −12)

check point: choose x = 2

$y = 3(2) - 12$

$y = 6 - 12$

$y = -6$

Plot (2, -6)

The graph of $y = 3x - 12$

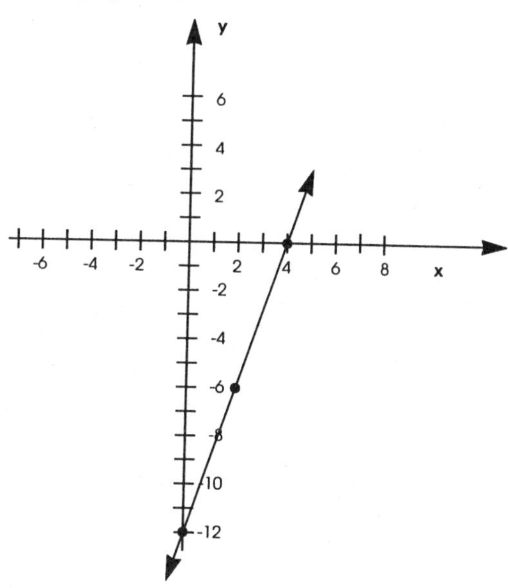

(c) $y + 2x = 0$

x-intercept:
$0 + 2x = 0$

$2x = 0$

$x = 0$

Plot (0, 0)

(This implies that y = 0 is the y-intercept.)

second point: choose x = 1

$y + 2(1) = 0$

$y + 2 = 0$

$y = -2$

Plot $(1, -2)$

check point: choose $x = -1$

$y + 2(-1) = 0$

$y - 2 = 0$

$y = 2$

Plot $(-1, 2)$

The graph of $y + 2x = 0$

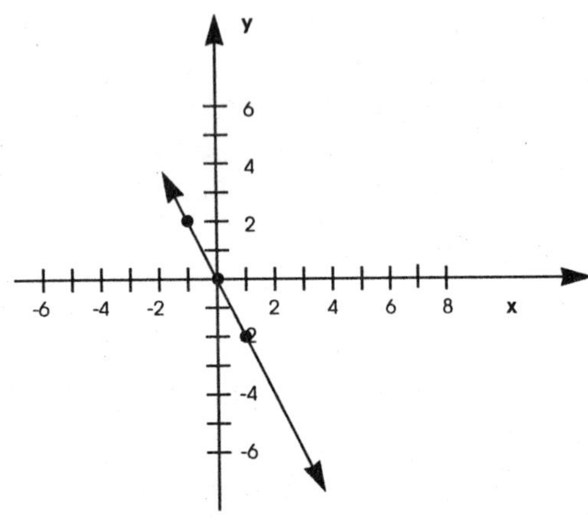

(d) $x = 4$. This is an equation of a line parallel to the y-axis and 4 units to the right of it.

The graph of $x = 4$

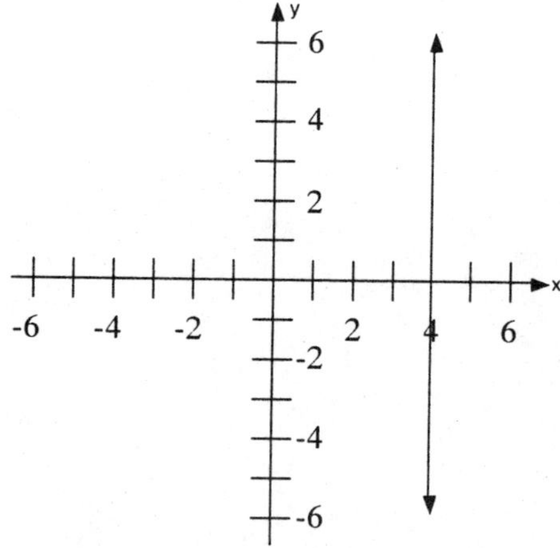

(e) $y = -3$. This is an equation of a line parallel to the x-axis and 3 units below it.

The graph of $y = -3$

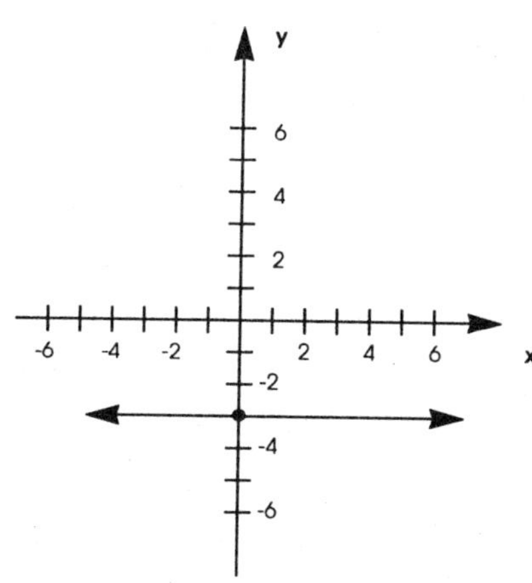

5. $m = \frac{4}{3}$, $(x_1, y_1) = (4, -1)$

$y - y_1 = m(x - x_1)$

$y - (-1) = \frac{4}{3}(x - 4)$

$y + 1 = \frac{4}{3}(x - 4)$

or $y = \frac{4}{3}x - \frac{19}{3}$

7. $y = 5$ is the horizontal line passing through (3, 5).

$x = 3$ is the vertical line passing through (3, 5).

9. The line whose equation is $y = 5x - 1$ has slope = 5. So the line in question has slope = 5, and it passes through (–4, 0).

$y - y_1 = m(x - x_1)$

$y - 0 = 5(x - (-4))$

$y - 0 = 5(x + 4)$

$y = 5x + 20$

CHAPTER 6
INTERPRETING GRAPHS AND SYSTEMS OF EQUATIONS

Exercises 6.1

1. (a) B

   (b) B

   (c) A

   (d) B

3. (a) 2,000 calls

   (b) 4:00 a.m.

   (c) 10,000 calls

   (d) Midnight to 4:00 a.m. and 4:00 p.m. to midnight

   (e) 10:00 a.m. to 4:00 p.m.

   (f) 10:00 a.m. to 4:00 p.m.

5. (a) This graph denies the belief that temperature drops steadily as altitude increases. Notice that the temperature increases as the altitude increases from 12 to 50 and again when the altitude is greater than 80.

   (b) The temperature decreases when the altitude is between 0 and 12 and also when it is between 50 and 80.

   (c) For the first 5 miles that the balloon rises, the temperature will increase; for the last 5 miles, it will decrease.

7. (a) Immediately after the students memorized the list of 20 words, the group remembered all twenty.

   (b) after 4 hours

   (c) during the first two hours

   (d) 16 words are forgotten during the first 4 hours; 2 more words are forgotten during the next 4 hours

9. (a) 150 deer

   (b) between 1996 and 1998

   (c) 800 deer

Exercises 6.2

1. $x + y = 4$

   Set $y = 0$ and solve for x to get an x-intercept of 4.

   Set $x = 0$ and solve for y to get a y-intercept of 4.

$x - y = 2$

Set $y = 0$ and solve for x to get an x-intercept of 2.

Set $x = 0$ and solve for y to get a y-intercept of – 2.

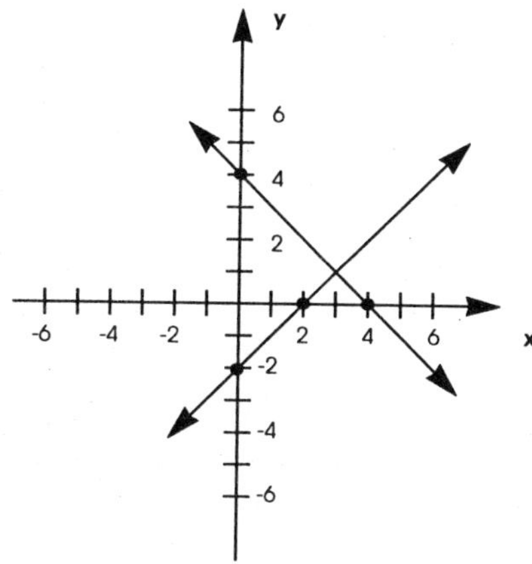

The lines cross at the point (3, 1). So the system

$$\begin{cases} x + y = 4 \\ x - y = 2 \end{cases}$$

is satisfied by the point (3,1).

5. $3x + y = 6$

Set $y = 0$ and solve for x to get an x-intercept of 2.

Set $x = 0$ and solve for y to get a y-intercept of 6.

$6x + 2y = 12$

Set $y = 0$ and solve for x to get an x-intercept of 2.

Set $x = 0$ and solve for y to get a y-intercept of 6.

Since these lines have the same x-intercept and the same y-intercept, the lines coincide.

Therefore there are infinitely many solutions to the system.

$$\begin{cases} 3x + y = 6 \\ 6x + 2y = 12 \end{cases} :\{(x, y) | 3x + y = 6\}$$

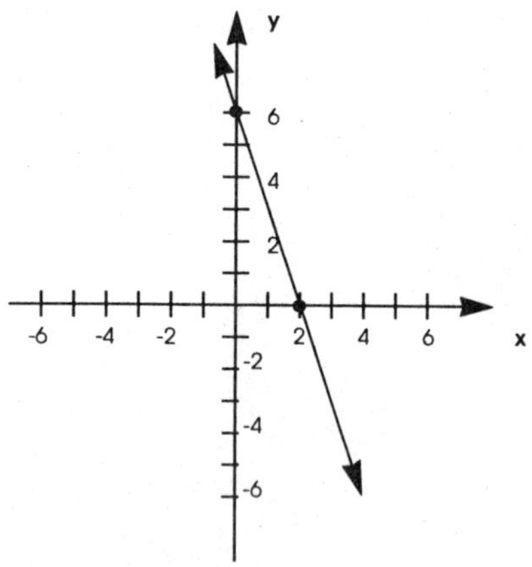

9. $3x - 2y = 6$

Set $y = 0$ and solve for x to get an x-intercept of 2.

Set $x = 0$ and solve for y to get a y-intercept of – 3.

$x + y = 2$

Set $y = 0$ and solve for x to get an x-intercept of 2.

Set $x = 0$ and solve for y to get a y-intercept of 2.

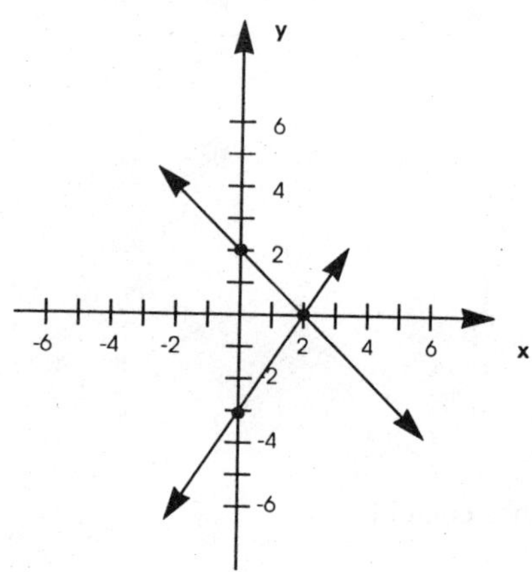

The lines cross at the point (2, 0). So the system

$$\begin{cases} 3x - 2y = 6 \\ x + y = 2 \end{cases}$$

is satisfied by the point (2, 0).

13. $5x - 3y = 15$

Set $y = 0$ and solve for x to get an x-intercept of 3.

Set $x = 0$ and solve for y to get a y-intercept of – 5.

$2x - y = 4$

Set $y = 0$ and solve for x to get an x-intercept of 2.

Set $x = 0$ and solve for y to get a y-intercept of – 4.

The lines cross at the point (–3, –10). So the system

$$\begin{cases} 5x - 3y = 15 \\ 2x - y = 4 \end{cases}$$

is satisfied by the point (–3, –10).

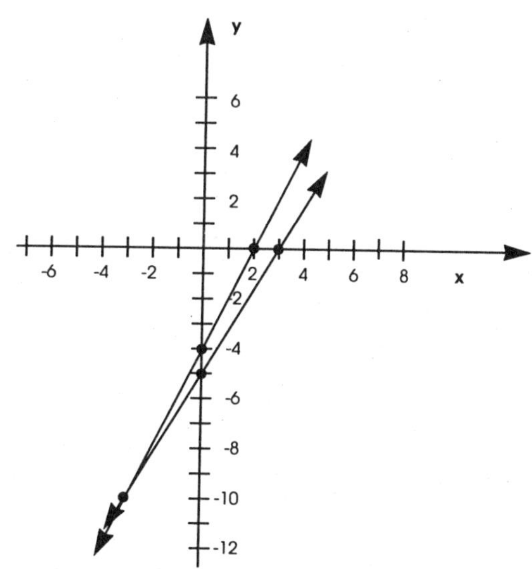

17. $y = x - 3$

Set $y = 0$ and solve for x to get an x-intercept of 3.

Set $x = 0$ and solve for y to get a y-intercept of –3.

$y = x + 4$

Set $y = 0$ and solve for x to get an x-intercept of – 4.

Set x = 0 and solve for y to get a y-intercept of 4.

These lines appear to be parallel. Therefore, the system

$\begin{cases} y = x - 3 \\ y = x + 4 \end{cases}$ has no solutions.

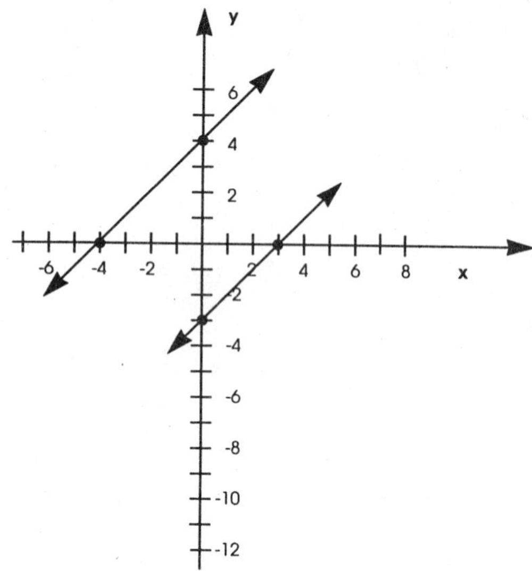

21. 3x + y = 3

Set y = 0 and solve for x to get an x-intercept of 1.

Set x = 0 and solve for y to get a y-intercept of 3.

y = x + 5

Set y = 0 and solve for x to get an x-intercept of – 5.

Set x = 0 and solve for y to get a y-intercept of 5.

The lines cross at the point

$\left(-\frac{1}{2}, \frac{9}{2}\right)$. So the system

$\begin{cases} 3x + y = 3 \\ \quad\;\; y = x + 5 \end{cases}$ is satisfied by the point $\left(-\frac{1}{2}, \frac{9}{2}\right)$.

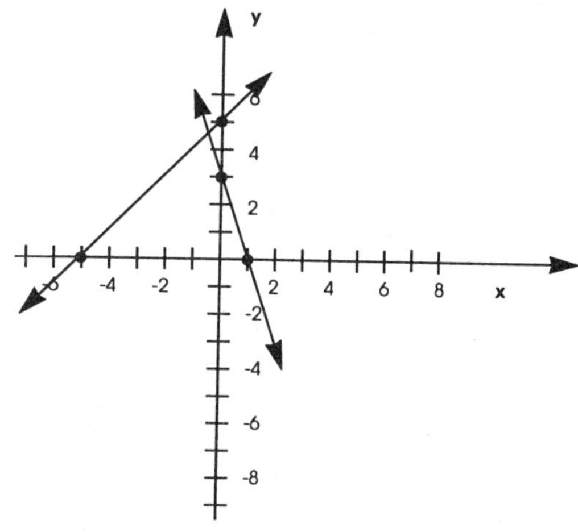

25. $2x + y = 10$

Set $y = 0$ and solve for x to get an x-intercept of 5.

Set $x = 0$ and solve for y to get a y-intercept of 10.

$2y = 20 - 4x$.

Set $y = 0$ and solve for x to get an x-intercept of 5.

Set $x = 0$ and solve for y to get a y-intercept of 10.

Since these lines have the same x-intercept and the same y-intercept, the lines coincide.

Therefore, there are infinitely many solutions to the system

$$\begin{cases} 2x + y = 10 \\ 2y = 20 - 4x \end{cases} : \{(x, y) | 2x + y = 10\}$$

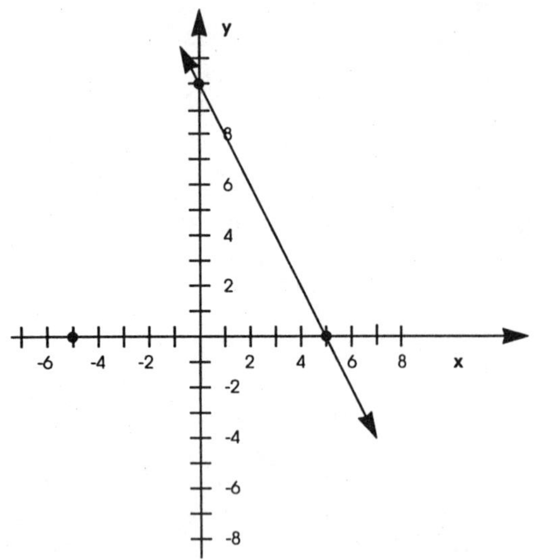

29. x + y = 6

Set y = 0 and solve for x to get an x-intercept of 6.

Set x = 0 and solve for y to get a y-intercept of 6.

y = – 2

This is a horizontal line two units below the x-axis.

The lines cross at the point (8, –2). So the system

$$\begin{cases} x + y = 6 \\ \quad\;\; y = -2 \end{cases}$$

is satisfied by the point (8, –2).

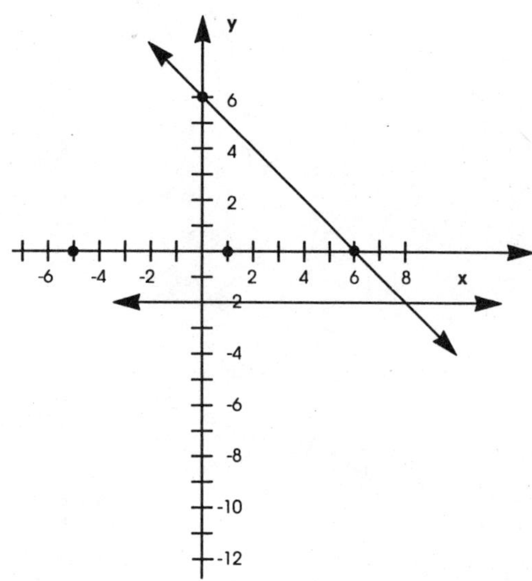

33. $y = 4$

This is a horizontal line four units above the x-axis.

$x = -1$

This is a vertical line one unit to the left of the y-axis.

The lines cross at the point $(-1, 4)$. So the system

$$\begin{cases} y = 4 \\ x = -1 \end{cases}$$

is satisfied by the point $(-1, 4)$.

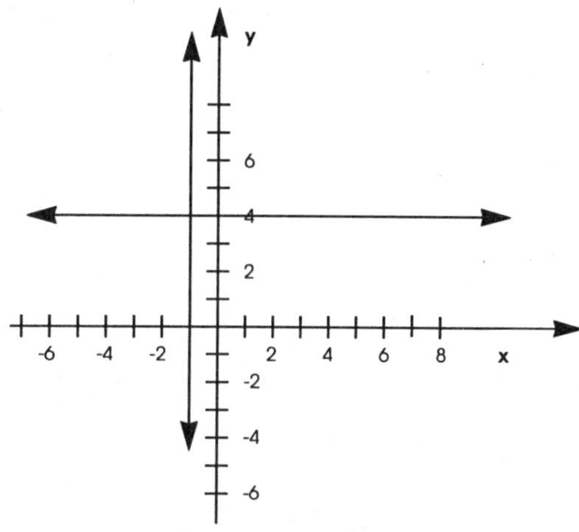

35. A solution to a system of equations is an ordered pair that satisfies both equations simultaneously.

36. It is impossible for a system of two linear equations to have exactly two solutions. It this were possible, then we would be able to draw two straight lines that intersect exactly twice. But two straight lines either do not intersect at all (when they are parallel), intersect in one point, or intersect in infinitely many points (when they coincide).

37. 1-b, 2-e, 3-a, 4-c, 5-d, 6-f

39. $9 - 5(x + 4) > 9$

$9 - 5x - 20 > 9$

$-5x - 11 > 9$

$-5x > 20$

$x < -4$

-4 -2 0 2 4

41.
$$\begin{aligned} 1 &\le 7-2x < 13 \\ -6 &\le -2x < 6 \\ 3 &\ge x > -3 \end{aligned}$$

or

$$-3 < x \le 3$$

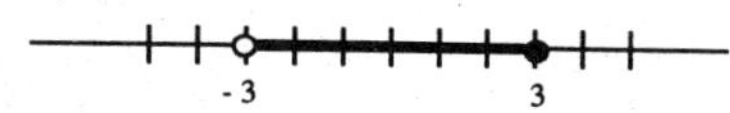

Exercises 6.3

3.
$$\begin{cases} 2x + y = 5 \\ x - y = 4 \end{cases}$$

Add: $3x = 9$

$x = 3$

$$\begin{aligned} 2x + y &= 5 \\ 2(3) + y &= 5 \\ 6 + y &= 5 \\ y &= -1 \end{aligned}$$

Solution: $(3, -1)$

CHECK: $x - y = 4$

$$\begin{aligned} (3) - (-1) &\stackrel{?}{=} 4 \\ 3 + 1 &\stackrel{?}{=} 4 \\ 4 &\stackrel{\checkmark}{=} 4 \end{aligned}$$

7.
$$\begin{cases} 2x + y = 15 \xrightarrow{\text{as is}} \\ x - 2y = 0 \xrightarrow{\text{multiply by } -2} \end{cases}$$

$$\begin{aligned} 2x + y &= 15 \\ -2x + 4y &= 0 \end{aligned}$$

Add: $5y = 15$

$y = 3$

$$\begin{aligned} 2x + y &= 15 \\ 2x + (3) &= 15 \\ 2x + 3 &= 15 \\ 2x &= 12 \\ x &= 6 \end{aligned}$$

Solution: $(6, 3)$

CHECK: $x - 2y = 0$

$$(6) - 2(3) \stackrel{?}{=} 0$$
$$6 - 6 \stackrel{?}{=} 0$$
$$0 \stackrel{\checkmark}{=} 0$$

11. $\begin{cases} x + 2y = 9 \\ \quad y = 3x + 1 \end{cases}$

Substitute the value of y given in the second equation into the first.

$$\begin{aligned} x + 2(3x + 1) &= 9 \\ x + 6x + 2 &= 9 \\ 7x + 2 &= 9 \\ 7x &= 7 \\ x &= 1 \end{aligned}$$

When $x = 1$,

$y = 3(1) + 1 = 3 + 1 = 4$

Solution: $(1, 4)$

CHECK: $x + 2y = 9$

$$(1) + 2(4) \stackrel{?}{=} 9$$
$$1 + 8 \stackrel{?}{=} 9$$
$$9 \stackrel{\checkmark}{=} 9$$

17. $\begin{cases} 2r - 5s = 9 \\ \quad s = 1 - r \end{cases}$

Substitute the value of s given in the second equation into the first.

$$\begin{aligned} 2r - 5(1 - r) &= 9 \\ 2r - 5 + 5r &= 9 \\ 7r - 5 &= 9 \\ 7r &= 14 \\ r &= 2 \end{aligned}$$

When $r = 2$, $s = 1 - 2 = -1$

Solution: $(2, -1)$

CHECK: $2r - 5s = 9$

$$2(2) - 5(-1) \stackrel{?}{=} 9$$
$$4 + 5 \stackrel{?}{=} 9$$
$$9 \stackrel{\checkmark}{=} 9$$

21. $\begin{cases} r + 2t = 10 \xrightarrow{\text{as is}} \\ 3r + t = -15 \xrightarrow{\text{multiply by } -2} \end{cases}$

$$\begin{aligned} r + 2t &= 10 \\ -6r - 2t &= 30 \\ \hline \text{Add: } -5r \quad &= 40 \\ r \quad &= 8 \end{aligned}$$

$$\begin{aligned} r + 2t &= 10 \\ (-8) + 2t &= 10 \\ -8 + 2t &= 10 \\ 2t &= 18 \\ t &= 9 \end{aligned}$$

Solution: (–8, 9)

CHECK: $3r + t = -15$

$$\begin{aligned} 3(-8) + (9) &\stackrel{?}{=} -15 \\ -24 + 9 &\stackrel{?}{=} -15 \\ -15 &\stackrel{\checkmark}{=} -15 \end{aligned}$$

23. $\begin{cases} 6x + y = 6 \\ 4x + 1 = y \end{cases}$

Rewrite the second equation in the system as $4x - y = -1$.

$$\begin{aligned} 6x + y &= 6 \\ \underline{4x - y} &\underline{= -1} \end{aligned}$$

Add: $10x = 5$

$$x = \frac{5}{10} = \frac{1}{2}$$

$$\begin{aligned} 6x + y &= 6 \\ 6\left(\frac{1}{2}\right) + y &= 6 \\ 3 + y &= 6 \\ y &= 3 \end{aligned}$$

Solution: $\left(\frac{1}{2}, 3\right)$

CHECK: $4x + 1 = y$

$$\begin{aligned} 4\left(\frac{1}{2}\right) + 1 &\stackrel{?}{=} (3) \\ 2 + 1 &\stackrel{?}{=} 3 \\ 3 &\stackrel{\checkmark}{=} 3 \end{aligned}$$

27. $\begin{cases} 11a - 2b = 30 \xrightarrow{\text{multiply by 3}} \\ 3a + 3b = -6 \xrightarrow{\text{multiply by 2}} \end{cases}$

$$\begin{aligned} 33a - 6b &= 90 \\ 6a + 6b &= -12 \\ \hline \text{Add: } 39a &= 78 \\ a &= 2 \end{aligned}$$

$$11a - 2b = 30$$
$$11(2) - 2b = 30$$
$$22 - 2b = 30$$
$$-2b = 8$$
$$b = -4$$

Solution: (2, –4)

CHECK: $3a + 3b = -6$

$$3(2) + 3(-4) \stackrel{?}{=} -6$$
$$6 + (-12) \stackrel{?}{=} -6$$
$$-6 \stackrel{\checkmark}{=} -6$$

31. $\begin{cases} 5x + 2y = 4y + 9 \xrightarrow{\text{subtract } 4y \text{ from both sides}} \\ 5x - 2y = 9 \\ y = x - 3 \xrightarrow{\text{subtract } x \text{ from both sides}} \end{cases}$

$$-x + y = -3$$

$$5x - 2y = 9 \xrightarrow{\text{as is}}$$
$$-x + y = -3 \xrightarrow{\text{multiply by 2}}$$

$$\begin{array}{r} 5x - 2y = 9 \\ -2x + 2y = -6 \\ \hline \text{Add: } 3x = 3 \end{array}$$
$$x = 1$$

$$y = x - 3$$
$$y = (1) - 3$$
$$y = 1 - 3$$
$$y = -2$$

Solution: (1, –2)

CHECK: $5x + 2y = 4y + 9$

$$5(1) + 2(-2) \stackrel{?}{=} 4(-2) + 9$$
$$5 + (-4) \stackrel{?}{=} -8 + 9$$
$$1 \stackrel{\checkmark}{=} 1$$

33. $\begin{cases} \dfrac{x}{2} + \dfrac{y}{3} = 1 \xrightarrow{\text{multiply by 6}} \\ \dfrac{x}{4} - y = 11 \xrightarrow{\text{multiply by 4}} \end{cases}$

$$3x + 2y = 6 \xrightarrow{\text{multiply by 2}}$$

$$x - 4y = 44 \xrightarrow{\text{as is}}$$

$$\begin{aligned} 6x + 4y &= 12 \\ x - 4y &= 44 \\ \hline \text{Add: } 7x &= 56 \\ x &= 8 \end{aligned}$$

$$\begin{aligned} \frac{x}{2} + \frac{y}{3} &= 1 \\ \frac{(8)}{2} + \frac{y}{3} &= 1 \\ 4 + \frac{y}{3} &= 1 \\ \frac{y}{3} &= -3 \\ y &= -9 \end{aligned}$$

Solution: (8, –9)

CHECK: $\dfrac{x}{4} - y = 11$

$$\begin{aligned} \frac{(8)}{4} - (-9) &\stackrel{?}{=} 11 \\ 2 + 9 &\stackrel{?}{=} 11 \\ 11 &\stackrel{\checkmark}{=} 11 \end{aligned}$$

37. $\begin{cases} .4x + .2y = 8 \xrightarrow{\text{multiply by 10}} \\ .7x - .3y = 1 \xrightarrow{\text{multiply by 10}} \end{cases}$

$$4x + 2y = 80 \xrightarrow{\text{multiply by 3}}$$

$$7x - 3y = 10 \xrightarrow{\text{multiply by 2}}$$

$$\begin{aligned} 12x + 6y &= 240 \\ 14x - 6y &= 20 \\ \hline \text{Add: } 26x &= 260 \\ x &= 10 \end{aligned}$$

$$\begin{aligned} .4x + .2y &= 8 \\ .4(10) + .2y &= 8 \\ 4 + .2y &= 8 \\ .2y &= 4 \\ 2y &= 40 \\ y &= 20 \end{aligned}$$

Solution: (10, 20)

CHECK: $.7x - .3y = 1$

$$.7(10) - .3(20) \stackrel{?}{=} 1$$

$$7 - 6 \stackrel{?}{=} 1$$

$$1 \stackrel{\surd}{=} 1$$

41. (5.7, 1.1)

43. (1.9, –.6)

45. The student forgot to multiply the right hand side of the second equation by 2.

46. The student mistakenly multiplied 0 times 2 in the second equation and got a product of 2 rather than 0.

Exercises 6.4

3. Let n = # of nickels that Sam has

Let q = # of quarters that Sam has

$$\begin{cases} n + q = 80 \xrightarrow{\text{multiply by } -5} \\ 5n + 25q = 1360 \xrightarrow{\text{as is}} \end{cases}$$

$$\begin{aligned} -5n - 5q &= -400 \\ 5n + 25q &= 1360 \\ \hline \text{Add: } 20q &= 960 \\ q &= 48 \end{aligned}$$

$$\begin{aligned} n + q &= 80 \\ n + 48 &= 80 \\ n &= 32 \end{aligned}$$

Thus, Sam has 32 nickels and 48 quarters.

CHECK: $32 + 48 \stackrel{\surd}{=} 80$

$32(5) + 48(25) = 160 + 1200 \stackrel{\surd}{=} 1360$

5. Let x = speed of the slower car.

Let y = speed of the faster car.

$$\begin{cases} y = x + 15 \xrightarrow{\text{subtract x from both sides}} \\ 5x + 5y = 275 \xrightarrow{\text{as is}} \end{cases}$$

$$-x + y = 15 \xrightarrow{\text{multiply by 5}}$$

$$5x + 5y = 275 \xrightarrow{\text{as is}}$$

$$\begin{aligned} -5x + 5y &= 75 \\ 5x + 5y &= 275 \\ \hline \text{Add: } 10y &= 350 \\ y &= 35 \end{aligned}$$

35 = x + 15

20 = x

Thus, the speed of the slower car is 20 kph and the speed of the faster car is 35 kph.

CHECK: The difference of the speeds is 35 – 20 = 15 kph. The slower car travels 5(20) = 100 km and the faster car travels 5(35) = 175 km. The distance between the cars after 5 hours is 100 + 175 = 275 km.

7. Let x = amount of money that Carmen invested at 6%.

Let y = amount of money that Carmen invested at 7%.

$$\begin{cases} x + y = 1700 \xrightarrow{\text{as is}} \\ .06x + .07y = 110 \xrightarrow{\text{multiply by 100}} \end{cases}$$

$$x + y = 1700 \xrightarrow{\text{multiply by } -6}$$
$$6x + 7y = 11000 \xrightarrow{\text{as is}}$$

$$\begin{aligned} -6x - 6y &= -10200 \\ 6x + 7y &= 11000 \\ \hline \text{Add: } y &= 800 \end{aligned}$$

$$\begin{aligned} x + y &= 1700 \\ x + 800 &= 1700 \\ x &= 900 \end{aligned}$$

Thus, Carmen invested \$900 at 6% and \$800 at 7%.

CHECK: $\$900 + \$800 \stackrel{\checkmark}{=} \$1700$

$.06(\$900) + .07(\$800) = \$54 + \$56 \stackrel{\checkmark}{=} \$110$

9. Let x = price of a cassette

Let y = price of a CD

$$\begin{cases} 4x + 6y = 107.66 \xrightarrow{\text{as is}} \\ 5x + 3y = 76.30 \xrightarrow{\text{multiply by } -2} \end{cases}$$

$$\begin{aligned} 4x + 6y &= 107.66 \\ -10x - 6y &= -152.60 \\ \hline \text{Add: } -6x &= -44.94 \\ x &= 7.49 \end{aligned}$$

$$\begin{aligned} 4x + 6y &= 107.66 \\ 4(7.49) + 6y &= 107.66 \\ 29.96 + 6y &= 107.66 \\ 6y &= 77.70 \\ y &= 12.95 \end{aligned}$$

Thus, a cassette costs \$7.49 and a CD costs \$12.95.

CHECK: $4(\$7.49) + 6(\$12.95) = \$29.96 + \$77.70 \stackrel{\checkmark}{=} \$107.66$

$5(\$7.49) + 3(\$12.95) = \$37.45 + \$38.85 \stackrel{\checkmark}{=} \$76.30$

. Let L = length of the rectangle

Let W = width of the rectangle

$$\begin{cases} L = 2W \\ 2L + 2W = 28 \end{cases}$$

Substitute the value of L from the first equation into the second to get

$$2(2W) + 2W = 28$$
$$4W + 2W = 28$$
$$6W = 28$$
$$W = \frac{28}{6} = \frac{14}{3}$$

$$\text{So } L = 2W = 2\left(\frac{14}{3}\right) = \frac{28}{3}$$

Thus, the width of the rectangle is

$\frac{14}{3}$ inches and the length is $\frac{28}{3}$ inches.

CHECK: $\frac{28}{3}$ is twice as large as $\frac{14}{3}$.

The perimeter of the rectangle is $2\left(\frac{28}{3}\right) + 2\left(\frac{14}{3}\right) = \frac{56}{3} + \frac{28}{3} = \frac{84}{3} = 28$, as required.

15. Let x = larger number

Let y = smaller number

$$\begin{cases} \frac{x}{y} = \frac{6}{5} \\ x - y = 8 \end{cases}$$

Solve the second equation for x, obtaining x = y + 8. Then substitute this result into the first equation:

$$\frac{y+8}{y} = \frac{6}{5}$$
$$\frac{5y}{1} \cdot \frac{y+8}{y} = \frac{5y}{1} \cdot \frac{6}{5}$$
$$5(y+8) = 6y$$
$$5y + 40 = 6y$$
$$40 = y$$

Then x = y + 8 = 40 + 8 = 48.

Thus, the numbers are 48 and 40.

CHECK: $\frac{48}{40} = \frac{6(8)}{5(8)} \stackrel{\checkmark}{=} \frac{6}{5}$ and

$48 - 40 \stackrel{\checkmark}{=} 8.$

19. Let x = cost of a receiver.

Let y = cost of a turntable.

$$\begin{cases} 8x + 4y = 2060 \xrightarrow{\text{multiply by 3}} \\ 5x + 6y = 1690 \xrightarrow{\text{multiply by -2}} \end{cases}$$

$$\begin{array}{r} 24x + 12y = \ \ 6180 \\ -10x - 12y = -3380 \\ \hline \text{Add: } 14x = 2800 \\ x = 200 \end{array}$$

$$\begin{aligned} 8(200) + 4y &= 2060 \\ 1600 + 4y &= 2060 \\ 4y &= 460 \\ y &= 115 \end{aligned}$$

Thus, a receiver costs \$200 and a turntable costs \$115.

CHECK: $8(\$200) + 4(\$115)$

$= \$1600 + \$460 = \ \$2060$

$5(\$200) + 6(\$115)$

$= \$1000 + \$690 = \ \$1690$

23. Let x = # of \$7 books bought.

Let y = # of \$9 books bought.

$$\begin{cases} x + \ \ y = \ 35 \xrightarrow{\text{multiply by -7}} \\ 7x + 9y = 271 \xrightarrow{\text{as is}} \end{cases}$$

$$\begin{array}{r} -7x - 7y = -245 \\ 7x + 9y = \ \ 271 \\ \hline \text{Add: } 2y = \ \ \ \ 26 \\ y = \ \ \ \ 13 \end{array}$$

$$\begin{aligned} x + y &= 35 \\ x + 13 &= 35 \\ x &= 22 \end{aligned}$$

Thus, the bookstore bought 22 books at \$7 each and 13 books at \$9 each.

CHECK: $22 + 13 \stackrel{\surd}{=} 35$

$22(\$7) + 13(\$9) = \$154 + \$117$

$\stackrel{\surd}{=} \ \$271$

7. Let p = speed (in mph) of the plane in still air.

Let w = speed (in mph) of the wind.

$$\begin{cases} \quad p + w = 150 \\ \underline{\quad p - w = \ \ 90} \\ \text{Add: } 2p = 240 \\ \qquad\quad p = 120 \end{cases}$$

$$\begin{aligned} p + w &= 150 \\ 120 + w &= 150 \\ w &= 30 \end{aligned}$$

Thus, the speed of the plane in still air is 120 mph and the speed of the wind is 30 mph.

CHECK: With the tailwind, the speed of the plane is increased by the speed of the wind, giving 120 + 30 = 150. With the headwind, the speed of the plane is decreased by the speed of the wind, giving 120 – 30 = 90.

Chapter 6 Review Exercises

1. $x + y = 4$

To find x-intercept, set y = 0 and solve for x. We get x = 4.

To find y-intercept, set x = 0 and solve for y. We get y = 4.

$x - y = 0$

Here, both the x and y intercepts are 0. To find a second point on this line, choose y = 1 and find x = 1. This gives (1, 1).

The lines cross at the point (2, 2). So the system

$$\begin{cases} x + y = 4 \\ x - y = 0 \end{cases}$$

is satisfied by the point (2, 2).

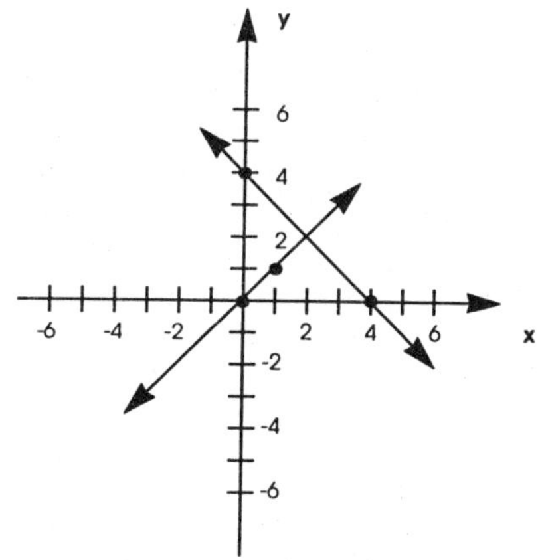

3. $x - 2y = 8$

To find x-intercept, set $y = 0$ and solve for x. We get $x = 8$.

To find y-intercept, set $x = 0$ and solve for y. We get $y = -4$.

$y = x - 5$

To find x-intercept, set $y = 0$ and solve for x. We get $x = 5$.

To find y-intercept, set $x = 0$ and solve for y. We get $y = -5$.

The lines cross at the point (2, –3). So the system

$$\begin{cases} x - 2y = 8 \\ \quad y = x - 5 \end{cases}$$

is satisfied by the point (2, –3).

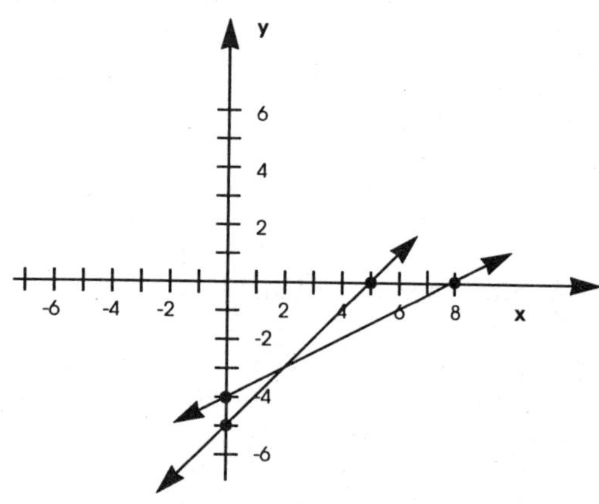

5. $y = x$

Here, both the x and y intercepts are 0. To find a second point on this line, choose $x = 1$ and find $y = 1$. This gives (1, 1).

$3x - 2y = 6$

To find the x-intercept, set $y = 0$ and solve for x. We get $x = 2$.

To find the y-intercept, set $x = 0$ and solve for y. We get $y = -3$.

The lines cross at the point (6, 6). So the system

$$\begin{cases} \quad y = x \\ 3x - 2y = 6 \end{cases}$$ is satisfied by the point (6, 6).

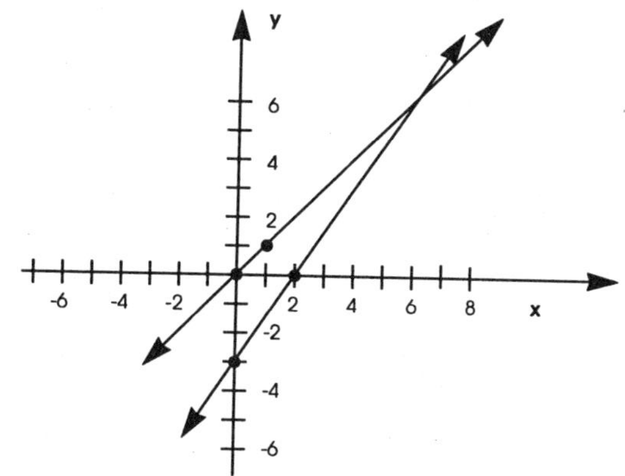

7. $\begin{cases} x + y = 4 \\ x - y = 6 \end{cases}$

$$\begin{aligned} \text{Add: } 2x &= 10 \\ x &= 5 \end{aligned}$$

$x + y = 4$

$5 + y = 4$

$y = -1$

Solution: $(5, -1)$

CHECK: $x - y = 6$

$(5) - (-1) \stackrel{?}{=} 6$

$5 + 1 \stackrel{?}{=} 6$

$6 \stackrel{\checkmark}{=} 6$

9. $\begin{cases} y = 2x - 3 \\ x = 3y - 2 \end{cases}$

Substitute the value of x given in the second equation into the first.

$y = 2(3y - 2) - 3$

$y = 6y - 4 - 3$

$y = 6y - 7$

$-5y = -7$

$y = \frac{7}{5}$

When $y = \frac{7}{5}, x = 3\left(\frac{7}{5}\right) - 2 = \frac{11}{5}$

Solution: $\left(\frac{11}{5}, \frac{7}{5}\right)$

CHECK: $y = 2x - 3$

$\frac{7}{5} \stackrel{?}{=} 2\left(\frac{11}{5}\right) - 3$

$\frac{7}{5} \stackrel{?}{=} \frac{22}{5} - 3$

$\frac{7}{5} \stackrel{\checkmark}{=} \frac{7}{5}$

11. $\begin{cases} 2x - y = 10 \xrightarrow{\text{multiply by 3}} \\ x + 3y = -16 \xrightarrow{\text{as is}} \end{cases}$

$$\begin{array}{r} 6x - 3y = 30 \\ x + 3y = -16 \\ \hline \text{Add: } 7x = 14 \\ x = 2 \end{array}$$

$$\begin{aligned} 2x - y &= 10 \\ 2(2) - y &= 10 \\ 4 - y &= 10 \\ -y &= 6 \\ y &= -6 \end{aligned}$$

Solution: (2, –6)

CHECK: $x + 3y = -16$

$(2) + 3(-6) \stackrel{?}{=} -16$

$2 - 18 \stackrel{?}{=} -16$

$-16 \stackrel{\checkmark}{=} -16$

13. $\begin{cases} x - 5y = 1 \xrightarrow{\text{multiply by -3}} \\ 3x - 2y = 3 \xrightarrow{\text{as is}} \end{cases}$

$$\begin{array}{r} -3x + 15y = -3 \\ 3x - 2y = 3 \\ \hline \text{Add: } 13y = 0 \\ y = 0 \end{array}$$

$$\begin{aligned} x - 5y &= 1 \\ x - 5(0) &= 1 \\ x - 0 &= 1 \\ x &= 1 \end{aligned}$$

Solution: (1, 0)

CHECK: $3x - 2y = 3$

$3(1) - 2(0) \stackrel{?}{=} 3$

$3 - 0 \stackrel{?}{=} 3$

$3 \stackrel{\checkmark}{=} 3$

15. $\begin{cases} 4x - 3y = 10 \xrightarrow{\text{multiply by 2}} \\ 9x + 2y = 5 \xrightarrow{\text{multiply by 3}} \end{cases}$

$$\begin{array}{r} 8x - 6y = 20 \\ \underline{27x + 6y = 15} \\ \text{Add: } 35x = 35 \\ x = 1 \end{array}$$

$$\begin{aligned} 4x - 3y &= 10 \\ 4(1) - 3y &= 10 \\ 4 - 3y &= 10 \\ -3y &= 6 \\ y &= -2 \end{aligned}$$

Solution: $(1, -2)$

CHECK: $9x + 2y = 5$

$$\begin{aligned} 9(1) + 2(-2) &\stackrel{?}{=} 5 \\ 9 + (-4) &\stackrel{?}{=} 5 \\ 9 - 4 &\stackrel{?}{=} 5 \\ 5 &\stackrel{\checkmark}{=} 5 \end{aligned}$$

17. $\begin{cases} \frac{x}{2} + y = 5 \xrightarrow{\text{multiply by 2}} \\ 2y = 8 - x \xrightarrow{\text{add x to both sides}} \end{cases}$

$$\begin{array}{l} x + 2y = 10 \\ \underline{x + 2y = \ 8} \end{array}$$

Subtract: $0 = 2$, a contradiction.

Therefore, the system of equations has no solution.

19. $\begin{cases} x + y - 8 = 2x - 4 \xrightarrow{\text{subtract 2x from both sides}} \\ 2(y - x) = 8 \xrightarrow{\text{divide both sides by 2}} \end{cases}$

$y - x - 8 = -4 \xrightarrow{\text{add 8 to both sides}}$

$y - x = 4 \xrightarrow{\text{as is}}$

$$\begin{array}{l} y - x = 4 \\ \underline{y - x = 4} \end{array}$$

Subtract: $0 = 0$, an identity.

Therefore, the system of equations has infinitely many solutions:
$\{(x, y) | y - x = 4\}$

21. Let x = # of gallons of pure water in the mixture

Let y = # of gallons of 30% alcohol solution in the mixture

$$\begin{cases} x + y = 30 \xrightarrow{\text{as is}} \\ 0x + .30y = .25(30) \xrightarrow{\text{multiply by 100}} \end{cases}$$

$x + y = 30$

$30y = 750$

From the second equation, find y = 25. Then 30 = x + y = x + 25, so x = 5. Thus, 5 gallons of water should be added to 25 gallons of a 30% alcohol solution to produce 30 gallons of a 25% alcohol solution.

CHECK: .30(25) = 7.5 gallons of pure alcohol.

.25(30) = 7.5 gallons of pure alcohol.

23. Let x = walking speed (in kph).

Let y = jogging speed (in kph).

$$\begin{cases} x + y = 17 \xrightarrow{\text{as is}} \\ 2x + \frac{1}{2}y = 16 \xrightarrow{\text{multiply by -2}} \end{cases}$$

$$\begin{array}{r} x + y = 17 \\ -4x - y = -32 \\ \hline \text{Add: } -3x = -15 \\ x = 5 \end{array}$$

$x + y = 17$

$5 + y = 17$

$y = 12$

Thus, their walking speed is 5 kph and their jogging speed is 12 kph.

CHECK: $1(5) + 1(12) \stackrel{\checkmark}{=} 17$

$2(5) + \frac{1}{2}(12) = 10 + 6 \stackrel{\checkmark}{=} 16$

25. (a) 7
(b) decreasing
(c) 0; occurs when w = 9
(d) 9; occurs when w = 7
(e) the value of R when w = 5

Chapter 6 Practice Test

1. (a) $32 per share
(b) between 10:00 a.m. and 11:00 a.m. and between 12:00 noon and 1:00 p.m.
(c) $30 per share
(d) Yes. This occurs between 11:00 a.m. and 12:00 noon
(e) rising

$$\begin{cases} 3x - 4 = y - 1 \\ 9 + 3y = x \end{cases}$$

Substitute the value of x given in the second equation into the first.

$$\begin{aligned} 3(9 + 3y) - 4 &= y - 1 \\ 27 + 9y - 4 &= y - 1 \\ 9y + 23 &= y - 1 \\ 8y + 23 &= -1 \\ 8y &= -24 \\ y &= -3 \end{aligned}$$

When $y = -3$, $x = 9 + 3(-3) = 9 - 9 = 0$

Solution: $(0, -3)$

CHECK: $3x - 4 = y - 1$

$$\begin{aligned} 3(0) - 4 &\stackrel{?}{=} (-3) - 1 \\ 0 - 4 &\stackrel{?}{=} -3 - 1 \\ -4 &\stackrel{\checkmark}{=} -4 \end{aligned}$$

5. $$\begin{cases} \dfrac{3x}{2} - y = 6 \xrightarrow{\text{multiply by 2}} \\ x - \dfrac{2y}{3} = 5 \xrightarrow{\text{multiply by 3}} \end{cases}$$

$3x - 2y = 12$

$\underline{3x - 2y = 15}$

Subtract: $0 = -3$, a contradiction.

Therefore, the system has no solution.

7. Let x = price of an orchestra ticket (in dollars).

Let y = price of a balcony ticket (in dollars).

$$\begin{cases} 5x + 3y = 227 \xrightarrow{\text{multiply by -2}} \\ 6x + 2y = 238 \xrightarrow{\text{multiply by 3}} \end{cases}$$

$$\begin{aligned} -10x - 6y &= -454 \\ 18x + 6y &= 714 \\ \hline \text{Add: } 8x &= 260 \\ x &= 32.50 \end{aligned}$$

$$\begin{aligned} 5x + 3y &= 227 \\ 5(32.50) + 3y &= 227 \\ 162.50 + 3y &= 227 \\ 3y &= 64.50 \\ y &= 21.50 \end{aligned}$$

Thus, an orchestra ticket costs \$32.50 and a balcony ticket costs \$21.50.

CHECK: $5(\$32.50) + 3(\$21.50) = \$162.50 + \$64.50 \stackrel{\checkmark}{=} \$227$

$6(\$32.50) + 2(\$21.50) = \$195.00 + \$43.00 \stackrel{\checkmark}{=} \$238$

CUMULATIVE REVIEW CHAPTERS 4–6

1. $\dfrac{-24}{42} = \dfrac{(-4)(6)}{(7)(6)} = \dfrac{-4}{7} = -\dfrac{4}{7}$

3. $\dfrac{36s^8y^9}{20s^9y^8} = \dfrac{9t}{5s}$

5. $\dfrac{6x}{25} \cdot \dfrac{10}{x} = \dfrac{12}{5}$

7. $\dfrac{6x}{25} + \dfrac{10}{x} = \dfrac{6x(x)}{25(x)} + \dfrac{10(25)}{x(25)}$

$= \dfrac{6x^2}{25x} + \dfrac{250}{25x}$

$= \dfrac{6x^2 + 250}{25x}$

9. $\dfrac{3t-5}{6t^2} + \dfrac{9t+5}{6t^2}$

$= \dfrac{3t - 5 + 9t + 5}{6t^2}$

$= \dfrac{12t}{6t^2} = \dfrac{2}{t}$

11. $\dfrac{12x^3y^2}{35z^2} \div \dfrac{20xy}{14z}$

$= \dfrac{12x^3y^2}{35z^2} \cdot \dfrac{14z}{20xy} = \dfrac{6x^2y}{25z}$

13. $\dfrac{5}{3x} - \dfrac{7}{2x} = \dfrac{5(2)}{3x(2)} - \dfrac{7(3)}{2x(3)}$

$= \dfrac{10}{6x} - \dfrac{21}{6x} = \dfrac{10-21}{6x}$

$= \dfrac{-11}{6x} = -\dfrac{11}{6x}$

15. $\dfrac{5}{6x^2y} - \dfrac{9}{10xy^3}$

$= \dfrac{5(5y^2)}{6x^2y(5y^2)} - \dfrac{9(3x)}{10xy^3(3x)}$

$= \dfrac{25y^2}{30x^2y^3} - \dfrac{27x}{30x^2y^3}$

$= \dfrac{25y^2 - 27x}{30x^2y^3}$

17. $\left(8 \cdot \frac{4}{x}\right) \div \frac{16}{x^2}$

$= \frac{32}{x} \div \frac{16}{x^2}$

$= \frac{\overset{2}{\cancel{32}}}{1\,\cancel{x}} \cdot \frac{\overset{x}{\cancel{x^2}}}{\cancel{16}_1} = 2x$

19. $2 + \frac{3}{x} - \frac{1}{x^2}$

$= \frac{2}{1} + \frac{3}{x} - \frac{1}{x^2}$

$= \frac{2(x^2)}{1(x^2)} + \frac{3(x)}{x(x)} - \frac{1}{x^2}$

$= \frac{2x^2}{x^2} + \frac{3x}{x^2} - \frac{1}{x^2}$

$= \frac{2x^2 + 3x - 1}{x^2}$

21. $\frac{x}{3} - \frac{x}{4} = \frac{x-4}{6}$ LCD = 12

$12(\frac{x}{3} - \frac{x}{4}) = 12 \cdot \frac{x-4}{6}$

$\frac{\overset{4}{\cancel{12}}}{1} \cdot \frac{x}{\cancel{3}_1} - \frac{\overset{3}{\cancel{12}}}{1} \cdot \frac{x}{\cancel{4}_1}$

$= \frac{\overset{2}{\cancel{12}}}{1} \cdot \frac{x-4}{\cancel{6}_1}$

$4x - 3x = 2(x - 4)$

$x = 2x - 8$

$-x = -8$

$x = 8$

23. $\frac{a}{5} - \frac{a}{6} = \frac{a}{30}$ LCD = 30

$30(\frac{a}{5} - \frac{a}{6}) = 30 \cdot \frac{a}{30}$

$\frac{\overset{6}{\cancel{30}}}{1} \cdot \frac{a}{\cancel{5}_1} - \frac{\overset{5}{\cancel{30}}}{1} \cdot \frac{a}{\cancel{6}_1}$

$= \frac{\overset{1}{\cancel{30}}}{1} \cdot \frac{a}{\cancel{30}_1}$

$6a - 5a = a$

$a = a$ Identity

25. $\frac{7-2y}{4} - \frac{5-4y}{6} = \frac{8y+5}{9}$

LCD = 36

$$36\left(\frac{7-2y}{4}-\frac{5-4y}{6}\right)$$

$$= 36\cdot\frac{8y+5}{9}$$

$$\frac{36}{1}\cdot\frac{7-2y}{4}-\frac{36}{1}\cdot\frac{5-4y}{6}$$

$$= \frac{36}{1}\cdot\frac{8y+5}{9}$$

$$9(7-2y)-6(5-4y)$$

$$= 4(8y+5)$$

$$63-18y-30+24y$$

$$= 32y+20$$

$$6y+33 = 32y+20$$

$$33 = 26y+20$$

$$13 = 26y$$

$$\frac{13}{26} = y$$

$$\frac{1}{2} = y$$

27. $.8x - .07(x-5) = 58.75$

LCD = 100

$$100\,(.8x - .07(x-5))$$

$$= 100\,(58.75)$$

$$100\,(.8x) - 100(.07(x-5))$$

$$= 100\,(58.75)$$

$$80x - 7(x-5) = 5875$$

$$80x - 7x + 35 = 5875$$

$$73x + 35 = 5875$$

$$73x = 5840$$

$$x = 80$$

29. $y = 2x - 6$

x- intercept:

$0 = 2x - 6$

$6 = 2x$

$3 = x$

Plot (3, 0)

y- intercept:

$y = 2(0) - 6$

$y = 0 - 6$

$y = -6$

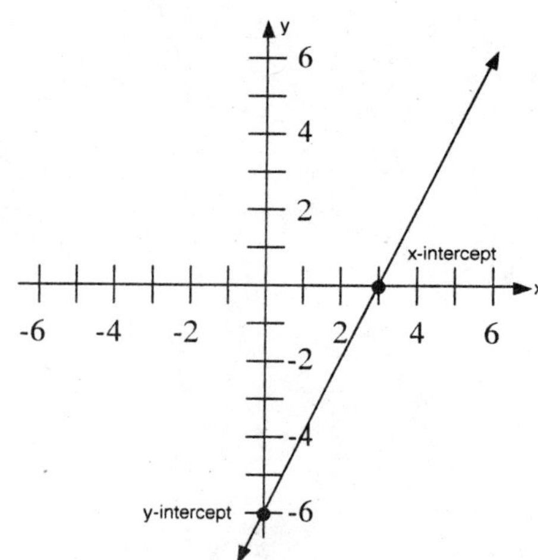

Plot (0, – 6)

check point: choose x = 2

$y = 2(2) - 6$

$y = 4 - 6$

$y = -2$

Plot (2, – 2)

The graph of $y = 2x - 6$

31. $3y - 6x = 12$

x-intercept:

$3(0) - 6x = 12$

$0 - 6x = 12$

$-6x = 12$

$x = -2$

Plot (–2, 0)

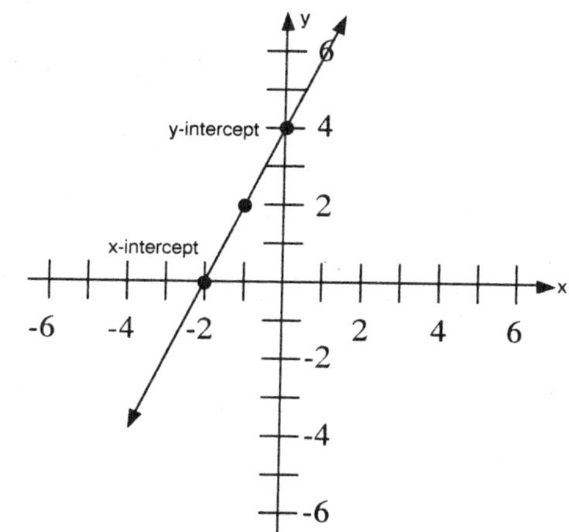

y-intercept:

$3y - 6(0) = 12$

$3y - 0 = 12$

$3y = 12$

$y = 4$

Plot (0, 4)

check point: choose y = 2

$3(2) - 6x = 12$

$6 - 6x = 12$

$-6x = 6$

$x = -1$

Plot $(-1, 2)$

The graph of $3y - 6x = 12$

33. $x - 2 = 0$

Add 2 to both sides of this equation to get the equivalent equation $x = 2$. The graph of this equation is a line parallel to the y-axis and 2 units to the right of it.

The graph of $x - 2 = 0$

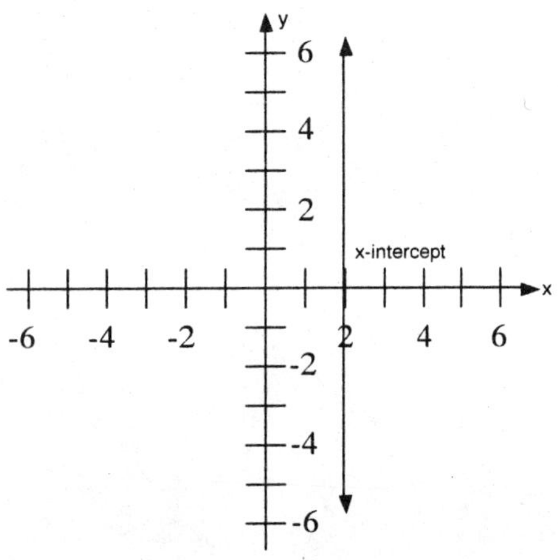

35. $y = 5x$

x-intercept:

$0 = 5x$

$0 = x$

Plot $(0, 0)$

(This implies that y-intercept is 0.)

second point: choose $x = 1$

$y = 5(1)$

$y = 5$

Plot (1, 5)

check point: choose $x = -1$

$y = 5(-1)$

$y = -5$

Plot (−1, −5)

The graph of $y = 5x$

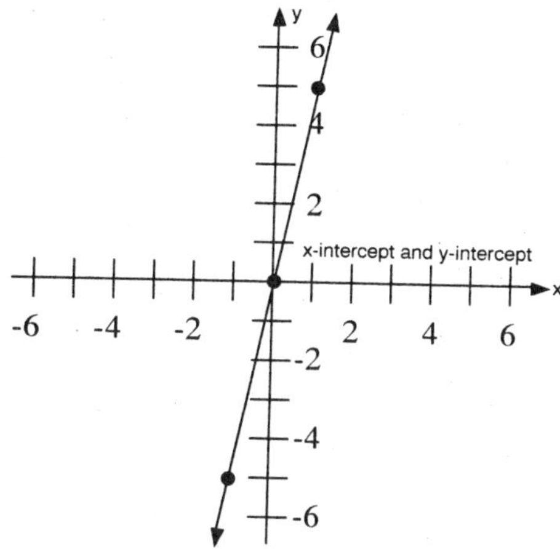

37. $m = \dfrac{y_2 - y_1}{x_2 - x_1} = \dfrac{4-(-1)}{-3-2} = \dfrac{4+1}{-3-2}$

$= \dfrac{5}{-5} = -1$

39. $m = \dfrac{y_2 - y_1}{x_2 - x_1} = \dfrac{4-4}{-1-2} = \dfrac{0}{-3} = 0$

41. $m = 4, (x_1, y_1) = (2, 3)$

$y - y_1 = m(x - x_1)$

$y - 3 = 4(x - 2)$ or $y = 4x - 5$

43. $m = -\dfrac{3}{4},\ b = 3$

$y = mx + b$

$y = -\dfrac{3}{4}x + 3$

45. $m = \dfrac{y_2 - y_1}{x_2 - x_1} = \dfrac{-2-5}{2-(-3)} = \dfrac{-2-5}{2+3}$

$= \dfrac{-7}{5} = -\dfrac{7}{5}$

$(x_1, y_1) = (-3, 5)$

$$y - y_1 = m(x - x_1)$$
$$y - 5 = -\frac{7}{5}(x - (-3))$$
$$y - 5 = -\frac{7}{5}(x + 3)$$
$$\text{or } y = -\frac{7}{5}x + \frac{4}{5}$$

47. $2x - y = 7$

To find x-intercept, set y = 0 and solve for x.

We get $x = \frac{7}{2}$.

To find y-intercept, set x = 0 and solve for y. We get y = –7.

$x + 2y = 6$

To find x-intercept, set y = 0 and solve for x. We get x = 6.

To find y-intercept, set x = 0 and solve for y. We get y = 3.

The lines cross at the point (4, 1). Therefore, the system

$$\begin{cases} 2x - y = 7 \\ x + 2y = 6 \end{cases}$$

is satisfied by the point (4, 1).

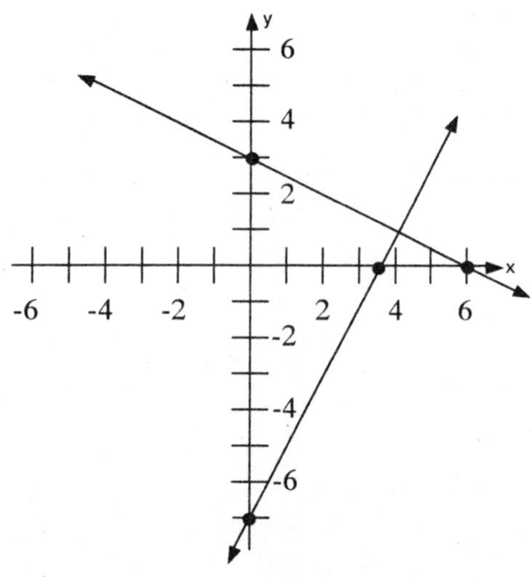

49. $\begin{cases} 2x - y = 7 \xrightarrow{\text{multiply by 2}} \\ x + 2y = 6 \xrightarrow{\text{as is}} \end{cases}$

$4x - 2y = 14$

$\underline{x + 2y = 6}$

Add: $5x = 20$

$x = 4$

$$2x - y = 7$$
$$2(4) - y = 7$$
$$8 - y = 7$$
$$-y = -1$$
$$y = 1$$

Solution: (4, 1)

CHECK: $x + 2y = 6$

$$(4) + (2)1 \stackrel{?}{=} 6$$
$$4 + 2 \stackrel{?}{=} 6$$
$$6 \stackrel{\surd}{=} 6$$

51. $\begin{cases} 4x - 3y = 0 \xrightarrow{\text{as is}} \\ 2x - y = \frac{1}{3} \xrightarrow{\text{multiply by } -3} \end{cases}$

$$4x - 3y = 0$$
$$\underline{-6x + 3y = -1}$$

Add: $-2x = -1$

$$x = \frac{1}{2}$$

$$4x - 3y = 0$$
$$4\left(\frac{1}{2}\right) - 3y = 0$$
$$2 - 3y = 0$$
$$2 = 3y$$
$$\frac{2}{3} = y$$

Solution: $\left(\frac{1}{2}, \frac{2}{3}\right)$

CHECK: $2x - y = \frac{1}{3}$

$$2\left(\frac{1}{2}\right) - \left(\frac{2}{3}\right) \stackrel{?}{=} \frac{1}{3}$$

$$1 - \frac{2}{3} \stackrel{?}{=} \frac{1}{3}$$

$$\frac{1}{3} \stackrel{\checkmark}{=} \frac{1}{3}$$

53. $\begin{cases} y = 5x - 4 \\ x = 3y + 12 \end{cases}$

Substitute the value of y given in the first equation into the second. We get

$x = 3(5x - 4) + 12$

$x = 15x - 12 + 12$

$x = 15x$

$0 = 14x$

$0 = x$

Then $y = 5x - 4 = 5(0) - 4$

$= 0 - 4 = -4$

Solution: $(0, -4)$

CHECK: $x = 3y + 12$

$(0) \stackrel{?}{=} 3(-4) + 12$

$0 \stackrel{?}{=} -12 + 12$

$0 \stackrel{\checkmark}{=} 0$

55. $\begin{cases} x + \frac{y}{2} = 5 \xrightarrow{\text{multiply by 2}} \\ 2x + y = 10 \xrightarrow{\text{as is}} \end{cases}$

$$\begin{array}{r} 2x + y = 10 \\ \underline{2x + y = 10} \end{array}$$

Subtract: $0 = 0$

an identity

Therefore, the system has infinitely many solutions:

$\{(x, y) \mid 2x + y = 10\}$

57. Let x = # of cheaper tickets sold

$360 - x$ = # of more expensive tickets sold

$$6.25x + 8.75(360 - x) = 2850$$
$$100(6.25x + 8.75(360 - x)) = 100 \cdot 2850$$
$$100(6.25x) + 100(8.75(360 - x)) = 100 \cdot 2850$$
$$625x + 875(360 - x) = 285000$$
$$625x + 315000 - 875x = 285000$$
$$-250x + 315000 = 285000$$
$$-250x = -30000$$
$$x = 120$$

Then 360 – x = 360 – 120 = 240. Thus, 120 tickets at \$6.25 each and 240 tickets at \$8.75 were sold.

59. Let x = # of votes that Party A received.

$$\frac{x}{15700} = \frac{8}{5}$$
$$15700 \cdot \frac{x}{15700} = 15700 \cdot \frac{8}{5}$$
$$\frac{15700}{1} \cdot \frac{x}{15700} = \frac{15700}{1} \cdot \frac{8}{5}$$
$$x = 25{,}120$$

Thus, Party A received 25,120 votes.

61. Let t = # of hours needed for the faster car to overtake the slower one.

$$80t = 65(t + \frac{1}{4})$$
$$80t = 65t + 65 \cdot \frac{1}{4}$$
$$15t = \frac{65}{4}$$
$$t = \frac{\frac{65}{4}}{15} = \frac{65}{4} \cdot \frac{1}{15}$$
$$= \frac{65}{60} = 1\frac{5}{60}$$

Thus, the faster car overtakes the slower one after 1 hour and 5 minutes.

Cumulative Practice Test: Chapters 4–6

1. $\frac{12s^2r^2}{5d^2} \cdot \frac{15d^5}{9st^4} = \frac{4sd^3}{t}$

3. $\frac{11a}{9x} - \frac{a}{9x} + \frac{5a}{9x}$

$= \frac{11a - a + 5a}{9x}$

$= \frac{15a}{9x}$

$= \frac{5a}{3x}$

5. $\frac{5}{6ab^2} + \frac{4}{9b} = \frac{5(3)}{6ab^2(3)}$

$+ \frac{4(2ab)}{9b(2ab)}$

$= \frac{15}{18ab^2} + \frac{8ab}{18ab^2}$

$= \frac{15 + 8ab}{18ab^2}$

7. $\frac{a}{6} - \frac{a}{9} = 18$

$18(\frac{a}{6} - \frac{a}{9}) = 18(18)$

$\frac{18}{1} \cdot \frac{a}{6} - \frac{18}{1} \cdot \frac{a}{9} = 324$

$3a - 2a = 324$

$a = 324$

9. $x - 3y = 0$

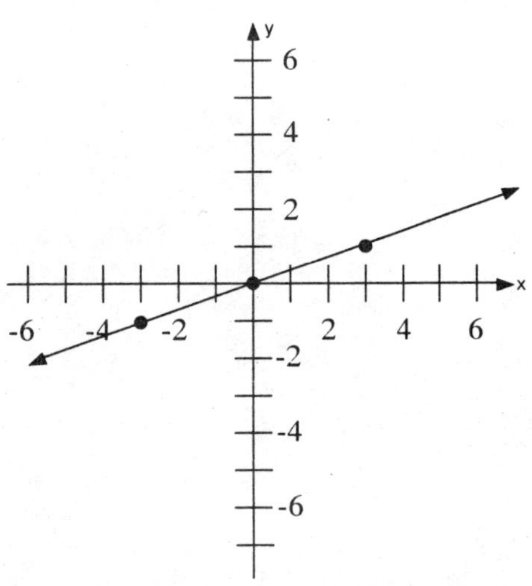

. $x - 3 = 6$

(This is equivalent to the equation $x = 9$.)

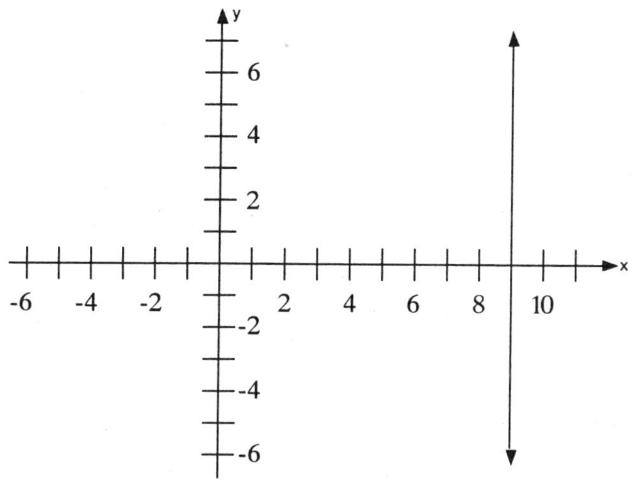

13. $m = \dfrac{y_2 - y_1}{x_2 - x_1} = \dfrac{3-(-4)}{-1-2} = \dfrac{7}{-3} = -\dfrac{7}{3}$

$(x_1, y_1) = (2, -4)$

$y - y_1 = m(x - x_1)$

$y - (-4) = \dfrac{-7}{3}(x - 2)$

$y + 4 = -\dfrac{7}{3}(x - 2)$ or

$y = -\dfrac{7}{3}x + \dfrac{2}{3}$

15. $\begin{cases} \dfrac{x}{2} - y = 5 \xrightarrow{\text{multiply by 2}} \\ -x + 2y = 8 \xrightarrow{\text{as is}} \end{cases}$

$x - 2y = 10$

$\underline{-x + 2y = \ \ 8}$

Add: $0 = 18$

a contradiction

So the system has no solution.

17. Let x = number of 22¢ stamps Jamie bought

y = number of 13¢ stamps Jamie bought

$$\begin{cases} x + y = 28 \xrightarrow{\text{multiply by } -13} \\ 22x + 13y = 535 \xrightarrow{\text{as is}} \end{cases}$$

$$\begin{aligned} -13x - 13y &= -364 \\ \underline{22x + 13y} &\underline{= 535} \end{aligned}$$

Add: $9x = 171$

$x = 19$

$x + y = 28$

$19 + y = 28$

$y = 9$

Thus, Jamie bought 19 22¢ satamps and 9 13¢ stamps.

CHECK: $19 + 9 \stackrel{\checkmark}{=} 28$

$22(19) + 13(9) = 418 + 117 \stackrel{\checkmark}{=} 535$

19. Let t = amount of time Terry walks (in hours)

$$6t + 8\left(t - \frac{1}{3}\right) = 9$$

$$6t + 8t - \frac{8}{3} = 9$$

$$14t - \frac{8}{3} = 9$$

$$14t = \frac{35}{3}$$

$$t = \frac{\frac{35}{3}}{14} = \frac{\overset{5}{\cancel{35}}}{3} \cdot \frac{1}{\underset{2}{\cancel{14}}} = \frac{5}{6}$$

Since Terry walks for $\frac{5}{6}$ hours or for 50 minutes, Terry and Tom will be 9 km apart at 11:50 a.m.

CHECK: From 11:00 a.m. to 11:50 a.m., Terry walks $6\left(\frac{5}{6}\right) = 5$ km. From 11:20 a.m. to 11:50 a.m., Tom walks $8\left(\frac{1}{2}\right) = 4$ km.

5 km + 4 km $\stackrel{\checkmark}{=}$ 9 km

21. Let r = regular price of a blouse

s = sale price of a blouse

$$\begin{cases} 4r + 3s = 55.30 \xrightarrow{\text{as is}} \\ 2r + 5s = 50.40 \xrightarrow{\text{multiply by } -2} \end{cases}$$

$$\begin{aligned} 4r + 3s &= 55.30 \\ \underline{-4r - 10s} &= \underline{-100.80} \end{aligned}$$

Add: $-7s = -45.50$

$$s = 6.50$$

$$\begin{aligned} 4r + 3s &= 55.30 \\ 4r + 3(6.50) &= 55.30 \\ 4r + 19.50 &= 55.30 \\ 4r &= 35.80 \\ r &= 8.95 \end{aligned}$$

So the regular price of a blouse is $8.95 and the sale price of a blouse is $6.50.

CHECK: 4 blouses at regular price and 3 at the sale price cost 4($8.95) + 3($6.50) = $35.80 + $19.50 = $55.30.

2 blouses at regular price and 5 at the sale price cost 2($8.95) + 5($6.50) = $17.90 + $32.50 = $50.40.

CHAPTER 7
EXPONENTS AND POLYNOMIALS

Exercises 7.1

1. $x^3x^2 = x^{3+2} = x^5$

3. $(x^3)^2 = x^{3\cdot2} = x^6$

5. $x^3xx^5 = x^{3+1+5} = x^9$

7. $10^4 10^5 = 10^{4+5} = 10^9$

9. $2^3 3^4$ cannot be simplified

11. $\dfrac{y^3y^5}{y^2y^4} = \dfrac{y^{3+5}}{y^{2+4}} = \dfrac{y^8}{y^6} = y^{8-6} = y^2$

13. $\dfrac{9u^9v^8}{3u^3v^4} = \dfrac{9}{3}u^{9-3}v^{8-4} = 3u^6v^4$

15. $\dfrac{(a^3)^5}{(a^4)^2} = \dfrac{a^{3\cdot5}}{a^{4\cdot2}} = \dfrac{a^{15}}{a^8} = a^{15-8} = a^7$

17. $(-x^2)^4 = (-1)^4(x^2)^4 = 1\cdot x^{2\cdot4} = x^8$

19. $(x^2y)^2 = (x^2)^2y^2 = x^{2\cdot2}y^2 = x^4y^2$

21. $(x^2y^3)^5 = (x^2)^5(y^3)^5 = x^{2\cdot5}y^{3\cdot5} = x^{10}y^{15}$

23. $(2r^3s^5)^4 = 2^4(r^3)^4(s^5)^4$

 $= 16r^{3\cdot4}s^{5\cdot4} = 16r^{12}s^{20}$

25. $(-x^3y)^3 = (-1)^3(x^3)^3y^3 = -1x^{3\cdot3}y^3$

 $= -x^9y^3$

27. $\left(\dfrac{x^3}{y^2}\right)^4 = \dfrac{(x^3)^4}{(y^2)^4} = \dfrac{x^{3\cdot4}}{y^{2\cdot4}} = \dfrac{x^{12}}{y^8}$

29. $(2x^3)^4(3x^2)^2 = 2^4(x^3)^4\cdot3^2(x^2)^2$

 $= 16x^{3\cdot4}\cdot9x^{2\cdot2} = (16\cdot9)x^{12}x^4$

 $= 144x^{12+4} = 144x^{16}$

31. $$\frac{(x^4y^2)^3}{x^5(y^3)^2} = \frac{(x^4)^3(y^2)^3}{x^5(y^3)^2} = \frac{x^{4\cdot 3}y^{2\cdot 3}}{x^5y^{3\cdot 2}}$$

$$= \frac{x^{12}\cancel{y^6}}{x^5\cancel{y^6}} = \frac{x^{12}}{x^5} = x^{12-5} = x^7$$

33. $$\frac{(3x^5y^4)^2}{9(x^3y)^3} = \frac{3^2(x^5)^2(y^4)^2}{9(x^3)^3y^3} = \frac{9x^{5\cdot 2}y^{4\cdot 2}}{9x^{3\cdot 3}y^3}$$

$$= \frac{\cancel{9}x^{10}y^8}{\cancel{9}x^9y^3} = \frac{x^{10}}{x^9}\cdot\frac{y^8}{y^3}$$

$$= x^{10-9}y^{8-3} = xy^5$$

35. $$\left(\frac{x^2y}{4u}\right)^4 = \frac{(x^2y)^4}{(4u)^4} = \frac{(x^2)^4y^4}{4^4u^4}$$

$$= \frac{x^{2\cdot 4}y^4}{256u^4} = \frac{x^8y^4}{256u^4}$$

37. $$\left(\frac{2x^3y^4}{xy^6}\right)^5 = \left(2\cdot\frac{x^3}{x}\cdot\frac{y^4}{y^6}\right)^5$$

$$= \left(2x^{3-1}\cdot\frac{1}{y^2}\right)^5 = \left(\frac{2x^2}{y^2}\right)^5$$

$$= \frac{(2x^2)^5}{(y^2)^5} = \frac{2^5(x^2)^5}{(y^2)^5} = \frac{32x^{2\cdot 5}}{y^{2\cdot 5}}$$

$$= \frac{32x^{10}}{y^{10}}$$

39. $$\left(\frac{-3a^2b^3}{2c}\right)^3 = \frac{(-3a^2b^3)^3}{(2c)^3}$$

$$= \frac{(-3)^3(a^2)^3(b^3)^3}{2^3c^3}$$

$$= \frac{-27a^{2\cdot 3}b^{3\cdot 3}}{8c^3} = \frac{-27a^6b^9}{8c^3}$$

41. $$\frac{-3^2}{(-3)^2} = \frac{-\cancel{9}^{\,1}}{\cancel{9}_{\,1}} = \frac{-1}{1} = -1$$

43. $$\frac{-x^2}{(-x)^2} = \frac{-\cancel{x^2}^{\,1}}{(-1)^2\cancel{x^2}_{\,1}} = \frac{-1}{1} = -1$$

45. $\frac{-u^3}{(-u)^3} = \frac{-u^3}{(-1)^3 u^3} = \frac{-u^3}{-u^3} = \frac{-1}{-1} = 1$

47. $\frac{(-x^2)^4}{-(x^3)^2} = \frac{(-1)^4(x^2)^4}{-(x^3)^2} = \frac{1 \cdot x^{2 \cdot 4}}{-x^{3 \cdot 2}}$

$= \frac{x^8}{-x^6} = -x^{8-6} = -x^2$

49. $\frac{-2^4 + 3^2}{(-4+3)^2} = \frac{-16+9}{(-1)^2} = \frac{-7}{1} = -7$

51. $\frac{-w^2}{(-w)^2} = \frac{-2^2}{(-2)^2} = \frac{-4}{4} = -1$

53. $\frac{-w^4 + x^2}{(-y+x)^2} = \frac{-2^4 + (-3)^2}{(-4+(-3))^2} = \frac{-16+9}{(-7)^2}$

$= \frac{-7}{49} = -\frac{1}{7}$

55. $(w - x + y - z)^2 = (2 - (-3) + 4 - (-5))^2$

$= (14)^2 = 196$

57. $\frac{x(-y)}{x-y} = \frac{-3(-4)}{-3-4} = \frac{12}{-7} = \frac{-12}{7}$

59. When we multiply two powers of the same base, we keep the base and add the exponents. When we raise a power to a power, we keep the base and multiply the exponents.

60. (a) $\frac{x^8}{x^8} = x^{8-8} = x^0$

(b) $\frac{x^4}{x^7} = x^{4-7} = x^{-3}$

61. (a) According to Exponent Rule 2, we must multiply the exponents, not add them.

(b) According to Exponent Rule 1, we must add the exponents, not multiply them.

(c) According to Exponent Rule 3, we must subtract the exponents, not divide them.

(d) Since x^2 and x^3 are unlike terms, we cannot combine them when they are added.

63. $\frac{2}{x} \div \frac{8}{y} = \frac{2}{x} \cdot \frac{y}{8} = \frac{y}{4x}$

65. $\frac{3}{2a}-\frac{4}{a^2}=\frac{3a}{2a^2}-\frac{8}{2a^2}$

$$=\frac{3a-8}{2a^2}$$

67. $\frac{x}{3}-\frac{x}{4}=3$

$$12\left(\frac{x}{3}-\frac{x}{4}\right)=12\cdot 3$$

$$\frac{12}{1}\cdot\frac{x}{3}-\frac{12}{1}\cdot\frac{x}{4}=36$$

$$4x-3x=36$$

$$x=36$$

69. Let d = number of dimes

40 – d = number of quarters

$$10d+25(40-d)=670$$

$$10d+1000-25d=670$$

$$-15d+1000=670$$

$$-15d=-330$$

$$d=22$$

Then 40 – d = 40 – 22 = 18, so there are 22 dimes and 18 quarters in the collection.

Exercises 7.2

1. (a) $-3(2)=-6$

(b) $x^2x^{-3}=x^{2+(-3)}=x^{-1}=\frac{1}{x}$

(c) $\left(x^2\right)^{-3}=x^{2(-3)}=x^{-6}=\frac{1}{x^6}$

(d) $2^{-3}=\frac{1}{2^3}=\frac{1}{8}$

3. (a) $-4(3)=-12$

(b) $x^3x^{-4}=x^{3+(-4)}=x^{-1}=\frac{1}{x}$

(c) $\left(x^3\right)^{-4}=x^{3(-4)}=x^{-12}=\frac{1}{x^{12}}$

(d) $3^{-4}=\frac{1}{3^4}=\frac{1}{81}$

5. $8^0=1$

7. $5\cdot 4^0=5\cdot 1=5$

9. $xy^0=x\cdot 1=x$

11. $5^{-2} = \frac{1}{5^2} = \frac{1}{25}$

13. $\frac{1}{5^{-2}} = \frac{1}{\frac{1}{5^2}} = \frac{1}{\frac{1}{25}} = 1 \cdot \frac{25}{1} = 25$

15. $x^{-4}x^4 = x^{-4+4} = x^0 = 1$

17. $x^{-4}x^{-6} = x^{-4+(-6)} = x^{-10} = \frac{1}{x^{10}}$

19. $a^2a^{-4}aa^{-7} = a^{2+(-4)+1+(-7)} = a^{-8} = \frac{1}{a^8}$

21. $10^{-3}10^8 = 10^{-3+8} = 10^5 = 100,000$

23. $10^6 10^{-5} 10^{-4} = 10^{6+(-5)+(-4)} = 10^{-3}$

$$= \frac{1}{10^3} = \frac{1}{1000}$$

25. $10^7 10^{-7} = 10^{7+(-7)} = 10^0 = 1$

27. $(xy)^4 = x^4y^4$

29. $2a^{-3} = \frac{2}{1} \cdot \frac{1}{a^3} = \frac{2}{a^3}$

31. $-3y^{-2} = \frac{-3}{1} \cdot \frac{1}{y^2} = -\frac{3}{y^2}$

33. $-\left(3y^{-2}\right) = -\left(\frac{3}{1} \cdot \frac{1}{y^2}\right) = -\left(\frac{3}{y^2}\right) = -\frac{3}{y^2}$

35. $xy^{-1} = \frac{x}{1} \cdot \frac{1}{y} = \frac{x}{y}$

37. $\left(a^4b^3\right)^{-2} = \left(a^4\right)^{-2}\left(b^3\right)^{-2} = a^{4(-2)}b^{3(-2)}$

$$= a^{-8}b^{-6} = \frac{1}{a^8} \cdot \frac{1}{b^6} = \frac{1}{a^8b^6}$$

39. $\left(a^{-4}b^3\right)^{-2} = \left(a^{-4}\right)^{-2}\left(b^3\right)^{-2} = a^{-4(-2)}b^{3(-2)}$

$$= a^8b^{-6} = \frac{a^8}{1} \cdot \frac{1}{b^6} = \frac{a^8}{b^6}$$

41. $\left(3x^{-2}y^3z^{-4}\right)^2 = 3^2\left(x^{-2}\right)^2\left(y^3\right)^2\left(z^{-4}\right)^2$

$$= 9x^{-2(2)}y^{3(2)}z^{-4(2)} = 9x^{-4}y^{6}z^{-8}$$

$$= \frac{9}{1}\cdot\frac{1}{x^4}\cdot\frac{y^6}{1}\cdot\frac{1}{z^8} = \frac{9y^6}{x^4z^8}$$

43. $4\left(x^{-1}y\right)^{-3} = 4\left(x^{-1}\right)^{-3}y^{-3} = 4x^{-1(-3)}y^{-3}$

$$= 4x^3y^{-3} = \frac{4}{1}\cdot\frac{x^3}{1}\cdot\frac{1}{y^3} = \frac{4x^3}{y^3}$$

45. $\frac{x^5}{x^2} = x^{5-2} = x^3$

47. $\frac{-3a^{-3}}{9a^9} = \frac{-3}{9}\cdot\frac{a^{-3}}{a^9} = -\frac{1}{3}a^{-3-9}$

$$= -\frac{1}{3}a^{-12} = -\frac{1}{3}\cdot\frac{1}{a^{12}} = -\frac{1}{3a^{12}}$$

49. $x^{-2} + y^{-1} = \frac{1}{x^2} + \frac{1}{y} = \frac{y}{x^2y} + \frac{x^2}{x^2y}$

$$= \frac{y + x^2}{x^2y}$$

51. $\frac{x^4x^{-10}}{x^{-2}x^{-5}} = \frac{x^{4+(-10)}}{x^{-2+(-5)}} = \frac{x^{-6}}{x^{-7}} = x^{-6-(-7)}$

$$= x^1 = x$$

53. $\frac{x^4y^{-10}}{x^{-2}y^{-5}} = \frac{x^4}{x^{-2}}\cdot\frac{y^{-10}}{y^{-5}} = x^{4-(-2)}y^{-10-(-5)}$

$$= x^6y^{-5} = \frac{x^6}{1}\cdot\frac{1}{y^5} = \frac{x^6}{y^5}$$

55. $\frac{10^{-3}10^5}{10^6 10^{-10}} = \frac{10^{-3+5}}{10^{6+(-10)}} = \frac{10^2}{10^{-4}} = 10^{2-(-4)}$

$$= 10^6 = 1{,}000{,}000$$

57. $\frac{12\left(10^{-3}\right)}{4\left(10^{-7}\right)} = \frac{12}{4}\cdot\frac{10^{-3}}{10^{-7}}$

$$= 3\cdot 10^{-3-(-7)} = 3\cdot 10^4$$

$$= 3\cdot 10000 = 30000$$

59. $\left(\frac{a^{-2}}{a^3}\right)^{-3} = \left(a^{-2-3}\right)^{-3} = \left(a^{-5}\right)^{-3} = a^{-5(-3)} = a^{15}$

61. $$\frac{(x^2y^{-1})^{-1}}{(x^3y^{-2})^2} = \frac{(x^2)^{-1}(y^{-1})^{-1}}{(x^3)^2(y^{-2})^2} = \frac{x^{2(-1)}y^{-1(-1)}}{x^{3(2)}y^{-2(2)}}$$
$$= \frac{x^{-2}y^1}{x^6y^{-4}} = \frac{x^{-2}}{x^6} \cdot \frac{y^1}{y^{-4}} = x^{-2-6}y^{1-(-4)}$$
$$= x^{-8}y^5 = \frac{1}{x^8} \cdot \frac{y^5}{1} = \frac{y^5}{x^8}$$

63. $$\left(\frac{2m^{-2}n^{-3}}{m^{-6}n^{-1}}\right)^{-2} = \left(\frac{2}{1} \cdot \frac{m^{-2}}{m^{-6}} \cdot \frac{n^{-3}}{n^{-1}}\right)^{-2}$$
$$= \left(2m^{-2-(-6)}n^{-3-(-1)}\right)^{-2} = \left(2m^4n^{-2}\right)^{-2}$$
$$= 2^{-2}\left(m^4\right)^{-2}\left(n^{-2}\right)^{-2} = \frac{1}{2^2} \cdot m^{4(-2)}n^{-2(-2)}$$
$$= \frac{1}{4}m^{-8}n^4 = \frac{1}{4} \cdot \frac{1}{m^8} \cdot \frac{n^4}{1} = \frac{n^4}{4m^8}$$

65. $$\left(\frac{x^{-1}y^{-2}}{3x^{-2}y^{-3}}\right)^{-1} = \left(\frac{1}{3} \cdot \frac{x^{-1}}{x^{-2}} \cdot \frac{y^{-2}}{y^{-3}}\right)^{-1}$$
$$= \left(\frac{1}{3}x^{-1-(-2)}y^{-2-(-3)}\right)^{-1} = \left(\frac{1}{3}xy\right)^{-1}$$
$$= \left(\frac{1}{3}\right)^{-1}x^{-1}y^{-1} = \frac{1}{\frac{1}{3}}x^{-1}y^{-1}$$
$$= \frac{3}{1} \cdot \frac{1}{x} \cdot \frac{1}{y} = \frac{3}{xy}$$

67. $$\frac{(2m^{-2}n^{-3})^{-4}}{(m^{-6}n^{-1})^{-2}} = \frac{2^{-4}(m^{-2})^{-4}(n^{-3})^{-4}}{(m^{-6})^{-2}(n^{-1})^{-2}}$$
$$= \frac{2^{-4}m^{-2(-4)}n^{-3(-4)}}{m^{-6(-2)}n^{-1(-2)}} = \frac{2^{-4}m^8n^{12}}{m^{12}n^2}$$
$$= \frac{1}{2^4} \cdot \frac{m^8}{m^{12}} \cdot \frac{n^{12}}{n^2} = \frac{1}{16}m^{8-12}n^{12-2}$$
$$= \frac{1}{16}m^{-4}n^{10} = \frac{1}{16} \cdot \frac{1}{m^4} \cdot \frac{n^{10}}{1}$$
$$= \frac{n^{10}}{16m^4}$$

69. $$\frac{(x^5y)^{-2}(x^{-2}y^3)^2}{(x^{-3}y^{-4})^{-2}} = \frac{(x^5)^{-2}y^{-2} \cdot (x^{-2})^2(y^3)^2}{(x^{-3})^{-2}(y^{-4})^{-2}}$$

$$= \frac{x^{5(-2)}y^{-2}x^{-2(2)}y^{3(2)}}{x^{-3(-2)}y^{-4(-2)}} = \frac{x^{-10}y^{-2}x^{-4}y^{6}}{x^{6}y^{8}}$$

$$= \frac{x^{-10+(-4)}y^{-2+6}}{x^{6}y^{8}} = \frac{x^{-14}y^{4}}{x^{6}y^{8}}$$

$$= \frac{x^{-14}}{x^{6}} \cdot \frac{y^{4}}{y^{8}} = x^{-14-6}y^{4-8} = x^{-20}y^{-4}$$

$$= \frac{1}{x^{20}} \cdot \frac{1}{y^{4}} = \frac{1}{x^{20}y^{4}}$$

71. $x^{-3} = 2^{-3} = \frac{1}{2^3} = \frac{1}{8}$

73. $8x^{-1} = 8 \cdot 2^{-1} = 8 \cdot \frac{1}{2} = 4$

75. $x^{-1} + y^{-1} = 2^{-1} + (-3)^{-1} = \frac{1}{2} + \frac{1}{-3}$

$= \frac{1}{2} - \frac{1}{3} = \frac{1}{6}$

77. $\frac{x^6}{x^4}$ requires us to divide x^6 by x^4, whereas $\frac{x^6}{x^{-4}} = \frac{x^6}{\frac{1}{x^4}} = x^6 \cdot \frac{x^4}{1}$ asks us to multiply x^6 by x^4.

78. When -1 appears in the exponent, it tells us to take the reciprocal of the base. Thus, $3^{-1} = \frac{1}{3}$.

When the minus sign appears in front of the 3, it tells us to take the opposite of 3. Put another way, 3^{-1} is the multiplicative inverse of 3, while -3 is the additive inverse of 3.

79. Let t = number of hours Maria works.

$t - 2$ = number of hours Francis works

$5t + 7(t - 2) = 70$. So
$5t + 7t - 14 = 70$ or
$12t - 14 = 70$. Then $12t = 84$,
so $t = 7$. Seven hours after Maria started, it is 4 P.M.

Exercises 7.3

1. $4{,}530 = 4.53 \times 10^3$

3. $.0453 = 4.53 \times 10^{-2}$

5. $.00007 = 7 \times 10^{-5}$

7. $7,000,000 = 7 \times 10^6$

9. $85,370 = 8.537 \times 10^4$

11. $.0085370 = 8.537 \times 10^{-3}$

13. $90 = 9 \times 10^1$

15. $9 = 9 \times 10^0$

17. $.9 = 9 \times 10^{-1}$

19. $.09 = 9 \times 10^{-2}$

21. $.00000003 = 3 \times 10^{-8}$

23. $28 = 2.8 \times 10^1$

25. $47.5 = 4.75 \times 10^1$

27. $9,727.3 = 9.727.3 \times 10^3$

29. $2.8 \times 10^4 = 28,000$

31. $2.8 \times 10^{-4} =.00028$

33. $4.29 \times 10^7 = 42,900,000$

35. $4.29 \times 10^{-7} =.000000429$

37. $3.52 \times 10^{-3} =.00352$

39. $3.5286 \times 10^5 = 352,860$

41. $.026 \times 10^{-3} =.000026$

43. $$\begin{aligned}(.004)(250) &= \left(4 \times 10^{-3}\right)\left(2.5 \times 10^2\right)\\ &= (4)(2.5) \times 10^{-3}10^2 = 10 \times 10^{-3+2}\\ &= 10 \times 10^{-1} = 1.0 \times 10^0 = 1 \times 1 = 1\end{aligned}$$

45. $\dfrac{.003}{6{,}000} = \dfrac{3 \times 10^{-3}}{6 \times 10^{3}} = \dfrac{3}{6} \times \dfrac{10^{-3}}{10^{3}}$

$= .5 \times 10^{-3-3} = .5 \times 10^{-6} = 5 \times 10^{-7}$

47. $\dfrac{(480)(.008)}{(.24)(4{,}000)} = \dfrac{(4.8 \times 10^{2})(8 \times 10^{-3})}{(2.4 \times 10^{-1})(4 \times 10^{3})}$

$= \dfrac{(4.8)(8)}{(2.4)(4)} \times \dfrac{10^{2}10^{-3}}{10^{-1}10^{3}} = 4 \times \dfrac{10^{2+(-3)}}{10^{-1+3}}$

$= 4 \times \dfrac{10^{-1}}{10^{2}} = 4 \times 10^{-1-2} = 4 \times 10^{-3}$

49. $\dfrac{(.0036)(.005)}{(.01)(.06)} = \dfrac{(3.6 \times 10^{-3})(5 \times 10^{-3})}{(1 \times 10^{-2})(6 \times 10^{-2})}$

$= \dfrac{(3.6)(5)}{(1)(6)} \times \dfrac{10^{-3}10^{-3}}{10^{-2}10^{-2}}$

$= 3 \times \dfrac{10^{-3+(-3)}}{10^{-2+(-2)}}$

$= 3 \times \dfrac{10^{-6}}{10^{-4}} = 3 \times 10^{-6-(-4)} = 3 \times 10^{-2}$

51. 5.98×10^{27} kg

53. $(80{,}000)(9.3 \times 10^{-23})$

$= (8 \times 10^{4})(9.3 \times 10^{-23})$

$= (8)(9.3) \times 10^{4}10^{-23}$

$= 74.4 \times 10^{4+(-23)} = 74.4 \times 10^{-19}$

$= 7.44 \times 10^{-18}$ grams

55. $.00000001 = 1 \times 10^{-8}$ cm

57. $(153)(1 \times 10^{-8}) = 153 \times 10^{-8}$

$= 1.53 \times 10^{-6}$ cm

59. $4{,}250$ million $= (4{,}250)(1{,}000{,}000)$

$= (4.25 \times 10^{3})(1 \times 10^{6})$

$= (4.25)(1) \times 10^{3}10^{6} = 4.25 \times 10^{3+6}$

$= 4.25 \times 10^{9}$ miles

61. Let w = weight of the Earth in tons.

$$\frac{5.98\times 10^{27}}{w}=\frac{888.9}{1}$$

$$5.98\times 10^{27}=\left(8.889\times 10^{2}\right)w$$

$$\frac{5.98\times 10^{27}}{8.889\times 10^{2}}=w$$

$$\left(\frac{5.98}{8.889}\right)\times\frac{10^{27}}{10^{2}}=w$$

$$.6727\times 10^{27-2}=w$$

$$.6727\times 10^{25}=6.727\times 10^{24}=w$$

Thus, the weight of the Earth is 6.727×10^{24} tons.

63. There are (365)(24)(60)(60) = 31,536,000 seconds in one year. Then one light year equals (186,000)(31,536,000) miles.

$$(186,000)(31,536,000)$$
$$=\left(1.86\times 10^{5}\right)\left(3.1536\times 10^{7}\right)$$
$$=(1.86)(3.1536)\times 10^{5}10^{7}$$
$$=5.865696\times 10^{5+7}$$
$$=5.865696\times 10^{12}\text{ miles}$$

65. $$\left(5\times 10^{9}\right)\left(5.865696\times 10^{12}\right)(1.6)$$
$$=(5)(5.865696)(1.6)\times 10^{9}10^{12}$$
$$=46.925568\times 10^{9+12}$$
$$=46.925568\times 10^{21}$$
$$=4.6925568\times 10^{22}\text{ kilometers}$$

67. First multiply 3.74 by 6.38; then multiply 10^{-5} by 10^{4}. Take the product of these two results and express this product in scientific notation.

$$(3.74)(6.38)=23.8612$$
$$10^{-5}10^{4}=10^{-5+4}=10^{-1}$$
$$23.8612\times 10^{-1}=2.38612\times 10^{0}$$
$$=2.38612$$

69. If the number is bigger than 1, the exponent cannot be negative; if the number is smaller than 1, the exponent must be negative.

Exercises 7.4

1. (a) one term: $3x^5$

   (b) degree of $3x^5$: 5

   (c) degree of polynomial: 5

3. (a) two terms: 3x, 4

(b) degree of 3x: 1

degree of 4: 0

(c) degree of polynomial: 1

5. (a) two terms: x^2, y^3

(b) degree of x^2: 2

degree of y^3: 3

(c) degree of polynomial: 3

7. (a) one term: x^2y^3

(b) degree of x^2y^3: 5 (= 2 + 3)

(c) degree of polynomial: 5

9. (a) one term: 8

(b) degree of 8: 0

(c) degree of polynomial: 0

11. (a) three terms: $2x^3$, $-5x^2$, x

(b) degree of $2x^3$: 3

degree of $-5x^2$: 2

degree of x: 1

(c) degree of polynomial: 3

13. (a) two terms: $2x^3$, y^5

(b) degree of $2x^3$: 3

degree of y^5: 5

(c) degree of polynomial: 5

15. (a) one term: $2x^3y^5$

(b) degree of $2x^3y^5$: 8 (= 3 + 5)

(c) degree of polynomial: 8

17. (a) four terms: x^5, $-x^3y^4$, $-2x^2y^3$, y^6

(b) degree of x^5: 5

degree of $-x^3y^4$: 7 (= 3 + 4)

degree of $-2x^2y^3$: 5 (= 2 + 3)

degree of y^6: 6

(c) degree of polynomial: 7

19. (a) degree of x^2: 2

degree of –5x: 1

degree of 6: 0

(b) degree of polynomial: 2

(c) The coefficient of x^2 is 1.

The coefficient of –5x is –5.

6 is both a term and a coefficient.

21. (a) degree of x^2: 2

degree of 4: 0

(b) degree of polynomial: 2

(c) Write the polynomial as $x^2 + 0x + 4$. Then the coefficient of x^2 is 1 and the coefficient of 0x is 0. 4 is both a term and a coefficient.

23. (a) degree of x^3: 3

degree of –1: 0

(b) degree of polynomial: 3

(c) Write the polynomial as $x^3 + 0x^2 + 0x - 1$. Then the coefficient of x^3 is 1, and the coefficients of $0x^2$ and 0x are 0. –1 is both a term and a coefficient.

25. (a) degree of 1: 0

degree of $-x^5$: 5

(b) degree of polynomial: 5

(c) Write the polynomial as $-x^5 + 0x^4 + 0x^3 + 0x^2 + 0x + 1$. Then the coefficient of $-x^5$ is –1, and the coefficients of $0x^4$, $0x^3$, $0x^2$, and $0x$ are all 0. 1 is both a term and a coefficient.

27. $\left(2x^2 - 5\right) + \left(3x^2 - 5\right)$

$= 2x^2 + 3x^2 - 5 - 5 = 5x^2 - 10$

29. $\left(3u^3 - 2u + 7\right) + \left(u^3 - u^2 + 7u\right)$

$= 3u^3 + u^3 - u^2 - 2u + 7u + 7$

$= 4u^3 - u^2 + 5u + 7$

31. $\left(3u^2 - 2u + 7\right) - \left(u^3 - u^2 + 7u\right)$

$= 3u^2 - 2u + 7 - u^3 + u^2 - 7u$

$= -u^3 + 3u^2 + u^2 - 2u - 7u + 7$

$= -u^3 + 4u^2 - 9u + 7$

33. $\left(4t^3 - t\right) + \left(t^2 + t\right) - \left(t^3 - t^2\right)$

$= 4t^3 - t + t^2 + t - t^3 + t^2$

$= 4t^3 - t^3 + t^2 + t^2 - t + t$

$= 3t^3 + 2t^2$

35. $\left(x^2y + 3xy - x^2y^2\right) + \left(x^2y - 5x^2y^2 - xy^2\right)$

$= x^2y + x^2y + 3xy - x^2y^2 - 5x^2y^2 - xy^2$

$= 2x^2y + 3xy - 6x^2y^2 - xy^2$

37. $\left(x^2y+3xy-x^2y^2\right)-\left(x^2y-5x^2y^2-xy^2\right)$
$=x^2y+3xy-x^2y^2-x^2y+5x^2y^2+xy^2$
$=x^2y-x^2y+3xy-x^2y^2+5x^2y^2+xy^2$
$=3xy+4x^2y^2+xy^2$

39. $2\left(y^2-4y+1\right)+3\left(2y^2-y-1\right)$
$=2y^2-8y+2+6y^2-3y-3$
$=2y^2+6y^2-8y-3y+2-3$
$=8y^2-11y-1$

41. $5\left(x^2-3x+2\right)-3\left(2x^2-5x-2\right)$
$=5x^2-15x+10-6x^2+15x+6$
$=5x^2-6x^2-15x+15x+10+6$
$=-x^2+16$

43. $\left(x^2+3x-7\right)+\left(5x-x^2\right)+\left(3x^2-x-2\right)$
$=x^2-x^2+3x^2+3x+5x-x-7-2$
$=3x^2+7x-9$

45. $\left(2x^2-3x+5\right)-\left(x^2-7x+3\right)$
$=2x^2-3x+5-x^2+7x-3$
$=2x^2-x^2-3x+7x+5-3$
$=x^2+4x+2$

47. $\left(a^3-b^2\right)+\left(a^2b+2b^2\right)-\left(a^3-a^2-b+b^2\right)$
$=a^3-b^2+a^2b+2b^2-a^3+a^2+b-b^2$
$=a^3-a^3-b^2+2b^2-b^2+a^2b+a^2+b$
$=a^2b+a^2+b$

49. $2x-1-\left((3x+6)+(5x-8)\right)$
$=2x-1-(8x-2)$
$=2x-1-8x+2$
$=2x-8x-1+2$
$=-6x+1$

51. $x^2-x+3=(-5)^2-(-5)+3$
$=25+5+3=33$

53. $y^4+y^3+y^2+y+1=(-3)^4+(-3)^3+(-3)^2+(-3)+1$
$=81-27+9-3+1=61$

55. $-3x^2y + 5xy^2 = -3(2)^2(-1) + 5(2)(-1)^2$
$= -3(4)(-1) + 5(2)(1) = 12 + 10$
$= 22$

57. $5x - 12 - (3x + 8) = 5x - 12 - 3x - 8$
$= 5x - 3x - 12 - 8 = 2x - 20$

59. In a sum, the expressions to be added are called terms; in a product, the expressions to be multiplied are called factors.

60. 3 is not a factor of the expression 6x + 8 because 3 does not exactly divide 8.

61. 2 is a factor of the expression 6x + 8 because 2 does exactly divide both 6x and 8. Here, we can write

$6x + 8 = 2(3x + 4)$.

63. $2(x + 3) - 5(x + 4) = x - 2$
$2x + 6 - 5x - 20 = x - 2$
$-3x - 14 = x - 2$
$-4x = 12$
$x = -3$

65. $4(x - 1) - 3x = x - 4$
$4x - 4 - 3x = x - 4$
$x - 4 = x - 4$

Identity

67. Let r = rate of interest on second investment

$.08(1,800) + r(1,400) = 284$
$144 + 1,400r = 284$
$1,400r = 140$
$r = .10$

The rate of interest on the second investment must be 10%.

Exercises 7.5

1. $3x(5x^3)(4x^2)$
$= (3)(5)(4)(x \cdot x^3 \cdot x^2)$
$= 60x^6$

3. $3x(5x^3 + 4x^2) = 3x \cdot 5x^3 + 3x \cdot 4x^2$
$= 15x^4 + 12x^3$

$4xy(3yz)(-5xz)$
$= (4)(3)(-5)(xx)(yy)(zz)$
$= -60x^2y^2z^2$

7. $4xy(3yz - 5xz)$
$= 4xy \cdot 3yz - 4xy \cdot 5xz$
$= 12xy^2z - 20x^2yz$

9. $3x^2(x + 3y) + 4xy(x - 3y)$
$= 3x^2 \cdot x + 3x^2 \cdot 3y + 4xy \cdot x - 4xy \cdot 3y$
$= 3x^3 + 9x^2y + 4x^2y - 12xy^2$
$= 3x^3 + 13x^2y - 12xy^2$

11. $5xy^2(xy - y) - 2y(x^2y^2 - xy^2)$
$= 5xy^2 \cdot xy - 5xy^2 \cdot y - 2y \cdot x^2y^2 + 2y \cdot xy^2$
$= 5x^2y^3 - 5xy^3 - 2x^2y^3 + 2xy^3$
$= 3x^2y^3 - 3xy^3$

13. $(x + 2)(x^2 - x + 3)$
$= x(x^2 - x + 3) + 2(x^2 - x + 3)$
$= x^3 - x^2 + 3x + 2x^2 - 2x + 6$
$= x^3 + x^2 + x + 6$

15. $(y - 5)(y^2 + 2y - 6)$
$= y(y^2 + 2y - 6) - 5(y^2 + 2y - 6)$
$= y^3 + 2y^2 - 6y - 5y^2 - 10y + 30$
$= y^3 - 3y^2 - 16y + 30$

17. $(3x - 2)(x^2 + 3x - 5)$
$= 3x(x^2 + 3x - 5) - 2(x^2 + 3x - 5)$
$= 3x^3 + 9x^2 - 15x - 2x^2 - 6x + 10$
$= 3x^3 + 7x^2 - 21x + 10$

19. $(5z + 2)(3z^2 + 2z + 8)$
$= 5z(3z^2 + 2z + 8) + 2(3z^2 + 2z + 8)$
$= 15z^3 + 10z^2 + 40z + 6z^2 + 4z + 16$
$= 15z^3 + 16z^2 + 44z + 16$

21. $(x+y)(x^2-xy+y^2)$
$= x(x^2-xy+y^2)+y(x^2-xy+y^2)$
$= x^3-x^2y+xy^2+x^2y-xy^2+y^3$
$= x^3+y^3$

23. $(x^2+x+1)(x^2+x-1)$
$= x^2(x^2+x-1)+x(x^2+x+1)+1(x^2+x-1)$
$= x^4+x^3-x^2+x^3+x^2-x+x^2+x-1$
$= x^4+2x^3+x^2-1$

25. $(x^3+xy-y^2)(x^3-3xy+y^2)$
$= x^3(x^3-3xy+y^2)+xy(x^3-3xy+y^2)-y^2(x^3-3xy+y^2)$
$= x^6-3x^4y+x^3y^2+x^4y-3x^2y^2+xy^3-x^3y^2+3xy^3-y^4$
$= x^6-2x^4y-3x^2y^2+4xy^3-y^4$

27. $(x+5)(x+3) = x^2+3x+5x+15$
$= x^2+8x+15$

29. $(x-5)(x-3) = x^2-3x-5x+15$
$= x^2-8x+15$

31. $(x-5)(x+3) = x^2+3x-5x-15$
$= x^2-2x-15$

33. $(x+5)(x-3) = x^2-3x+5x-15$
$= x^2+2x-15$

35. $(x+2y)(x+3y)$
$= x^2+3xy+2xy+6y^2$
$= x^2+5xy+6y^2$

37. $(a+8b)(a-5b)$
$= a^2-5ab+8ab-40b^2$
$= a^2+3ab-40b^2$

39. $(3x-4)(4x-1)$
$= 12x^2-3x-16x+4$
$= 12x^2-19x+4$

1. $(2r-s)(r+3s)$
$= 2r^2 + 6rs - rs - 3s^2$
$= 2r^2 + 5rs - 3s^2$

43. $\left(x^2+3\right)\left(x^2+2\right)$
$= x^4 + 2x^2 + 3x^2 + 6$
$= x^4 + 5x^2 + 6$

45. $(x+7)(x+7)$
$= x^2 + 7x + 7x + 49$
$= x^2 + 14x + 49$

47. $(x+7)(x-7)$
$= x^2 - 7x + 7x - 49$
$= x^2 - 49$

49. $(x-4)^2 = (x-4)(x-4)$
$= x^2 - 4x - 4x + 16$
$= x^2 - 8x + 16$

51. $(x+2)^3$
$= (x+2)(x+2)(x+2)$
$= (x+2)\left(x^2 + 2x + 2x + 4\right)$
$= (x+2)\left(x^2 + 4x + 4\right)$
$= x\left(x^2 + 4x + 4\right) + 2\left(x^2 + 4x + 4\right)$
$= x^3 + 4x^2 + 4x + 2x^2 + 8x + 8$
$= x^3 + 6x^2 + 12x + 8$

53. $(3x-5)^2 = (3x-5)(3x-5)$
$= 9x^2 - 15x - 15x + 25$
$= 9x^2 - 30x + 25$

55. $(2a-9b)^2$
$= (2a-9b)(2a-9b)$
$= 4a^2 - 18ab - 18ab + 81b^2$
$= 4a^2 - 36ab + 81b^2$

57. $2x^2(x+4)(x-8)$

$= 2x^2(x^2 - 8x + 4x - 32)$

$= 2x^2(x^2 - 4x - 32)$

$= 2x^4 - 8x^3 - 64x^2$

59. $3x(5x - 6)(3x - 2)$

$= 3x(15x^2 - 10x - 18x + 12)$

$= 3x(15x^2 - 28x + 12)$

$= 45x^3 - 84x^2 + 36x$

61. $(x + 4)(x - 3) + (x - 6)(x - 2)$

$= x^2 - 3x + 4x - 12 + x^2 - 2x - 6x + 12$

$= 2x^2 - 7x$

63. $(a - 5)(a - 4) - (a - 3)(a - 2)$

$= a^2 - 4a - 5a + 20 - (a^2 - 2a - 3a + 6)$

$= a^2 - 9a + 20 - (a^2 - 5a + 6)$

$= a^2 - 9a + 20 - a^2 + 5a - 6$

$= -4a + 14$

65. $(x - 6)^2 - (x + 6)^2$

$= (x - 6)(x - 6) - (x + 6)(x + 6)$

$= x^2 - 6x - 6x + 36 - (x^2 + 6x + 6x + 36)$

$= (x^2 - 12x + 36) - (x^2 + 12x + 36)$

$= x^2 - 12x + 36 - x^2 - 12x - 36$

$= -24x$

67. $(x + 2)^3 - (x + 2)^2 - (x + 2) + 2$

$= (x + 2)(x + 2)(x + 2) - (x + 2)(x + 2) - (x + 2) + 2$

$= x^3 + 6x^2 + 12x + 8 - (x^2 + 4x + 4) - (x + 2) + 2$

(see Exercise (51) for details)

$= x^3 + 6x^2 + 12x + 8 - x^2 - 4x - 4 - x - 2 + 2$

$= x^3 + 5x^2 + 7x + 4$

69. Let W = width of the rectangle

2W + 3 = length of the rectangle

Then P = 2W + 2(2W + 3) = 2W + 4W + 6 = 6W + 6 is the perimeter of the rectangle and A = W(2W + 3) = $2W^2 + 3W$ is the area of the rectangle.

1. Let s = length of the side of the original square

s + 4 = length of the side of the new square

Then $s \cdot s = s^2$ is the area of the original square and $(s+4)(s+4) = s^2 + 4s + 4s + 16 = s^2 + 8s + 16$ is the area of the new square. Therefore, the change in area is $s^2 + 8s + 16 - s^2 = 8s + 16$.

73. Let a = width of original rectangle

5a – 8 = length of original rectangle

a + 3 = width of new rectangle

Then $a(5a - 8) = 5a^2 - 8a$ is the area of the original rectangle and $(a+3)(5a-8) = 5a^2 - 8a + 15a - 24 = 5a^2 + 7a - 24$ is the area of the new rectangle. Therefore, the change in area is

$5a^2 + 7a - 24 - (5a^2 - 8a)$

$= 5a^2 + 7a - 24 - 5a^2 + 8a$
$= 15a - 24$

75. Let x = length of side of square to be cut out. The original rectangle has dimensions 8 by 10, so has an area of 80. The area of the square to be cut out is $x \cdot x = x^2$. Therefore, the remaining area is $80 - x^2$.

77. $1.92x^3 - 2.4x^2 + 4.88x$

79. $.06x^2 + .01x - .4$

81. $.007x^2 - 1.876x + 31.5$

83. (a) Both examples require us to find the square of an expression involving x and y.

(b) The first example asks us to square a product; the second asks us to square a sum.

(c) $(xy)^2 = x^2y^2$, by Exponent Rule 4. $(x+y)^2 = x^2 + 2xy + y^2$ by the FOIL method.

84. $(x+y)^n = x^n + y^n$ only when n = 1.

85. (a) $-4x^2(2x-7) = -8x^3 + 28x^2$

$$-4x^2(2x-7)(3x+1) = (-8x^3 + 28x^2)(3x+1)$$
$$= -24x^4 - 8x^3 + 84x^3 + 28x^2$$
$$= -24x^4 + 76x^3 + 28x^2$$

(b) $-4x^2(3x+1) = -12x^3 - 4x^2$

$$-4x^2(2x-7)(3x+1) = -4x^2(3x+1)(2x-7)$$
$$= (-12x^3 - 4x^2)(2x-7)$$
$$= -24x^4 + 84x^3 - 8x^3 + 28x^2$$
$$= -24x^4 + 76x^3 + 28x^2$$

(c) $$(2x-7)(3x+1)=6x^2+2x-21x-7$$
$$=6x^2-19x-7$$
$$-4x^2(2x-7)(3x+1)=-4x^2\left(6x^2-19x-7\right)$$
$$=-24x^4+76x^3+28x^2$$

The three answers are the same, and should be because of the commutative and associative laws of multiplication.

87. $$8-3(x+1)<32$$
$$8-3x-3<32$$
$$-3x+5<32$$
$$-3x<27$$
$$x>-9$$

89. $$-3\le 9-x<1$$
$$\underline{-9\quad -9\quad -9}$$
$$-12\le -x<-8$$
$$12\ge x>8$$
or
$$8<x\le 12$$

91. Let x = number of ounces of the 55% solution

40 – x = number of ounces of the 35% solution

$$.55x+.35(40-x)=.51(40)$$
$$.55x+14-.35x=20.4$$
$$.20x+14=20.4$$
$$.20x=6.4$$
$$2.0x=64$$
$$x=32$$

Then 40 – x = 40 – 32 = 8. So 32 ounces of the 55% solution should be mixed with 8 ounces of the 35% solution.

Chapter 7 Review Exercises

1. $$3^{-4}=\frac{1}{3^4}=\frac{1}{81}$$

3. $$\left(3^{-1}+2^{-2}\right)^2=\left(\frac{1}{3^1}+\frac{1}{2^2}\right)^2$$
$$=\left(\frac{1}{3}+\frac{1}{4}\right)^2=\left(\frac{7}{12}\right)^2$$
$$=\frac{7^2}{12^2}=\frac{49}{144}$$

5. $$\frac{\left(xy^2\right)^3}{\left(x^2y\right)^4}=\frac{x^3\left(y^2\right)^3}{\left(x^2\right)^4y^4}=\frac{x^3y^6}{x^8y^4}$$

$$= \frac{x^3}{x^8} \cdot \frac{y^6}{y^4} = x^{-5}y^2$$

$$= \frac{1}{x^5} \cdot \frac{y^2}{1} = \frac{y^2}{x^5}$$

7. $$\frac{(3x^3y^2)^4}{9(x^2y^4)^3} = \frac{3^4(x^3)^4(y^2)^4}{9(x^2)^3(y^4)^3}$$

$$= \frac{81x^{12}y^8}{9x^6y^{12}} = \frac{81}{9} \cdot \frac{x^{12}}{x^6} \cdot \frac{y^8}{y^{12}}$$

$$= 9 \cdot x^6 \cdot y^{-4}$$

$$= \frac{9}{1} \cdot \frac{x^6}{1} \cdot \frac{1}{y^4} = \frac{9x^6}{y^4}$$

9. $(x^{-2})^{-3} = x^6$

11. $$\left(\frac{2x^{-2}x^3}{x^{-3}}\right)^{-2} = \left(\frac{2x}{x^{-3}}\right)^{-2}$$

$$= (2x^4)^{-2} = 2^{-2}(x^4)^{-2}$$

$$= \frac{1}{2^2}x^{-8} = \frac{1}{4} \cdot \frac{1}{x^8} = \frac{1}{4x^8}$$

13. $58,700,000 = 5.87 \times 10^7$

15. $.000002 = 2 \times 10^{-6}$

17. $2.56 \times 10^{-3} = .00256$

19. $5.773 \times 10^8 = 577,300,000$

21. $(.008)(250,000)$

$= (8 \times 10^{-3})(2.5 \times 10^5)$

$= (8)(2.5) \times 10^{-3}10^5$

$= 20 \times 10^2 = 2 \times 10^3 = 2,000$

23. $$\frac{.001}{.000025} = \frac{1 \times 10^{-3}}{2.5 \times 10^{-5}}$$

$$= \frac{1}{2.5} \times \frac{10^{-3}}{10^{-5}} = .4 \times 10^2$$

$$= 4 \times 10^1 = 40$$

25. (a) three terms: x^2, $3x$, -7

(b) degree of x^2: 2

degree of $3x$: 1

degree of -7: 0

(c) degree of polynomial: 2

27. (a) three terms: $3x^3y$, $-5y^2$, $6xy$

(b) degree of $3x^3y$: 4 (= 3 + 1)

degree of $-5y^2$: 2

degree of 6xy: 2 (= 1 + 1)

(c) degree of polynomial: 4

29. (a) two terms: 8x, –5

(b) degree of 8x: 1

degree of –5: 0

(c) degree of polynomial: 1

31. (a) one term

(b) degree of 9: 0

(c) degree of polynomial: 0

33. (a) one term

(b) degree of $\left(3x^5\right)\left(2x^3\right) = 6x^8$: 8

(c) degree of polynomial: 8

35. $2x^3 - 7x^2 + 0x + 4$

37. $y^5 + 0y^4 + 0y^3 + y^2 - 2y - 1$

39. $\left(3x^2 - 5x + 7\right) + \left(5x - x^2 - 5\right)$
$= 3x^2 - 5x + 7 + 5x - x^2 - 5$
$= 2x^2 + 2$

41. $\left(3x^2 - 5x + 7\right) - \left(5x - x^2 - 5\right)$
$= 3x^2 - 5x + 7 - 5x + x^2 + 5$
$= 4x^2 - 10x + 12$

43. $2\left(x^2y - xy^2 - 5x^2y^2\right) + 3\left(xy^2 + x^2y + x^2y^2\right)$
$= 2x^2y - 2xy^2 - 10x^2y^2 + 3xy^2 + 3x^2y + 3x^2y^2$
$= 5x^2y + xy^2 - 7x^2y^2$

45. $2\left(x^2y - xy^2\right) - 5x^2y^2 - 3\left(xy^2 - x^2y + x^2y^2\right)$
$= 2x^2y - 2xy^2 - 5x^2y^2 - 3xy^2 + 3x^2y - 3x^2y^2$
$= 5x^2y - 5xy^2 - 8x^2y^2$

47. $2a^2(a - 3b) + 4a\left(a^2 + ab\right) - 2\left(a^3 - a^2b\right)$
$= 2a^3 - 6a^2b + 4a^3 + 4a^2b - 2a^3 + 2a^2b = 4a^3$

49. $x^2 + 4x - (x^2 - 4x)$

$= x^2 + 4x - x^2 + 4x = 8x$

51. $(x^2 + 4x - 3) + (2x^2 - x - 2) - (3x - 5)$

$= x^2 + 4x - 3 + 2x^2 - x - 2 - 3x + 5$

$= 3x^2$

53. $(x + 4)(x - 7)$

$= x^2 - 7x + 4x - 28$

$= x^2 - 3x - 28$

55. $(2x - 3)(4x - 5)$

$= 8x^2 - 10x - 12x + 15$

$= 8x^2 - 22x + 15$

57. $(3a - 4b)(2a + 5b)$

$= 6a^2 + 15ab - 8ab - 20b^2$

$= 6a^2 + 7ab - 20b^2$

59. $(x + 2)(x - 3)(x + 1)$

$= (x + 2)(x^2 + x - 3x - 3)$

$= (x + 2)(x^2 - 2x - 3)$

$= x(x^2 - 2x - 3) + 2(x^2 - 2x - 3)$

$= x^3 - 2x^2 - 3x + 2x^2 - 4x - 6$

$= x^3 - 7x - 6$

61. $(x + 6)^2 = (x + 6)(x + 6)$

$= x^2 + 6x + 6x + 36$

$= x^2 + 12x + 36$

63. $(x - 5)^3 = (x - 5)(x - 5)(x - 5)$

$= (x - 5)(x^2 - 5x - 5x + 25)$

$= (x - 5)(x^2 - 10x + 25)$

$= x(x^2 - 10x + 25) - 5(x^2 - 10x + 25)$

$= x^3 - 10x^2 + 25x - 5x^2 + 50x - 125$

$= x^3 - 15x^2 + 75x - 125$

65. $3x^2(x-4)(x+2)$

$= 3x^2\left(x^2 + 2x - 4x - 8\right)$

$= 3x^2\left(x^2 - 2x - 8\right)$

$= 3x^4 - 6x^3 - 24x^2$

67. $(x-5)(x+5)$

$= x^2 + 5x - 5x - 25$

$= x^2 - 25$

69. $(x+2)\left(x^2 - 3x + 4\right)$

$= x\left(x^2 - 3x + 4\right) + 2\left(x^2 - 3x + 4\right)$

$= x^3 - 3x^2 + 4x + 2x^2 - 6x + 8$

$= x^3 - x^2 - 2x + 8$

71. $\left(x^2 + 2x - 1\right)\left(x^2 + 2x + 1\right)$

$= x^2\left(x^2 + 2x + 1\right) + 2x\left(x^2 + 2x + 1\right) - 1\left(x^2 + 2x + 1\right)$

$= x^4 + 2x^3 + x^2 + 2x^3 + 4x^2 + 2x - x^2 - 2x - 1$

$= x^4 + 4x^3 + 4x^2 - 1$

73. $(2x-3)(x+4) - (x-2)(x-1)$

$= 2x^2 + 8x - 3x - 12 - \left(x^2 - x - 2x + 2\right)$

$= 2x^2 + 5x - 12 - \left(x^2 - 3x + 2\right)$

$= 2x^2 + 5x - 12 - x^2 + 3x - 2$

$= x^2 + 8x - 14$

75. Let w = width of the rectagle.

Then 3w – 5 = length of the rectangle. Since the area of a rectangle, A, is the product of its length and width, we find $A = (3w-5)w = 3w^2 - 5w$.

Chapter 7 Practice Test

1. $5^0 + 2^{-2} + 4^{-1}$

$= 1 + \frac{1}{2^2} + \frac{1}{4^1}$

$= 1 + \frac{1}{4} + \frac{1}{4}$

$= \frac{4}{4} + \frac{1}{4} + \frac{1}{4} = \frac{6}{4} = \frac{3}{2}$

3. $x^{-4}x^{-5} = x^{-4+(-5)} = x^{-9} = \frac{1}{x^9}$

5. $\dfrac{\left(2x^{-3}y^{4}\right)^{4}}{4\left(x^{-2}y^{-1}\right)^{3}} = \dfrac{2^{4}\left(x^{-3}\right)^{4}\left(y^{4}\right)^{4}}{4\left(x^{-2}\right)^{3}\left(y^{-1}\right)^{3}}$

$= \dfrac{16x^{-3(4)}y^{4(4)}}{4x^{-2(3)}y^{-1(3)}}$

$= \dfrac{16x^{-12}y^{16}}{4x^{-6}y^{-3}}$

$= \left(\dfrac{16}{4}\right)\left(\dfrac{x^{-12}}{x^{-6}}\right)\left(\dfrac{y^{16}}{y^{-3}}\right)$

$= 4x^{-12-(-6)}y^{16-(-3)}$

$= 4x^{-6}y^{19}$

$= \dfrac{4}{1}\cdot\dfrac{1}{x^{6}}\cdot\dfrac{y^{19}}{1} = \dfrac{4y^{19}}{x^{6}}$

7. $-3x^{2}y\left(4x^{2}y\right)\left(-2x^{3}\right)$

$= -3(4)(-2)x^{2}x^{2}x^{3}yy$

$= 24x^{2+2+3}y^{1+1}$

$= 24x^{7}y^{2}$

9. $2x\left(x^{2} - y\right) - 3\left(x - xy\right) - \left(2x^{3} - 3x\right)$

$= 2x^{3} - 2xy - 3x + 3xy - 2x^{3} + 3x$

$= 2x^{3} - 2x^{3} - 2xy + 3xy - 3x + 3x$

$= xy$

11. $3x^{2}\left(2x - y\right) - xy\left(x + y\right)$

$= 6x^{3} - 3x^{2}y - x^{2}y - xy^{2}$

$= 6x^{3} - 4x^{2}y - xy^{2}$

13. $(a - 1)^{2} - (a + 1)^{2}$

$= (a - 1)(a - 1) - (a + 1)(a + 1)$

$= a^{2} - 2a + 1 - \left(a^{2} + 2a + 1\right)$

$= a^{2} - 2a + 1 - a^{2} - 2a - 1$

$= a^{2} - a^{2} - 2a - 2a + 1 - 1$

$= -4a$

15. $$\frac{(.24)(5,000)}{.006} = \frac{\left(2.4 \times 10^{-1}\right)\left(5 \times 10^{3}\right)}{6 \times 10^{-3}}$$

$$= \frac{(2.4)(5)}{6} \times \frac{10^{-1}10^{3}}{10^{-3}}$$

$$= \frac{12}{6} \times \frac{10^{2}}{10^{-3}}$$

$$= 2 \times 10^{2-(-3)}$$

$$= 2 \times 10^{5} = 200,000$$

CHAPTER 8
FACTORING

Exercises 8.1

1. $(x+4)(x+3) = x^2 + 3x + 4x + 12$
$= x^2 + 7x + 12$

3. $(x-4)(x-3) = x^2 - 3x - 4x + 12$
$= x^2 - 7x + 12$

5. $(x+4)(x-3) = x^2 - 3x + 4x - 12$
$= x^2 + x - 12$

7. $(x-4)(x+3) = x^2 + 3x - 4x - 12$
$= x^2 - x - 12$

9. $(x+6)(x+2) = x^2 + 2x + 6x + 12$
$= x^2 + 8x + 12$

11. $(x-6)(x-2) = x^2 - 2x - 6x + 12$
$= x^2 - 8x + 12$

13. $(x+6)(x-2) = x^2 - 2x + 6x - 12$
$= x^2 + 4x - 12$

15. $(x-6)(x+2) = x^2 + 2x - 6x - 12$
$= x^2 - 4x - 12$

17. $(x+12)(x+1) = x^2 + x + 12x + 12$
$= x^2 + 13x + 12$

19. $(x-12)(x-1) = x^2 - x - 12x + 12$
$= x^2 - 13x + 12$

21. $(x+12)(x-1) = x^2 - x + 12x - 12$
$= x^2 + 11x - 12$

23. $(x-12)(x+1) = x^2 + x - 12x - 12$
$= x^2 - 11x - 12$

25. $(a+8)(a+8) = a^2 + 8a + 8a + 64$
$= a^2 + 16a + 64$

(Perfect square)

27. $(a-8)(a-8) = a^2 - 8a - 8a + 64$
$= a^2 - 16a + 64$

(Perfect square)

29. $(a+8)(a-8) = a^2 - 8a + 8a - 64$
$= a^2 - 64$

(Difference of two squares)

31. $(c-4)^2 = (c-4)(c-4)$
$= c^2 - 4c - 4c + 16$
$= c^2 - 8c + 16$

(Perfect square)

33. $(c+4)^2 = (c+4)(c+4)$
$= c^2 + 4c + 4c + 16$
$= c^2 + 8c + 16$

(Perfect square)

35. $(c+4)(c-4) = c^2 - 4c + 4c - 16$
$= c^2 - 16$

(Difference of two squares)

37. $(3x+4)(x+7) = 3x^2 + 21x + 4x + 28$
$= 3x^2 + 25x + 28$

39. $(3x+7)(x+4) = 3x^2 + 12x + 7x + 28$
$= 3x^2 + 19x + 28$

41. $(3x+4)(x-7) = 3x^2 - 21x + 4x - 28$
$= 3x^2 - 17x - 28$

43. $(3x-4)(x+7) = 3x^2 + 21x - 4x - 28$
$= 3x^2 + 17x - 28$

45. $(3x+4)(5x+7)$
$= 15x^2 + 21x + 20x + 28$
$= 15x^2 + 41x + 28$

47. $(3x+7)(5x+4)$
$= 15x^2 + 12x + 35x + 28$
$= 15x^2 + 47x + 28$

49. $(3x+4)(5x-7)$
$= 15x^2 - 21x + 20x - 28$
$= 15x^2 - x - 28$

51. $(3x-4)(5x+7)$
$= 15x^2 + 21x - 20x - 28$
$= 15x^2 + x - 28$

$(2a+5)^2 = (2a+5)(2a+5)$
$= 4a^2 + 10a + 10a + 25$
$= 4a^2 + 20a + 25$

(Perfect square)

55. $(2a+5)(2a-5)$
$= 4a^2 - 10a + 10a - 25$
$= 4a^2 - 25$

(Difference of two squares)

57. $(3xy)^2 = 3^2x^2y^2 = 9x^2y^2$

59. $(x^3+y^2)^2 = (x^3+y^2)(x^3+y^2)$
$= x^6 + x^3y^2 + x^3y^2 + y^4$
$= x^6 + 2x^3y^2 + y^4$

(Perfect square)

61. $(2a+5y)^2 = (2a+5y)(2a+5y)$
$= 4a^2 + 10ay + 10ay + 25y^2$
$= 4a^2 + 20ay + 25y^2$

(Perfect square)

63. $(2a+5y)(2a-5y)$
$= 4a^2 - 10ay + 10ay - 25y^2$
$= 4a^2 - 25y^2$

(Difference of two squares)

65. $(x+6)(x+4) = x^2 + 4x + 6x + 24$
$= x^2 + 10x + 24$
$(x-6)(x-4) = x^2 - 4x - 6x + 24$
$= x^2 - 10x + 24$

The effect of switching both + signs to – signs is to change the sign of the cross term from + to –.

66. $(x+6)(x-4) = x^2 - 4x + 6x - 24$
$= x^2 + 2x - 24$
$(x-6)(x+4) = x^2 + 4x - 6x - 24$
$= x^2 - 2x - 24$

The middle terms of the resulting trinomials have opposite signs.

67. (5) and (7) (6) and (8)

(13) and (15) (14) and (16)

(21) and (23) (22) and (24)

(41) and (43) (42) and (44)

(49) and (51) (50) and (52)

The middle terms will always have opposite signs, since

$(x + a)(x - b) = x^2 + (a - b)x - ab$

$(x - a)(x + b) = x^2 + (-a + b)x - ab$

and (a – b) and (–a + b) are always opposites of one another.

69. $$\frac{20xy}{9z} \div 18xyz = \frac{20xy}{9z} \cdot \frac{1}{18xyz}$$
$$= \frac{10}{81z^2}$$

71. $\frac{a}{3} - \frac{a}{2} = \frac{a}{5}$ LCD = 30

$$30\left(\frac{a}{3} - \frac{a}{2}\right) = 30\left(\frac{a}{5}\right)$$
$$10a - 15a = 6a$$
$$-5a = 6a$$
$$0 = 11a$$
$$0 = a$$

73. Let r = rate of interest on second investment

$$.09(1400) + r(1100) = 203$$
$$126 + 1100r = 203$$
$$1100r = 77$$
$$r = .07$$

So the rate of interest on the second investment is 7%.

Exercises 8.2

1. $5x = 20 = 5 \cdot x + 5 \cdot 4 = 5(x + 4)$

3. $6x - 18 = 6 \cdot x - 6 \cdot 3 = 6(x - 3)$

5. $4y + 27$ is not factorable.

7. $8a - 12 = 4 \cdot 2a - 4 \cdot 3 = 4(2a - 3)$

9. $12a + 9 = 3 \cdot 4a + 3 \cdot 3$
$= 3(4a + 3)$

11. $3a + 6b - 8c$ is not factorable.

13. $x^2 + 3x = x \cdot x + x \cdot 3 = x(x + 3)$

15. $2t^2 + 8t = 2 \cdot t \cdot t + 2 \cdot t \cdot 4$
$= 2t(t + 4)$

17. $6y^3 - 9y^2 = 3y^2 \cdot 2y - 3y^2 \cdot 3$
$= 3y^2(2y - 3)$

19. $4x^5 + 2x^2 - 8x$
$= 2x \cdot 2x^4 + 2x \cdot x - 2x \cdot 4$
$= 2x(2x^4 + x - 4)$

21. $a^2 + a = a \cdot a + a \cdot 1 = a(a + 1)$

23. $x^2 - 5x + xy = x \cdot x - x \cdot 5 + x \cdot y$
$= x(x - 5 + y)$

25. $3c^6 - 6c^3 = 3c^3 \cdot c^3 - 3c^3 \cdot 2$
$= 3c^3(c^3 - 2)$

27. $x^2y - xy^2 = xy \cdot x - xy \cdot y$
$= xy(x - y)$

29. $6x^2 + 3x = 3x \cdot 2x + 3x \cdot 1$
$= 3x(2x + 1)$

31. $8x^3y^2 - 25z^4$ is not factorable.

33. $12c^3d^5 + 4c^2d^3$
$= 4c^2d^3 \cdot 3cd^2 + 4c^2d^3 \cdot 1$
$= 4c^2d^3(3cd^2 + 1)$

35. $x^2y^3 - y^2z^4 + x^3z^2$ is not factorable.

37. $2x^2yz^3 + 8xyz^2 - 10x^2y^2z^2$
$= 2xyz^2 \cdot xz + 2xyz^2 \cdot 4$
$- 2xyz^2 \cdot 5xy$
$= 2xyz^2(xz + 4 - 5xy)$

39. $6u^3v^2 + 18u^3v^3 - 12u^3v^5$
$= 6u^3v^2 \cdot 1 + 6u^3v^2 \cdot 3v$
$- 6u^3v^2 \cdot 2v^3$
$= 6u^3v^2(1 + 3v - 2v^3)$

41. $x(x - 5) + 4(x - 5) = (x - 5)(x + 4)$

43. $y(y + 6) - 3(y + 6) = (y + 6)(y - 3)$

45. $x^2 + 8x + xy + 8y$

$= (x^2 + 8x) + (xy + 8y)$

$= x(x + 8) + y(x + 8)$

$= (x + 8)(x + y)$

47. $m^2 + mn + 9m + 9n$

$= (m^2 + mn) + (9m + 9n)$

$= m(m + n) + 9(m + n)$

$= (m + n)(m + 9)$

49. $x^2 - xy - 4x + 4y$

$= (x^2 - xy) + (-4x + 4y)$

$= x(x - y) - 4(x - y)$

$= (x - y)(x - 4)$

51. $3x^2y + 6xy - 5x - 10$

$= (3x^2y + 6xy) + (-5x - 10)$

$= 3xy(x + 2) - 5(x + 2)$

$= (x + 2)(3xy - 5)$

53. (a) $(x + 2)(x + 3)$

(b) $x^2y^2(x + y)$

54. Step 1: $-x$ is replaced by $-5x + 4x$.

Step 2: The first two terms are grouped and the last two terms are grouped. The common factor is then removed from each group.

Step 3: The common factor of $(x - 5)$ is removed from the sum.

55. When we factor out -5 from the third and fourth terms, the remaining factor should be $x - 3$, not $x + 3$. This means that there is no common factor to remove.

57. $(x^{-3})^{-4} = x^{(-3)(-4)} = x^{12}$

59. $\frac{(x^{-2}y^3)^{-1}}{(x^3)^{-2}} = \frac{(x^{-2})^{-1}(y^3)^{-1}}{x^{-6}} = \frac{x^2y^{-3}}{x^{-6}}$

$= x^{2-(-6)}y^{-3} = \frac{x^8}{y^3}$

61. Let x = number of liters of pure water to be added.

$.18(36) = .12(36 + x)$

$6.48 = 4.32 + .12x$

$2.16 = .12x$

$216 = 12x$

$18 = x$

So 18 liters of pure water must be added.

Exercises 8.3

1. $x^2 + 3x = x(x + 3)$

3. $x^2 + 3x + 2 = (x + 1)(x + 2)$

5. $x^2 - 3x + 2 = (x - 1)(x - 2)$

7. $x^2 + 3x - 2$ is not factorable.

9. $x^2 + x - 2 = (x - 1)(x + 2)$

11. $x^2 - x - 2 = (x + 1)(x - 2)$

13. $a^2 + 8a + 12 = (a + 2)(a + 6)$

15. $a^2 - a - 12 = (a - 4)(a + 3)$

17. $a^2 - a + 12$ is not factorable.

19. $a^2 - 12a = a(a - 12)$

21. $a - 12 + a^2 = a^2 + a - 12$

    $= (a + 4)(a - 3)$

23. $y^2 + 11y + 28 = (y + 4)(y + 7)$

25. $x^2 - 5x - 36 = (x - 9)(x + 4)$

27. $a^2 + 6a - 40 = (a + 10)(a - 4)$

29. $x^2 - 17x + 30 = (x - 2)(x - 15)$

31. $x^2 - 9x = x(x - 9)$

33. $x^2 - 9 = (x - 3)(x + 3)$

35. $x^2 - 9x - 10 = (x - 10)(x + 1)$

37. $x^2 - 3xy + 2y^2 = (x - y)(x - 2y)$

39. $a^2 + 10a + 24 = (a + 4)(a + 6)$

41. $y^2 + 12y + 36 = (y + 6)(y + 6)$

    or $(y + 6)^2$

43. $y^2 - 36 = (y - 6)(y + 6)$

45. $x^2 - 7x - 18 = (x - 9)(x + 2)$

47. $r^2 - 3rs - 10s^2 = (r - 5s)(r + 2s)$

49. $c^2 - 6c + 5 = (c-1)(c-5)$

51. $4x^2 + 8x + 4 = 4(x^2 + 2x + 1)$
$= 4(x+1)(x+1)$
or $4(x+1)^2$

53. $x^2 - 30 + x = x^2 + x - 30$
$= (x+6)(x-5)$

55. $2x^2 - 50 = 2(x^2 - 25)$
$= 2(x-5)(x+5)$

57. $x^2 - x - 20 = (x-5)(x+4)$

59. $x^2 - x + 20$ is not factorable.

61. $y^2 + 11y + 28 = (y+4)(y+7)$

63. $2y^2 + 2y - 84 = 2(y^2 + y - 42)$
$= 2(y+7)(y-6)$

65. $49 - d^2 = (7-d)(7+d)$

67. $49 + d^2$ is not factorable.

69. $10x^2 - 40xy - 120y^2$
$= 10(x^2 - 4xy - 12y^2)$
$= 10(x-6y)(x+2y)$

71. $a^2 + 13 + 14a = a^2 + 14a + 13$
$= (a+13)(a+1)$

73. $6s^2 - 6s - 72 = 6(s^2 - s - 12)$
$= 6(s-4)(s+3)$

75. $4x^2 - 64 = 4(x^2 - 16)$
$= 4(x-4)(x+4)$

77. The factors of 10 are 1 and 10 and 2 and 5. Since the last term is positive, the signs in the parentheses must be the same.

Possibilities:
$(x+1)(x+10) = x^2 \underline{+11}x + 10$
$(x-1)(x-10) = x^2 \underline{-11}x + 10$
$(x+2)(x+5) = x^2 \underline{+7}x + 10$
$(x-2)(x-5) = x^2 \underline{-7}x + 10$

Therefore, the possible values of k are 11, –11, 7, and –7.

[illegible]. As in (71), the factors of 10 are 1 and 10 and 2 and 5. Since the last term is negative, the signs in the parentheses must be opposite. Possibilities:

$$(x+1)(x-10) = x^2 \underline{-9}x - 10$$
$$(x-1)(x+10) = x^2 \underline{+9}x - 10$$
$$(x+2)(x-5) = x^2 \underline{-3}x - 10$$
$$(x-2)(x+5) = x^2 \underline{+3}x - 10$$

Therefore, the possible values of b are –9, 9, –3, and 3.

79. There are infinitely many such integers. We can always find such a "c" by choosing two integers that differ by 5 and forming the product.

(x + larger integer)(x – smaller integer)

For example, since 9 and 4 differ by 5, $(x+9)(x-4) = x^2 + 5x - 36$ will give the value c = –36. Clearly, we can find two integers that differ by 5 in infinitely many ways. (Interestingly, if we ask for all <u>positive</u> integers c with this property, there are only two answers: c = 4 and c = 6.)

81.
$$\begin{aligned}(5x^2+y)^2 &= (5x^2+y)(5x^2+y)\\ &= 25x^4 + 5x^2y + 5x^2y + y^2\\ &= 25x^4 + 10x^2y + y^2\end{aligned}$$

83.
$$\begin{aligned}(2x-7)(3x+4) &= 6x^2 + 8x - 2(x-28)\\ &= 6x^2 - 13x - 28\end{aligned}$$

85. $28,700 = 2.87 \times 10^4$

87.
$$\begin{aligned}\frac{(2,400)(.003)}{(.02)(.004)} &= \frac{(2.4\times 10^3)(3\times 10^{-3})}{(2\times 10^2)(4\times 10^{-3})}\\ &= \frac{(2.4)(3)}{(2)(4)} \times \frac{10^3\cancel{10^{-3}}}{10^{-2}\cancel{10^{-3}}}\\ &= .9\times 10^5\\ &= 90,000\end{aligned}$$

Exercises 8.4

1. $x^2 + 5x = x(x+5)$

3. $x^2 + 5x + 4 = (x+1)(x+4)$

5. $x^2 + 5x - 4$ is not factorable.

7. $3x^2 + 8x + 4 = (3x+2)(x+2)$

9. $2x^2 + 11x + 12 = (2x+3)(x+4)$

11.
$$\begin{aligned}2x^2 + 10x + 12 &= 2(x^2 + 5x + 6)\\ &= 2(x+2)(x+3)\end{aligned}$$

13. $5x^2 - 27x + 10 = (5x - 2)(x - 5)$

15. $5x^2 - 15x + 10 = 5(x^2 - 3x + 2)$
$= 5(x - 1)(x - 2)$

17. $2y^2 - y - 6 = (2y + 3)(y - 2)$

19. $5a^2 + 9a - 18 = (5a - 6)(a + 3)$

21. $2t^2 + 7t + 6 = (2t + 3)(t + 2)$

23. $2t^2 + 6t + 6 = 2(t^2 + 3t + 3)$

25. $3w^2 - 6w - 30 = 3(w^2 - 2w - 10)$

27. $3x^2 - 4x + 2$ is not factorable.

29. $3x^2 - 14xy + 15y^2$

$= (3x - 5y)(x - 3y)$

31. $6a^2 + 17a + 10 = (6a + 5)(a + 2)$

33. $6a^2 + 17a - 10 = (3a + 10)(2a - 1)$

35. $6a^2 - 18a - 24 = 6(a^2 - 3a - 4)$
$= 6(a - 4)(a + 1)$

37. $x^2 - 36y^2 = (x - 6y)(x + 6y)$

39. $4x^2 - 36y^2 = 4(x^2 - 9y^2)$
$= 4(x - 3y)(x + 3y)$

41. $x^3 + 5x^2 - 24x = x(x^2 + 5x - 24)$
$= x(x + 8)(x - 3)$

43. $x^2 + 5x^2 - 24x = 6x^2 - 24x$
$= 6x(x - 4)$

45. $4x^4 - 24x^3 + 32x^2 = 4x^2(x^2 - 6x + 8)$
$= 4x^2(x - 2)(x - 4)$

47. $6x^2y - 8xy^2 + 12xy = 2xy(3x - 4y + 6)$

49. $3x^2 - 7x - 48 = (3x - 16)(x + 3)$

51. $8x^2 - 32x = 8x(x - 4)$

53. $2x - x^2 + 15 = -x^2 + 2x + 15$
$= -(x^2 - 2x - 15)$
$= -(x - 5)(x + 3)$

55. $84xy - 16x^2y - 4x^3y$
$$= 4xy(21 - 4x - x^2)$$
$$= -4xy(x^2 + 4x - 21)$$
$$= -4xy(x + 7)(x - 3)$$

57. $-x^2 + 25 = 25 - x^2 = (5 - x)(5 + x)$

59. The proposed factor 2x + 2 has a common factor of 2, which would imply that the original trinomial $6x^2 - 5x - 4$ has a common factor of 2. This is not the case. We can eliminate eight possible factorizations of $6x^2 - 5x - 4$ in this way. In addition to $(3x - 2)(2x + 2)$, we can eliminate

$(3x - 1)(2x + 4)$ $\quad (3x + 1)(2x - 4)$

$(3x + 2)(2x - 2)$ $\quad (6x - 2)(x + 2)$

$(6x + 2)(x - 2)$ $\quad (6x - 4)(x + 1)$

$(6x + 4)(x - 1)$

This leaves four possibilities:

$(6x - 1)(x + 4)$, $(6x + 1)(x - 4)$,
$(3x + 4)(2x - 1)$, and $(3x - 4)(2x + 1)$,

the last of which is the correct one.

61. Let f = number of franks sold

f + 20 = number of knishes sold

$$1.25 + .80(f + 20) = 169.75$$
$$1.25 + .80f + 16 = 169.75$$
$$2.05f + 16 = 169.75$$
$$2.05f = 153.75$$
$$f = 75$$

Then $f + 20 = 75 + 20 = 95$. So the vendor sold 75 franks and 95 knishes on the day in question.

Exercises 8.5

1. $$\frac{3x + 12}{6} = \frac{3x}{6} + \frac{12}{6} = \frac{x}{2} + 2$$
$$= \frac{x + 4}{2}$$

3. $$\frac{t^2 - 6t}{6t} = \frac{t^2}{6t} - \frac{6t}{6t} = \frac{t}{6} - 1$$
$$= \frac{t - 6}{6}$$

5. $$\frac{3x^2y - 9xy^2}{3xy} = \frac{3x^2y}{3xy} - \frac{9xy^2}{3xy}$$
$$= x - 3y$$

7. $\frac{3x^2y - 9xy^2}{6x^2y^2} = \frac{3x^2y}{6x^2y^2} - \frac{9xy^2}{6x^2y^2}$

$= \frac{1}{2y} - \frac{3}{2x}$

$= \frac{x - 3y}{2xy}$

9. $\frac{10a^2b^3c - 15ab^2c^2 - 20a^3b^2c^3}{5ab^2c}$

$= \frac{10a^2b^3c}{5ab^2c} - \frac{15ab^2c^2}{5ab^2c} - \frac{20a^3b^2c^3}{5ab^2c}$

$= 2ab - 3c - 4a^2c^2$

11.
$$\begin{array}{r} x - 5 \\ x + 2 \overline{) x^2 - 3x + 2} \\ \underline{-(x^2 + 2x)} \\ -5x + 2 \\ \underline{-(-5x - 10)} \\ 12 \end{array}$$

13.
$$\begin{array}{r} t + 2 \\ t - 5 \overline{) t^2 - 3t - 10} \\ \underline{-(t^2 - 5t)} \\ 2t - 10 \\ \underline{-(2t - 10)} \\ 0 \end{array}$$

15.
$$\begin{array}{r} w + 1 \\ w + 3 \overline{) w^2 + 4w - 21} \\ \underline{-(w^2 + 3w)} \\ w - 21 \\ \underline{-(w + 3)} \\ -24 \end{array}$$

17.
$$\begin{array}{r} 2x - 1 \\ x - 1 \overline{) 2x^2 - 3x + 7} \\ \underline{-(2x^2 - 2x)} \\ -x + 7 \\ \underline{-(-x + 1)} \\ 6 \end{array}$$

19.

$$\begin{array}{r}
y^2 + 3y + 7 \\
y - 2 \overline{\big) y^3 + y^2 + y - 14} \\
\underline{-(y^3 - 2y^2)} \\
3y^2 + y \\
\underline{-(3y^2 - 6y)} \\
7y - 14 \\
\underline{-(7y - 14)} \\
0
\end{array}$$

21.

$$\begin{array}{r}
2a^2 - a - 2 \\
a + 1 \overline{\big) 2a^3 + a^2 - 3a + 2} \\
\underline{-(2a^3 + 2a^2)} \\
-a^2 - 3a \\
\underline{-(-a^2 - a)} \\
-2a + 2 \\
\underline{-(-2a - 2)} \\
4
\end{array}$$

23.

$$\begin{array}{r}
x^2 - 4x + 12 \\
x + 3 \overline{\big) x^3 - x^2 + 0x + 36} \\
\underline{-(x^3 + 3x^2)} \\
-4x^2 + 0x \\
\underline{-(-4x^2 - 12x)} \\
12x + 36 \\
\underline{-(12x + 36)} \\
0
\end{array}$$

25.

$$\begin{array}{r}
x^3 + 2x^2 + 4x + 8 \\
x - 2 \overline{\big) x^4 + 0x^3 + 0x^2 + 0x - 16} \\
\underline{-(x^4 - 2x^3)} \\
2x^3 + 0x^2 \\
\underline{-(2x^3 - 4x^2)} \\
4x^2 + 0x \\
\underline{-(4x^2 - 8x)} \\
8x - 16 \\
\underline{-(8x - 16)} \\
0
\end{array}$$

27. $$\begin{array}{r} x^2 + 4x - 2 \\ 3x+2 \overline{\smash{)}\, 3x^3 + 14x^2 + 2x - 4} \\ \underline{-(3x^3 + 2x^2)} \\ 12x^2 + 2x \\ \underline{-(12x^2 + 8x)} \\ -6x - 4 \\ \underline{-(-6x - 4)} \\ 0 \end{array}$$

29. $$\begin{array}{r} 2t^2 + 5t - 4 \\ 2t-5 \overline{\smash{)}\, 4t^3 + 0t^2 - 33t + 24} \\ \underline{-(4t^3 - 10t^2)} \\ 10t^2 - 33t \\ \underline{-(10t^2 - 25t)} \\ -8t + 24 \\ \underline{-(-8t + 20)} \\ 4 \end{array}$$

31. To divide $5x^4 - 8x^3 + 7x^2 - 2x + 1$ by $x^2 + x - 1$, we would begin by asking what x^2 must be multiplied by in order to give a product of $5x^4$. When we determine this to be $5x^2$, we multiply $5x^2$ and $x^2 + x - 1$ and subtract the resulting product from $5x^4 - 8x^3 + 7x^2 - 2x + 1$. We continue this process, as shown below.

$$\begin{array}{r} 5x^2 - 13x + 25 \\ x^2 + x - 1 \overline{\smash{)}\, 5x^4 - 8x^3 + 7x^2 - 2x + 1} \\ \underline{-(5x^4 + 5x^3 - 5x^2)} \\ -13x^3 + 12x^2 - 2x \\ \underline{-(-13x^3 - 13x^2 + 13x)} \\ 25x^2 - 15x + 1 \\ \underline{-(25x^2 + 25x - 25)} \\ -40x + 26 \end{array}$$

The long division process ends when the degree of the remainder (in this case, 1) is less than the degree of the divisor (in this case, 2).

33. $-x^2 = -(-3)^2 = -9$

35. $(xy)^2 = (-3 \cdot 5)^2 = (-15)^2 = 225$

37. $x^{-2} = (-3)^{-2} = \dfrac{1}{(-3)^2} = \dfrac{1}{9}$

39. $(xy)^{-1} = (-3 \cdot 5)^{-1} = (-15)^{-1}$
$= \dfrac{1}{-15} = -\dfrac{1}{15}$

Chapter 8 Review Exercises

1. $(x+5)(x+7) = x^2 + 7x + 5x + 35$
$= x^2 + 12x + 35$

3. $(x-5)(x-7) = x^2 - 7x - 5x + 35$
$= x^2 - 12x + 35$

5. $(x+5)(x-7) = x^2 - 7x + 5x - 35$
$= x^2 - 2x - 35$

7. $(x-5)(x+7) = x^2 + 7x - 5x - 35$
$= x^2 + 2x - 35$

9. $(x-5)(x-5) = x^2 - 5x - 5x + 25$
$= x^2 - 10x + 25$

11. $(x-5)(x+5) = x^2 + 5x - 5x - 25$
$= x^2 - 25$

13. $(x+9y)(x-9y)$
$= x^2 - 9xy + 9xy - 81y^2$
$= x^2 - 81y^2$

15. $(2x+3)(x-7)$
$= 2x^2 - 14x + 3x - 21$
$= 2x^2 - 11x - 21$

17. $(5x-2)(3x+4)$
$= 15x^2 + 20x - 6x - 8$
$= 15x^2 + 14x - 8$

19. $x^2 + 7x + 12 = (x+3)(x+4)$

21. $x^2 + 7x = x(x+7)$

23. $x^2 - 13x + 12 = (x-1)(x-12)$

25. $x^2 - 6xy - 27y^2 = (x+3y)(x-9y)$

27. $x^2 - 64 = (x-8)(x+8)$

29. $2x^2 + 9x + 10 = (2x+5)(x+2)$

31. $3x^2 - 6x - 24 = 3(x^2 - 2x - 8)$
$= 3(x-4)(x+2)$

33. $6a^2 + 36a + 48 = 6(a^2 + 6a + 8)$
$= 6(a+2)(a+4)$

35. $5x^3y - 80xy^3 = 5xy(x^2 - 16y^2)$
$= 5xy(x - 4y)(x + 4y)$

37. $x^2 + 9x = x(x + 9)$

39. $25t^2 - 1 = (5t - 1)(5t + 1)$

41. $30 - x^2 + x = -x^2 + x + 30$
$= -(x^2 - x - 30)$
$= -(x - 6)(x + 5)$

43. $12x - 3x^2 - 9 = -3x^2 + 12x - 9$
$= -3(x^2 - 4x + 3)$
$= -3(x - 1)(x - 3)$

45. $x(x - 7) + 3(x - 7) = (x - 7)(x + 3)$

47. $m^2 + 2mn + 9m + 18n$
$= m(m + 2n) + 9(m + 2n)$
$= (m + 2n)(m + 9)$

49. $\frac{x^2y - xy^2}{xy} = \frac{x^2y}{xy} - \frac{xy^2}{xy} = x - y$

51.
$$\begin{array}{r} x - 3 \\ x - 1\overline{)x^2 - 4x - 5} \\ \underline{-(x^2 - x)} \\ -3x - 5 \\ \underline{-(-3x + 3)} \\ -8 \end{array}$$

53.
$$\begin{array}{r} 2x^2 + 6x + 14 \\ x - 3\overline{)2x^3 + 0x^2 - 4x - 4} \\ \underline{-(2x^3 - 6x^2)} \\ 6x^2 - 4x \\ \underline{-(6x^2 - 18x)} \\ 14x - 4 \\ \underline{-14x - 42} \\ 38 \end{array}$$

55. $$\begin{array}{r} x^2 - 2x + 4 \\ x + 2 \overline{)x^3 + 0x^2 + 0x + 8} \\ \underline{x^3 + 2x^2} \\ -2x^2 + 0x \\ \underline{-(-2x^2 - 4x)} \\ 4x + 8 \\ \underline{-(4x + 8)} \\ 0 \end{array}$$

Chapter 8 Practice Test

1. $6x^2 + 12x^2 - 15x = 3x\left(2x^2 + 4x - 5\right)$

3. $x^2 + 9x + 8 = (x + 1)(x + 8)$

5. $4x^2 - 20x = 4x(x - 5)$

7. $6x^2 + 24x + 18 = 6(x^2 + 4x + 3)$
$= 6(x + 1)(x + 3)$

9. $x^2 + 4x + 4 = (x + 2)(x + 2)$

11. $x^2y^2 - 9 = (xy - 3)(xy + 3)$

13. $$\frac{12r^3t^2 - 18r^2t^2 + 20r^3t^4}{4r^3t^3}$$
$$= \frac{12r^3t^2}{4r^3t^3} - \frac{18r^2t^2}{4r^3t^3} + \frac{20r^3t^4}{4r^3t^3}$$
$$= \frac{3}{t} - \frac{9}{2rt} + 5t$$
$$= \frac{6r}{2rt} - \frac{9}{2rt} + \frac{10rt^2}{2rt}$$
$$= \frac{6r - 9 + 10rt^2}{2rt}$$

CHAPTER 9
MORE RATIONAL EXPRESSIONS

Exercises 9.1

1. $\frac{8x^3y^{10}}{10x^6y^5} = \frac{4y^5}{5x^3}$

3. $\frac{5x - 7x}{x^2 - 7x^2} = \frac{-2x}{-6x^2} = \frac{1}{3x}$

5. $\frac{6x^2(x+4)^5}{9x^3(x+4)} = \frac{2(x+4)^4}{3x}$

7. $\frac{12a^2b + 6c^3}{8a^2b + 4c^3} = \frac{6(2a^2b + c^3)}{4(2a^2b + c^3)} = \frac{3}{2}$

9. $\frac{3x - 6}{5x - 10} = \frac{3(x-2)}{5(x-2)} = \frac{3}{5}$

11. $\frac{3x - 6}{6x - 12} = \frac{3(x-2)}{6(x-2)} = \frac{1}{2}$

13. $\frac{3x - 6}{6x + 12} = \frac{3(x-2)}{6(x+2)} = \frac{x-2}{2(x+2)}$

15. $\frac{5y}{10y + 20} = \frac{5y}{10(y+2)} = \frac{y}{2(y+2)}$

17. $\frac{6x + 18}{x^2 - 9} = \frac{6(x+3)}{(x-3)(x+3)}$

$= \frac{6}{x - 3}$

19. $\frac{6x^2 + 18}{x^2 - 9} = \frac{6(x^2 + 3)}{x^2 - 9}$

(cannot be reduced)

21. $\frac{t^2 + 3t}{t^2 + 3t - 10} = \frac{t(t+3)}{(t+5)(t-2)}$

(cannot be reduced)

23. $\frac{2x^2 + x + x}{4x^3 + 4x^2} = \frac{2x^2 + 2x}{4x^3 + 4x^2}$

$= \frac{2x(x+1)}{4x^2(x+1)} = \frac{1}{2x}$

25. $\dfrac{y^2-5y-6}{y^2-12y+36}=\dfrac{\cancel{(y-6)}(y+1)}{\cancel{(y-6)}(y-6)}$

$=\dfrac{y+1}{y-6}$

27. $\dfrac{s^2-2s-15}{s^2-6s+5}=\dfrac{\cancel{(s-5)}(s+3)}{\cancel{(s-5)}(s-1)}$

$=\dfrac{s+3}{s-1}$

29. $\dfrac{x^2(x+3)(x-4)}{x^4-4x^3}=\dfrac{\cancel{x^2}(x+3)\cancel{(x-4)}}{\cancel{x^3}^{\,x}\cancel{(x-4)}}$

$=\dfrac{x+3}{x}$

31. $\dfrac{3a^2+a-2}{a^2-a-2}=\dfrac{(3a-2)\cancel{(a+1)}}{(a-2)\cancel{(a+1)}}$

$=\dfrac{3a-2}{a-2}$

33. $\dfrac{4x^2+7x-2}{x^2+4x+4}=\dfrac{(4x-1)\cancel{(x+2)}}{(x+2)\cancel{(x+2)}}$

$=\dfrac{4x-1}{x+2}$

35. $\dfrac{x^2-x+3x-8}{x^2+4x}=\dfrac{x^2+2x-8}{x^2+4x}$

$=\dfrac{(x-2)\cancel{(x+4)}}{x\cancel{(x+4)}}$

$=\dfrac{x-2}{x}$

37. $\dfrac{6x^2-12x-18}{3x^2-9x-30}=\dfrac{6(x^2-2x-3)}{3(x^2-3x-10)}$

$=\dfrac{\overset{2}{\cancel{6}}(x-3)(x+1)}{\underset{1}{\cancel{3}}(x-5)(x+2)}$

$=\dfrac{2(x-3)(x+1)}{(x-5)(x+2)}$

39. $\dfrac{6x^2-5x^2-4}{x^2-6x+8}=\dfrac{x^2-4}{x^2-6x+8}$

$=\dfrac{\cancel{(x-2)}(x+2)}{\cancel{(x-2)}(x-4)}$

$=\dfrac{x+2}{x-4}$

41. $\dfrac{x^2-7x+10}{x^2-7x+12}=\dfrac{(x-2)(x-5)}{(x-3)(x-4)}$

(cannot be reduced)

43. $\dfrac{y^3-y^2-2y}{6y^2-24}=\dfrac{y(y^2-y-2)}{6(y^2-4)}$

$=\dfrac{y(y-2)(y+1)}{6(y-2)(y+2)}$

$=\dfrac{y(y+1)}{6(y+2)}$

45. $\dfrac{c^2-9c}{c^3-9c^2}=\dfrac{c(c-9)}{c^2(c-9)}=\dfrac{1}{c}$

47. The student was incorrect in each case, because he or she tried to cancel terms instead of factors. Neither of the fractions can be reduced because neither contains a common factor in its numerator and denominator. (In part (a), we could write

$\dfrac{x^2+5}{x}=\dfrac{x^2}{x}+\dfrac{5}{x}=\dfrac{x}{1}+\dfrac{5}{x}$ if we wished.)

49. $\dfrac{9x^2}{25y^6}\div\dfrac{xy}{15}=\dfrac{9x^2}{25y^6}\cdot\dfrac{15}{xy}$

$=\dfrac{27x}{5y^7}$

51. $\dfrac{5}{x}\div 10=\dfrac{5}{x}\cdot\dfrac{1}{10}=\dfrac{1}{2x}$

53. $\dfrac{2a}{5}=\dfrac{a-2}{3}$ LCD = 15

$15\left(\dfrac{2a}{5}\right)=15\left(\dfrac{a-2}{3}\right)$

$6a=5a-10$

$a=-10$

Exercises 9.2

1. $\dfrac{8x^2y^2}{9yz^3}\cdot\dfrac{12a^2x}{2ax^2}=\dfrac{16ay^2}{3z^2}$

3. $\dfrac{3st^2}{5p}\div\dfrac{15t^2}{s^2}=\dfrac{3st^2}{5p}\cdot\dfrac{s^2}{15t^2}$

$=\dfrac{s^3}{25p}$

5. $\dfrac{12x^2y^3}{5z}\cdot 30xyz=\dfrac{12x^2y^3}{5z}\cdot\dfrac{30xyz}{1}$

$=72x^3y^4$

7. $28a^2b^3z \div \frac{4a}{7b} = \frac{28a^2b^3z^4}{1} \cdot \frac{7b}{4a}$
$= 49ab^4z^4$

9. $\frac{x^2+4x}{x^2} \cdot \frac{x}{x^2+6x+5}$
$= \frac{x(x+4)}{x^2} \cdot \frac{x}{(x+1)(x+5)}$
$= \frac{x+4}{(x+1)(x+5)}$

11. $\frac{x^2+4x}{x^2+4x+4} \cdot \frac{x^2-4}{x^2-4x+4}$
$= \frac{x(x+4)}{(x+2)(x+2)} \cdot \frac{(x-2)(x+2)}{(x-2)(x-2)}$
$= \frac{x(x+4)}{(x+2)(x-2)}$

13. $\frac{r^2-4r-5}{2r-10} \div \frac{r^2-3r+2}{4r^2}$
$= \frac{r^2-4r-5}{2r-10} \cdot \frac{4r^2}{r^2-3r+2}$
$= \frac{(r-5)(r+1)}{2(r-5)} \cdot \frac{4r^2}{(r-1)(r-2)}$
$= \frac{2r^2(r+1)}{(r-1)(r-2)}$

15. $\frac{m^2}{m^2+3m} \div \frac{3m^2}{m^2+6m}$
$= \frac{m^2}{m^2+3m} \cdot \frac{m^2+6m}{3m^2}$
$= \frac{m^2}{m(m+3)} \cdot \frac{m(m+6)}{3m^2}$
$= \frac{m+6}{3(m+3)}$

17. $\frac{x^2+3x+2}{x^2+2x} \cdot \frac{x}{x^2+2}$
$= \frac{(x+1)(x+2)}{x(x+2)} \cdot \frac{x}{x^2+2}$
$= \frac{x+1}{x^2+2}$

19. $\frac{y^2-3y-4}{y^2-2y-8}\cdot\frac{y^2+4y+4}{y^2-8y+16}$

$=\frac{(y-4)(y+1)}{(y-4)(y+2)}\cdot\frac{(y+2)(y+2)}{(y-4)(y-4)}$

$=\frac{(y+1)(y+2)}{(y-4)(y-4)}$

21. $\frac{x^2+2x}{x^2-x-2}\cdot\frac{x-2}{x}$

$=\frac{x(x+2)}{(x-2)(x+1)}\cdot\frac{x-2}{x}$

$=\frac{x+2}{x+1}$

23. $\frac{2x^2+x-15}{x^2-9}\cdot\frac{6x^2+7x+1}{2x^2-3x-5}$

$=\frac{(2x-5)(x+3)}{(x-3)(x+3)}\cdot\frac{(6x+1)(x+1)}{(2x-5)(x+1)}$

$=\frac{6x+1}{x-3}$

25. $\frac{4a}{a+4}\cdot\frac{a+5}{5a}\div\frac{a^2+6a+5}{a^2+5a+4}$

$=\left(\frac{4a}{a+4}\cdot\frac{a+5}{5a}\right)\div\frac{a^2+6a+5}{a^2+5a+4}$

$=\frac{4(a+5)}{5(a+4)}\cdot\frac{a^2+5a+4}{a^2+6a+5}$

$=\frac{4(a+5)}{5(a+4)}\cdot\frac{(a+1)(a+4)}{(a+1)(a+5)}$

$=\frac{4}{5}$

27. $\frac{x}{2}\div\frac{2}{x}\cdot\frac{x^2-16}{x^2-4x}$

$=\left(\frac{x}{2}\div\frac{2}{x}\right)\cdot\frac{x^2-16}{x^2-4x}$

$=\left(\frac{x}{2}\cdot\frac{x}{2}\right)\cdot\frac{x^2-16}{x^2-4x}$

$=\frac{x^2}{4}\cdot\frac{(x-4)(x+4)}{x(x-4)}$

$=\frac{x(x+4)}{4}$

29. $\frac{t^2+2t}{t^2+2t+1} \div \frac{2t^2+7t+6}{t^2+t}$

$= \frac{t^2+2t}{t^2+2t+1} \cdot \frac{t^2+t}{2t^2+7t+6}$

$= \frac{t(t+2)}{(t+1)(t+1)} \cdot \frac{t(t+1)}{(2t+3)(t+2)}$

$= \frac{t^2}{(t+1)(2t+3)}$

31. $\frac{2x^2+6x+4x}{x^2-25} \cdot \frac{(x+5)^2}{4x-2x}$

$= \frac{2x^2+10x}{x^2-25} \cdot \frac{(x+5)^2}{2x}$

$= \frac{2x(x+5)}{(x-5)(x+5)} \cdot \frac{(x+5)(x+5)}{2x}$

$= \frac{(x+5)(x+5)}{x-5}$

33. $\frac{x^3y-xy^3}{8x^2y+4xy^2} \div \frac{(x-y)^2}{2x^2+3xy+y^2}$

$= \frac{x^3y-xy^3}{8x^2y+4xy^2} \cdot \frac{2x^2+3xy+y^2}{(x-y)^2}$

$= \frac{xy(x^2-y^2)}{4xy(2x+y)} \cdot \frac{(2x+y)(x+y)}{(x-y)^2}$

$= \frac{xy(x-y)(x+y)}{4xy(2x+y)} \cdot \frac{(2x+y)(x+y)}{(x-y)(x-y)}$

$= \frac{(x+y)(x+y)}{4(x-y)}$

35. $\left(\frac{x}{3} \cdot \frac{x}{4}\right) \cdot \frac{12}{x^2-3x}$

$= \frac{x^2}{12} \cdot \frac{12}{x^2-3x}$

$= \frac{x^2}{12} \cdot \frac{12}{x(x-3)}$

$= \frac{x}{x-3}$

37. $\left(\frac{c}{2}+\frac{c}{5}\right) \div \frac{c^2+7c}{10}$

$$= \left(\frac{c(5)}{2(5)}+\frac{c(2)}{5(2)}\right) \div \frac{c^2+7c}{10}$$

$$= \left(\frac{5c}{10}+\frac{2c}{10}\right) \div \frac{c^2+7c}{10}$$

$$= \frac{7c}{10} \div \frac{c^2+7c}{10}$$

$$= \frac{7c}{10} \cdot \frac{10}{c^2+7c}$$

$$= \frac{7\cancel{c}}{\cancel{10}_1} \cdot \frac{\cancel{10}^1}{\cancel{c}(c+7)} = \frac{7}{c+7}$$

39. $\left(\frac{w}{4}-\frac{9}{w}\right) \cdot \left(\frac{w+10}{2w}-\frac{5}{w}\right)$

$$= \left(\frac{w(w)}{4(w)}-\frac{9(4)}{w(4)}\right) \cdot \left(\frac{w+10}{2w}-\frac{5(2)}{w(2)}\right)$$

$$= \left(\frac{w^2}{4w}-\frac{36}{4w}\right) \cdot \left(\frac{w+10}{2w}-\frac{10}{2w}\right)$$

$$= \left(\frac{w^2-36}{4w}\right) \cdot \left(\frac{w+10-10}{2w}\right)$$

$$= \frac{(w-6)(w+6)}{4w} \cdot \frac{\cancel{w}^1}{2\cancel{w}}$$

$$= \frac{(w-6)(w+6)}{8w}$$

41. Two cars leave from the same point and travel in opposite directions. The slower car, traveling at the rate of 20 kph, leaves two hours later than the faster one, which travels at 30 kph. How long has the slower car been traveling when the two cars are 310 km apart? (Let t = time traveled by slower car.)

42. Wilma invested a sum of money at an 8% interest rate and a second sum, \$500 more than the first, at a 12% interest rate. How much money did she invest at each rate if her total interest on the two investments is \$260? (Let x = amount Wilma invested at 8%.)

43. $\frac{5}{6x}+\frac{3}{8x^2}$ LCD $= 24x^2$

$$\frac{5}{6x} = \frac{5(4x)}{6x(4x)} = \frac{20x}{24x^2}$$

$$\frac{3}{8x^2} = \frac{3(3)}{8x^2(3)} = \frac{9}{24x^2}$$

Then $\frac{5}{6x}+\frac{3}{8x^2} = \frac{20x}{24x^2}+\frac{9}{24x^2}$

$$= \frac{20x+9}{24x^2}$$

45. $9-\frac{6}{x} = \frac{9x}{x}-\frac{6}{x} = \frac{9x-6}{x}$

47. Let x = cost of an orchestra seat

x – 4 = cost of a balcony seat

$$\begin{aligned} 250x + 150(x - 4) &= 3400 \\ 250x + 150x - 600 &= 3400 \\ 400x - 600 &= 3400 \\ 400x &= 4000 \\ x &= 10 \end{aligned}$$

They should charge $10 for an orchestra seat.

Exercises 9.3

1. $$\frac{5x-2}{4x+8} + \frac{3x+2}{4x+8} = \frac{5x-2+3x+2}{4x+8}$$
$$= \frac{8x}{4x+8} = \frac{8x}{4(x+2)} = \frac{2x}{x+2}$$

3. $$\frac{5x-2}{4x+8} - \frac{3x+2}{4x+8}$$
$$= \frac{5x-2-(3x+2)}{4x+8}$$
$$= \frac{5x-2-3x-2}{4x+8}$$
$$= \frac{2x-4}{4x+8} = \frac{2(x-2)}{4(x+2)}$$
$$= \frac{x-2}{2(x+2)}$$

7. $$\frac{x+7}{x+2} - \frac{x+3}{x+2} + \frac{x-2}{x+2}$$
$$= \frac{(x+7)-(x+3)+(x-2)}{x+2}$$
$$= \frac{x+7-x-3+x-2}{x+2}$$
$$= \frac{x+2}{x+2} = \frac{1}{1} = 1$$

11. $$\frac{5}{2x} + \frac{4}{x+2} = \frac{5(x+2)}{2x(x+2)} + \frac{4(2x)}{(x+2)(2x)}$$
$$= \frac{5x+10}{2x(x+2)} + \frac{8x}{2x(x+2)}$$
$$= \frac{5x+10+8x}{2x(x+2)} = \frac{13x+10}{2x(x+2)}$$

13. $$\frac{5}{2x} \cdot \frac{4}{x+2} = \frac{10}{x(x+2)}$$

17. $$\frac{7}{a+7}-\frac{5}{a+5}=\frac{7(a+5)}{(a+7)(a+5)}-\frac{5(a+7)}{(a+5)(a+7)}$$
$$=\frac{7a+35}{(a+7)(a+5)}-\frac{5a+35}{(a+7)(a+5)}$$
$$=\frac{7a+35-(5a+35)}{(a+7)(a+5)}$$
$$=\frac{7a+35-5a-35}{(a+7)(a+5)}$$
$$=\frac{2a}{(a+7)(a+5)}$$

19. $$\frac{4}{3x^2}-\frac{2}{x^2+3x}=\frac{4}{3x^2}-\frac{2}{x(x+3)}$$
$$=\frac{4(x+3)}{3x^2(x+3)}-\frac{2(3x)}{x(x+3)(3x)}$$
$$=\frac{4x+12}{3x^2(x+3)}-\frac{6x}{3x^2(x+3)}$$
$$=\frac{4x+12-6x}{3x^2(x+3)}$$
$$=\frac{-2x+12}{3x^2(x+3)}=\frac{2(-x+6)}{3x^2(x+3)}$$

23. $$\frac{4}{x^2+4x}-\frac{2}{x^2-4x}=\frac{4}{x(x+4)}-\frac{2}{x(x-4)}$$
$$=\frac{4(x-4)}{x(x+4)(x-4)}-\frac{2(x+4)}{x(x-4)(x+4)}$$
$$=\frac{4x-16}{x(x+4)(x-4)}-\frac{2x+8}{x(x+4)(x-4)}$$
$$=\frac{4x-16-(2x+8)}{x(x+4)(x-4)}$$
$$=\frac{4x-16-2x-8}{x(x+4)(x-4)}$$
$$=\frac{2x-24}{x(x+4)(x-4)}=\frac{2(x-12)}{x(x+4)(x-4)}$$

25. $$2-\frac{x}{x-1}=\frac{2}{1}-\frac{x}{x-1}$$
$$=\frac{2(x-1)}{1(x-1)}-\frac{x}{x-1}$$
$$=\frac{2x-2}{x-1}-\frac{x}{x-1}$$
$$=\frac{2x-2-x}{x-1}=\frac{x-2}{x-1}$$

29. $\dfrac{3}{x^2-16}-\dfrac{3}{2x^2+8x}$

$= \dfrac{3}{(x-4)(x+4)} - \dfrac{3}{2x(x+4)}$

$= \dfrac{3(2x)}{(x-4)(x+4)(2x)}$

$- \dfrac{3(x-4)}{2x(x+4)(x-4)}$

$= \dfrac{6x}{2x(x+4)(x-4)}$

$- \dfrac{3x-12}{2x(x+4)(x-4)}$

$= \dfrac{6x-(3x-12)}{2x(x+4)(x-4)}$

$= \dfrac{6x-3x+12}{2x(x+4)(x-4)}$

$= \dfrac{3x+12}{2x(x+4)(x-4)}$

$= \dfrac{3\cancel{(x+4)}}{2x\cancel{(x+4)}(x-4)} = \dfrac{3}{2x(x-4)}$

33. $\dfrac{x}{x^2+6x+9}+\dfrac{1}{x^2+4x+3}$

$= \dfrac{x}{(x+3)(x+3)} + \dfrac{1}{(x+1)(x+3)}$

$= \dfrac{x(x+1)}{(x+3)(x+3)(x+1)}$

$+ \dfrac{1(x+3)}{(x+1)(x+3)(x+3)}$

$= \dfrac{x^2+x}{(x+1)(x+3)(x+3)}$

$+ \dfrac{x+3}{(x+1)(x+3)(x+3)}$

$= \dfrac{x^2+x+x+3}{(x+1)(x+3)(x+3)}$

$= \dfrac{x^2+2x+3}{(x+1)(x+3)(x+3)}$

37. $5 - \frac{1}{3x-6} + \frac{3}{x^2-2x}$

$$= \frac{5}{1} - \frac{1}{3(x-2)} + \frac{3}{x(x-2)}$$

$$= \frac{5(3x(x-2))}{1(3x(x-2))} - \frac{1(x)}{3(x-2)(x)}$$

$$+ \frac{3(3)}{x(x-2)(3)}$$

$$= \frac{15x^2-30x}{3x(x-2)} - \frac{x}{3x(x-2)}$$

$$+ \frac{9}{3x(x-2)}$$

$$= \frac{15x^2-30x-x+9}{3x(x-2)}$$

$$= \frac{15x^2-31x+9}{3x(x-2)}$$

39. $\frac{5}{x^2+x+6} - \frac{3}{x^2+3x} + \frac{2}{x^2-2x}$

$$= \frac{5}{(x+3)(x-2)} - \frac{3}{x(x+3)}$$

$$+ \frac{2}{x(x-2)}$$

$$= \frac{5(x)}{(x+3)(x-2)(x)} - \frac{3(x-2)}{x(x+3)(x-2)}$$

$$+ \frac{2(x+3)}{x(x-2)(x+3)}$$

$$= \frac{5x}{x(x+3)(x-2)} - \frac{3x-6}{x(x+3)(x-2)}$$

$$+ \frac{2x+6}{x(x+3)(x-2)}$$

$$= \frac{5x-(3x-6)+2x+6}{x(x+3)(x-2)}$$

$$= \frac{5x-3x+6+2x+6}{x(x+3)(x-2)}$$

$$= \frac{4x+12}{x(x+3)(x-2)} = \frac{4\cancel{(x+3)}}{x\cancel{(x+3)}(x-2)}$$

$$= \frac{4}{x(x-2)}$$

45. $\dfrac{1-\frac{1}{a}}{\frac{1}{a}-\frac{1}{a^2}} = \dfrac{a^2\left(1-\frac{1}{a}\right)}{a^2\left(\frac{1}{a}-\frac{1}{a^2}\right)}$

$= \dfrac{a^2 \cdot 1 - \frac{a^2}{1}\cdot\frac{1}{a}}{\frac{a^2}{1}\cdot\frac{1}{a} - \frac{a^2}{1}\cdot\frac{1}{a^2}}$

$= \dfrac{a^2 - a}{a-1} = \dfrac{a(a-1)}{a-1} = \dfrac{a}{1}$

$= a$

47. $\dfrac{\frac{a}{2}-\frac{b}{4}}{\frac{4}{b^2}-\frac{1}{a^2}} = \dfrac{4a^2b^2\left(\frac{a}{2}-\frac{b}{4}\right)}{4a^2b^2\left(\frac{4}{b^2}-\frac{1}{a^2}\right)}$

$= \dfrac{\frac{4a^2b^2}{1}\cdot\frac{a}{2} - \frac{4a^2b^2}{1}\cdot\frac{b}{4}}{\frac{4a^2b^2}{1}\cdot\frac{4}{b^2} - \frac{4a^2b^2}{1}\cdot\frac{1}{a^2}}$

$= \dfrac{2a^3b^2 - a^2b^3}{16a^2 - 4b^2} = \dfrac{a^2b^2(2a-b)}{4(4a^2-b^2)}$

$= \dfrac{a^2b^2(2a-b)}{4(2a-b)(2a+b)}$

$= \dfrac{a^2b^2}{4(2a+b)}$

49. (a) $\left(x+\frac{2}{x}\right)\left(x-\frac{3}{x}\right)$

$= x^2 - \frac{x}{1}\left(\frac{3}{x}\right) + \frac{x}{1}\left(\frac{2}{x}\right)$

$- \left(\frac{2}{x}\right)\left(\frac{3}{x}\right)$

$= x^2 - 3 + 2 - \dfrac{6}{x^2}$

$= x^2 - 1 - \dfrac{6}{x^2}$

(b) $x + \frac{2}{x} = \frac{x}{1} + \frac{2}{x}$

$= \frac{x(x)}{1(x)} + \frac{2}{x}$

$= \frac{x^2}{x} + \frac{2}{x} = \frac{x^2 + 2}{x}$

$x - \frac{3}{x} = \frac{x}{1} - \frac{3}{x}$

$= \frac{x(x)}{1(x)} - \frac{3}{x}$

$= \frac{x^2}{x} - \frac{3}{x} = \frac{x^2 - 3}{x}$

Then $\left(x + \frac{2}{x}\right)\left(x - \frac{3}{x}\right)$

$= \frac{x^2 + 2}{x} \cdot \frac{x^2 - 3}{x}$

$= \frac{(x^2 + 2)(x^2 - 3)}{x^2}$

The second method is easier. If the instructions ask for the answer in the form of a single fraction, we get such an answer directly in part (b). In part (a), there would be further work to do.

51. $\frac{x}{2} - 1 = \frac{x}{3}$ LCD = 6

$6\left(\frac{x}{2} - 1\right) = 6\left(\frac{x}{3}\right)$

$\overset{3}{\cancel{6}}\left(\frac{x}{\cancel{2}_1}\right) - 6 = \overset{2}{\cancel{6}}\left(\frac{x}{\cancel{3}_1}\right)$

$3x - 6 = 2x$

$-6 = -x$

$6 = x$

53. $.25m + .1m = 56$

$.35m = 56$

$35m = 5600$

$m = \frac{5600}{35}$

$m = 160$

Exercises 9.4

1.
$$\frac{x}{2}+\frac{x}{3}=10$$
$$6\left(\frac{x}{2}+\frac{x}{3}\right)=6\cdot 10$$
$$\frac{6}{1}\cdot\frac{x}{2}+\frac{6}{1}\cdot\frac{x}{3}=60$$
$$3x+2x=60$$
$$5x=60$$
$$x=12$$

CHECK x = 12:
$$\frac{(12)}{2}+\frac{(12)}{3}\stackrel{?}{=}10$$
$$6+4\stackrel{?}{=}10$$
$$10\stackrel{\checkmark}{=}10$$

5.
$$\frac{a+1}{3}+\frac{a+3}{4}=4$$
$$12\left(\frac{a+1}{3}+\frac{a+3}{4}\right)=12\cdot 4$$
$$\frac{12}{1}\cdot\frac{a+1}{3}+\frac{12}{1}\cdot\frac{a+3}{4}=48$$
$$4(a+1)+3(a+3)=48$$
$$4a+4+3a+9=48$$
$$7a+13=48$$
$$7a=35$$
$$a=5$$

CHECK a = 5:
$$\frac{(5)+1}{3}+\frac{(5)+3}{4}\stackrel{?}{=}4$$
$$\frac{6}{3}+\frac{8}{4}\stackrel{?}{=}4$$
$$2+2\stackrel{?}{=}4$$
$$4\stackrel{\checkmark}{=}4$$

9. $\frac{x}{2}+\frac{x}{3}+\frac{x}{4}$ LCD = 12
$$=\frac{x(6)}{2(6)}+\frac{x(4)}{3(4)}+\frac{x(3)}{4(3)}$$
$$=\frac{6x}{12}+\frac{4x}{12}+\frac{3x}{12}$$
$$=\frac{6x+4x+3x}{12}=\frac{13x}{12}$$

13.

$$\frac{x+1}{2} - \frac{3}{x} = \frac{x}{2}$$

$$2x\left(\frac{x+1}{2} - \frac{3}{x}\right) = 2x\left(\frac{x}{2}\right)$$

$$\frac{\cancel{2}x}{1} \cdot \frac{x+1}{\cancel{2}_1} - \frac{2\cancel{x}}{1} \cdot \frac{3}{\cancel{x}_1} = \frac{\cancel{2}x}{1} \cdot \frac{x}{\cancel{2}_1}$$

$$x(x+1) - 6 = x^2$$

$$x^2 + x - 6 = x^2$$

$$x - 6 = 0$$

$$x = 6$$

CHECK x = 6:

$$\frac{(6)+1}{2} - \frac{3}{(6)} \stackrel{?}{=} \frac{(6)}{2}$$

$$\frac{7}{2} - \frac{3}{6} \stackrel{?}{=} \frac{6}{2}$$

$$\frac{7}{2} - \frac{1}{2} \stackrel{?}{=} \frac{6}{2}$$

$$\frac{6}{2} \stackrel{\checkmark}{=} \frac{6}{2}$$

17.

$$\frac{x}{5} + \frac{x-1}{4} + \frac{x-3}{2} = 3$$

$$20\left(\frac{x}{5} + \frac{x-1}{4} + \frac{x-3}{2}\right) = 20 \cdot 3$$

$$\frac{\overset{4}{\cancel{20}}}{1} \cdot \frac{x}{\cancel{5}_1} + \frac{\overset{5}{\cancel{20}}}{1} \cdot \frac{x-1}{\cancel{4}_1} + \frac{\overset{10}{\cancel{20}}}{1} \cdot \frac{x-3}{\cancel{2}_1} = 60$$

$$4x + 5(x-1) + 10(x-3) = 60$$

$$4x + 5x - 5 + 10x - 30 = 60$$

$$19x - 35 = 60$$

$$19x = 95$$

$$x = 5$$

CHECK x = 5:

$$\frac{(5)}{5} + \frac{(5)-1}{4} + \frac{(5)-3}{2} \stackrel{?}{=} 3$$

$$\frac{5}{5} + \frac{4}{4} + \frac{2}{2} \stackrel{?}{=} 3$$

$$1 + 1 + 1 \stackrel{?}{=} 3$$

$$3 \stackrel{\checkmark}{=} 3$$

23.

$$\frac{2r+1}{3} - \frac{r+1}{5} = \frac{r+8}{6}$$

$$30\left(\frac{2r+1}{3} - \frac{r+1}{5}\right) = 30\left(\frac{r+8}{6}\right)$$

$$\frac{\overset{10}{\cancel{30}}}{1} \cdot \frac{2r+1}{\cancel{3}_1} - \frac{\overset{6}{\cancel{30}}}{1} \cdot \frac{r+1}{\cancel{5}_1} = \frac{\overset{5}{\cancel{30}}}{1} \cdot \frac{r+8}{\cancel{6}_1}$$

$$10(2r+1)-6(r+1)=5(r+8)$$
$$20r+10-6r-6=5r+40$$
$$14r+4=5r+40$$
$$9r+4=40$$
$$9r=36$$
$$r=4$$

CHECK $r=4$:

$$\frac{2(4)+1}{3}-\frac{(4)+1}{5}\stackrel{?}{=}\frac{(4)+8}{6}$$
$$\frac{9}{3}-\frac{5}{5}\stackrel{?}{=}\frac{12}{6}$$
$$3-1\stackrel{?}{=}2$$
$$2\stackrel{\checkmark}{=}2$$

25. $$\frac{3}{x}-\frac{2}{3}=\frac{2}{x}$$
$$3x\left(\frac{3}{x}-\frac{2}{3}\right)=3x\left(\frac{2}{x}\right)$$
$$\frac{3\cancel{x}}{1}\cdot\frac{3}{\cancel{x}_1}-\frac{\cancel{3}x}{1}\cdot\frac{2}{\cancel{3}_1}=\frac{3\cancel{x}}{1}\cdot\frac{2}{\cancel{x}_1}$$
$$9-2x=6$$
$$-2x=-3$$
$$x=\frac{3}{2}$$

CHECK $x=\frac{3}{2}$:

$$\frac{3}{\left(\frac{3}{2}\right)}-\frac{2}{3}\stackrel{?}{=}\frac{2}{\left(\frac{3}{2}\right)}$$
$$\frac{3}{1}\cdot\frac{2}{3}-\frac{2}{3}\stackrel{?}{=}\frac{2}{1}\cdot\frac{2}{3}$$
$$2-\frac{2}{3}\stackrel{?}{=}\frac{4}{3}$$
$$\frac{4}{3}\stackrel{\checkmark}{=}\frac{4}{3}$$

29. $\frac{5}{2x}-\frac{8}{x+2}$ LCD $=2x(x+2)$

$$=\frac{5(x+2)}{2x(x+2)}-\frac{8(2x)}{2x(x+2)}$$
$$=\frac{5x+10}{2x(x+2)}-\frac{16x}{2x(x+2)}$$
$$=\frac{5x+10-16x}{2x(x+2)}$$
$$=\frac{-11x+10}{2x(x+2)}$$

31.
$$\frac{4}{x-1}-\frac{5}{8}=\frac{3}{2x-2}$$
$$\frac{4}{x-1}-\frac{5}{8}=\frac{3}{2(x-1)}$$
$$8(x-1)\left(\frac{4}{x-1}-\frac{5}{8}\right)=8(x-1)\left(\frac{3}{2(x-1)}\right)$$
$$\frac{8(x-1)}{1}\cdot\frac{4}{x-1}-\frac{8(x-1)}{1}\cdot\frac{5}{8}=\frac{8(x-1)}{1}\cdot\frac{3}{2(x-1)}$$
$$32-5(x-1)=12$$
$$32-5x+5=12$$
$$-5x+37=12$$
$$-5x=-25$$
$$x=5$$

CHECK x = 5:
$$\frac{4}{(5)-1}-\frac{5}{8}\overset{?}{=}\frac{3}{2(5)-2}$$
$$\frac{4}{4}-\frac{5}{8}\overset{?}{=}\frac{3}{8}$$
$$\frac{8}{8}-\frac{5}{8}\overset{?}{=}\frac{3}{8}$$
$$\frac{3}{8}\overset{\checkmark}{=}\frac{3}{8}$$

33.
$$\frac{8}{x-2}+3=\frac{x+6}{x-2}$$
$$(x-2)\left(\frac{8}{x-2}+3\right)=(x-2)\left(\frac{x+6}{x-2}\right)$$
$$\frac{x-2}{1}\cdot\frac{8}{x-2}+\frac{x-2}{1}\cdot\frac{3}{1}=\frac{x-2}{1}\cdot\frac{x+6}{x-2}$$
$$8+3(x-2)=x+6$$
$$8+3x-6=x+6$$
$$3x+2=x+6$$
$$2x+2=6$$
$$2x=4$$
$$x=2$$

CHECK x = 2:
$$\frac{8}{(2)-2}+3\overset{?}{=}\frac{(2)+6}{(2)-2}$$
$$\frac{8}{0}+3\overset{?}{=}\frac{8}{0}$$

$\frac{8}{0}$ is not defined, so $x = 2$ is not a solution. Since $x = 2$ does not satisfy the original equation, no solution can be found.

5. $\dfrac{4}{x+2} - \dfrac{3}{x+6}$ LCD = $(x+2)(x+6)$

$$= \frac{4(x+6)}{(x+2)(x+6)} - \frac{3(x+2}{(x+2)(x+6)}$$

$$= \frac{4x+24}{(x+2)(x+6)} - \frac{3x+6}{(x+2)(x+6)}$$

$$= \frac{4x+24-(3x+6)}{(x+2)(x+6)}$$

$$= \frac{4x+24-3x-6}{(x+2)(x+6)}$$

$$= \frac{x+18}{(x+2)(x+6)}$$

39.

$$\frac{5}{x^2-x} - \frac{1}{2x-2} = \frac{1}{x}$$

$$\frac{5}{x(x-1)} - \frac{1}{2(x-1)} = \frac{1}{x}$$

$$2x(x-1)\left(\frac{5}{x(x-1)} - \frac{1}{2(x-1)}\right) = 2x(x-1)\left(\frac{1}{x}\right)$$

$$\frac{2\cancel{x}\cancel{(x-1)}}{1} \cdot \frac{5}{\cancel{x}\cancel{(x-1)}_1} - \frac{\cancel{2}x\cancel{(x-1)}}{1} \cdot \frac{1}{\cancel{2}\cancel{(x-1)}_1} = \frac{2\cancel{x}(x-1)}{1} \cdot \frac{1}{\cancel{x}_1}$$

$$10 - x = 2(x-1)$$

$$10 - x = 2x - 2$$

$$10 = 3x - 2$$

$$12 = 3x$$

$$4 = x$$

CHECK x = 4:

$$\frac{5}{(4)^2-(4)} - \frac{1}{2(4)-2} \stackrel{?}{=} \frac{1}{(4)}$$

$$\frac{5}{16-4} - \frac{1}{8-2} \stackrel{?}{=} \frac{1}{4}$$

$$\frac{5}{12} - \frac{1}{6} \stackrel{?}{=} \frac{1}{4}$$

$$\frac{5}{12} - \frac{2}{12} \stackrel{?}{=} \frac{1}{4}$$

$$\frac{3}{12} \stackrel{?}{=} \frac{1}{4}$$

$$\frac{1}{4} \stackrel{\checkmark}{=} \frac{1}{4}$$

45. $\frac{2}{x^2 - x} + \frac{8}{x^2 - 1}$ LCD = x(x − 1)(x + 1)

$$= \frac{2}{x(x-1)} + \frac{8}{(x-1)(x+1)}$$
$$= \frac{2(x+1)}{x(x-1)(x+1)} + \frac{8(x)}{x(x-1)(x+1)}$$
$$= \frac{2x+2}{x(x-1)(x+1)} + \frac{8x}{x(x-1)(x+1)}$$
$$= \frac{2x+2+8x}{x(x-1)(x+1)}$$
$$= \frac{10x+2}{x(x-1)(x+1)}$$

49.

$$\frac{5}{x^2 - 2x - 3} = \frac{4}{x^2 - 3x - 4}$$
$$\frac{5}{(x-3)(x+1)} = \frac{4}{(x-4)(x+1)}$$
$$\frac{\cancel{(x-3)}(x-4)\cancel{(x+1)}}{1}\left(\frac{5}{\cancel{(x-3)}\cancel{(x+1)}}\right) = \frac{(x-3)\cancel{(x-4)}\cancel{(x+1)}}{1}\left(\frac{4}{\cancel{(x-4)}\cancel{(x+1)}}\right)$$
$$5(x-4) = 4(x-3)$$
$$5x - 20 = 4x - 12$$
$$x - 20 = -12$$
$$x = 8$$

CHECK x = 8:

$$\frac{5}{(8)^2 - 2(8) - 3} \stackrel{?}{=} \frac{4}{(8)^2 - 3(8) - 4}$$
$$\frac{5}{64 - 16 - 3} \stackrel{?}{=} \frac{4}{64 - 24 - 4}$$
$$\frac{5}{45} \stackrel{?}{=} \frac{4}{36}$$
$$\frac{1}{9} \stackrel{\checkmark}{=} \frac{1}{9}$$

51.

$$\frac{9}{x^2 - 3x + 2} - \frac{2}{x-1} = \frac{1}{x-2}$$
$$\frac{9}{(x-1)(x-2)} - \frac{2}{x-1} = \frac{1}{x-2}$$
$$(x-1)(x-2)\cdot\left(\frac{9}{(x-1)(x-2)} - \frac{2}{x-1}\right) = (x-1)(x-2)\left(\frac{1}{x-2}\right)$$
$$\frac{\cancel{(x-1)}\cancel{(x-2)}}{1}\cdot\frac{9}{\cancel{(x-1)}\cancel{(x-2)}} - \frac{\cancel{(x-1)}(x-2)}{1}\cdot\frac{2}{\cancel{x-1}} = \frac{(x-1)\cancel{(x-2)}}{1}\cdot\frac{1}{\cancel{x-2}}$$

$$
\begin{aligned}
9-2(x-2) &= x-1\\
9-2x+4 &= x-1\\
13-2x &= x-1\\
14 &= 3x\\
\frac{14}{3} &= x
\end{aligned}
$$

53. The length of a rectangle is five less than three times its width. Find the dimensions of the rectangle if its perimeter is 62 cm. (Let x = width of the rectangle.)

55. Let x = number of students who take longer than four years to graduate.

2600 + x = number of graduates altogether.

$$
\begin{aligned}
\frac{2600}{x} &= \frac{13}{3}\\
\frac{3x}{1}\left(\frac{2600}{x}\right) &= \frac{3x}{1}\left(\frac{13}{3}\right)\\
7800 &= 13x\\
600 &= x
\end{aligned}
$$

Then 2600 + x = 2600 + 600 = 3200. So there were 3200 graduates altogether during the last five years.

Exercises 9.5

3.

$$
\begin{aligned}
3x+6y &= 18\\
-3x \qquad &\quad -3x\\
6y &= 18-3x\\
\frac{6y}{6} &= \frac{18-3x}{6}\\
y &= \frac{18-3x}{6} = \frac{3(6-x)}{6}\\
&= \frac{6-x}{2}
\end{aligned}
$$

7.

$$
\begin{aligned}
3(m+2p) &= 4(p-m)\\
3m+6p &= 4p-4m\\
+4m \qquad &\quad +4m\\
7m+6p &= 4p\\
-6p &= -6p\\
7m &= -2p\\
\frac{7m}{7} &= \frac{-2p}{7}\\
m &= \frac{-2p}{7} = -\frac{2p}{7}
\end{aligned}
$$

11. $2r + 3(r-5) = r - 5s + 1$

$$2r + 3r - 15 = r - 5s + 1$$
$$5r - 15 = r - 5s + 1$$
$$\underline{-r \qquad -r}$$
$$4r - 15 = -5s + 1$$
$$\underline{+15 \qquad +15}$$
$$4r = -5s + 16$$
$$\frac{\not{4}r}{\not{4}} = \frac{-5s+16}{4}$$
$$r = \frac{-5s+16}{4}$$

15. $\frac{x}{2} + \frac{y}{3} = \frac{x}{3} + \frac{y}{4} - \frac{1}{6}$

$$12\left(\frac{x}{2} + \frac{y}{3}\right) = 12\left(\frac{x}{3} + \frac{y}{4} - \frac{1}{6}\right)$$
$$\frac{\overset{6}{\not{12}}}{1} \cdot \frac{x}{\not{2}_1} + \frac{\overset{4}{\not{12}}}{1} \cdot \frac{y}{\not{3}_1}$$
$$= \frac{\overset{4}{\not{12}}}{1} \cdot \frac{x}{\not{3}_1} + \frac{\overset{3}{\not{12}}}{1} \cdot \frac{y}{\not{4}_1} - \frac{\overset{2}{\not{12}}}{1} \cdot \frac{1}{\not{6}_1}$$
$$6x + 4y = 4x + 3y - 2$$
$$\underline{-4x \qquad -4x}$$
$$2x + 4y = 3y - 2$$
$$\underline{-4y \qquad -4y}$$
$$2x = -y - 2$$
$$\frac{\not{2}x}{\not{2}} = \frac{-y-2}{2}$$
$$x = \frac{-y-2}{2}$$

21.

$$2pn - \frac{y}{3} = 3n + ay + p$$

$$3\left(2pn - \frac{y}{3}\right) = 3(3n + ay + p)$$

$$6pn - \frac{\cancel{3}}{1} \cdot \frac{y}{\cancel{3}_1} = 9n + 3ay + 3p$$

$$6pn - y = 9n + 3ay + 3p$$

$$\underline{\quad + y \qquad\qquad\qquad + y}$$

$$6pn = 9n + 3ay + 3p + y$$

$$\underline{\quad - 9n \qquad - 9n \qquad\qquad}$$

$$6pn - 9n = 3ay + 3p + y$$

$$\underline{\quad - 3p \qquad\qquad - 3p}$$

$$6pn - 9n - 3p = 3ay \quad + y$$

$$6pn - 9n - 3p = (3a + 1)y$$

$$\frac{6pn - 9n - 3p}{3a + 1} = \frac{(\cancel{3a + 1})y}{\cancel{3a + 1}}$$

$$\frac{6pn - 9n - 3p}{3a + 1} = y$$

23.

$$y = \frac{u + 1}{u - 1}$$

$$(u - 1)y = \frac{(\cancel{u - 1})}{1} \frac{(u + 1)}{(\cancel{u - 1})_1}$$

$$(u - 1)y = u + 1$$

$$uy - y = u + 1$$

$$\underline{\quad - u = -u}$$

$$uy - y - u = 1$$

$$\underline{\quad + y \qquad + y}$$

$$uy - u = 1 + y$$

$$u(y - 1) = 1 + y$$

$$\frac{u(\cancel{y - 1})}{\cancel{y - 1}_1} = \frac{1 + y}{y - 1}$$

$$u = \frac{1 + y}{y - 1}$$

29.

$$A = \frac{1}{2}h(b_1 + b_2)$$

$$2A = 2\left(\frac{1}{2}h(b_1 + b_2)\right)$$

$$2A = h(b_1 + b_2)$$

$$\frac{2A}{b_1 + b_2} = h$$

33. $C = \frac{5}{9}(F - 32)$

$$9C = 9\left(\frac{5}{9}(F - 32)\right)$$

$$9C = 5(F - 32)$$

$$\frac{9C}{5} = F - 32$$

$$\frac{9C}{5} + 32 = F$$

37. $\frac{x - \mu}{s} < 2$

$$s\left(\frac{x - \mu}{s}\right) < s(2)$$

(since s is positive)

$$x - \mu < 2s$$

$$x < 2s + \mu$$

39. $\frac{1}{f} = \frac{1}{f_1} + \frac{1}{f_2}$

$$ff_1f_2\left(\frac{1}{f}\right) = ff_1f_2\left(\frac{1}{f_1} + \frac{1}{f_2}\right)$$

$$\frac{ff_1f_2}{1} \cdot \frac{1}{f} = \frac{ff_1f_2}{1} \cdot \frac{1}{f_1} + \frac{ff_1f_2}{1} \cdot \frac{1}{f_2}$$

$$f_1f_2 = ff_2 + ff_1$$

$$f_1f_2 = f(f_2 + f_1)$$

$$\frac{f_1f_2}{f_2 + f_1} = f$$

41. The ratio of men to women on the faculty of a local college is 9 to 10. If there are 600 women on the faculty, how many men are on the faculty? (Let x = number of men.)

Exercises 9.6

3. Let x = # of people who preferred Brand X

200 – x = # of people who did not

$$\frac{x}{200 - x} = \frac{13}{12}$$

$$\frac{12(200 - x)}{1} \cdot \frac{x}{200 - x} = \frac{12(200 - x)}{1} \cdot \frac{13}{12}$$

$$12x = 13(200 - x)$$

$$12x = 2600 - 13x$$

$$25x = 2600$$

$$x = 104$$

Then 200 – x = 200 – 104 = 96.

Thus, 104 people preferred Brand X.

CHECK: $\frac{104}{96} = \frac{\not{8}(13)}{\not{8}(12)} = \frac{13}{12}$,

and $104 + 96 = 200$.

5. Let x = # of hits Joe must get in his next 50 at-bats

$$\frac{120 + x}{400 + 50} = .400 = \frac{400}{1000} = \frac{2}{5}$$

$$\frac{120 + x}{450} = \frac{2}{5}$$

$$\frac{\not{450}}{1} \cdot \frac{120 + x}{\not{450}\,1} = \frac{\overset{90}{\not{450}}}{1} \cdot \frac{2}{\not{5}\,1}$$

$$120 + x = 180$$

$$x = 60$$

This is impossible, since the largest possible value of x is 50. Therefore, Joe cannot raise his average to .400 in his next 50 at-bats.

9. Let x = # of hours that the electrician needs to complete the job working alone

2x = # of hours that the apprentice needs to complete the job working alone

$$6\left(\frac{1}{x}\right) + 6\left(\frac{1}{2x}\right) = 1$$

$$\frac{6}{1} \cdot \frac{1}{x} + \frac{\overset{3}{\not{6}}}{1} \cdot \frac{1}{\not{2}x} = 1$$

$$\frac{6}{x} + \frac{3}{x} = 1$$

$$\frac{9}{x} = 1$$

$$\frac{\not{x}}{1} \cdot \frac{9}{\not{x}} = x \cdot 1$$

$$9 = x$$

Thus, the electrician can complete the job in 9 hours, working alone.

CHECK: In one hour, the electrician completes $\frac{1}{9}$ of the job, so in 6 hours, he completes $\frac{6}{9} = \frac{2}{3}$ of the job. In one hour, the apprentice completes $\frac{1}{18}$ of the job, so in 6 hours, he completes $\frac{6}{18} = \frac{1}{3}$ of the job. $\frac{2}{3} + \frac{1}{3} = 1$ (the entire job).

11. Let r = rate of the train (in kph)

r + 100 = rate of the plane (in kph)

$$\frac{300}{r} = \frac{500}{r + 100}$$

$$\frac{\not{r}(r + 100)}{1} \cdot \frac{300}{\not{r}\,1} = \frac{r(\not{r + 100})}{1} \cdot \frac{500}{\not{r + 100}\,1}$$

$$300(r+100) = 500r$$
$$300r + 30000 = 500r$$
$$30000 = 200r$$
$$150 = r$$

Then $250 = r + 100$. Thus the train travels at the rate of 150 kph and the plane travels at the rate of 250 kph.

CHECK: At the rate of 150 kph, the train covers 300 kilometers in $\frac{300}{150} = 2$ hours. At the rate of 250 kph, the plane covers 500 kilometers in $\frac{500}{250} = 2$ hours, the same amount of time.

15. Let t = # of hours that Ronnie walks

$3 - t$ = # of hours that Ronnie jogs

$$6t = 14(3-t)$$
$$6t = 42 - 14t$$
$$20t = 42$$
$$t = \frac{42}{20} = \frac{20}{10} = 2.1$$

Then the distance to Ronnie's friend's house is 6(2.1) = 12.6 kilometers.

CHECK: Ronnie walks 6(2.1) = 12.6 kilometers to his friend's house and jogs 14(3 – 2.1) = 14(.9) = 12.6 kilometers back, the same distance.

17. Let x = smaller number

3x = larger number

$$\frac{1}{x} + \frac{1}{3x} = \frac{5}{3}$$
$$3x\left(\frac{1}{x} + \frac{1}{3x}\right) = 3x\left(\frac{5}{3}\right)$$
$$\frac{3x}{1}\cdot\frac{1}{x} + \frac{3x}{1}\cdot\frac{1}{3x} = \frac{3x}{1}\cdot\frac{5}{3}$$
$$3 + 1 = 5x$$
$$4 = 5x$$
$$\frac{4}{5} = x$$

Then, $3x = 3\left(\frac{4}{5}\right) = \frac{12}{5}$. Thus, the numbers are $\frac{4}{5}$ and $\frac{12}{5}$.

CHECK: $\frac{12}{5}$ is three times $\frac{4}{5}$, and $\frac{1}{\left(\frac{12}{5}\right)} + \frac{1}{\left(\frac{4}{5}\right)} = \frac{5}{12} + \frac{5}{4}$
$$= \frac{5}{12} + \frac{15}{12} = \frac{20}{12} = \frac{5}{3}.$$

19. Let x = number to be added

$$\frac{3+x}{5+x}=\frac{5}{6}$$

$$\frac{6\cancel{(5+x)}}{1}\left(\frac{3+x}{\cancel{5+x}}\right)=\frac{\cancel{6}(5+x)}{1}\cdot\frac{5}{\cancel{6}}$$

$$6(3+x)=5(5+x)$$

$$18+6x=25+5x$$

$$18+x=25$$

$$x=7$$

Thus, the number to be added to both the numerator and denominator is 7.

CHECK: $\frac{3+(7)}{5+(7)}=\frac{10}{12}=\frac{5}{6}$

21. Let x = numerator of the fraction

x +2 = denominator of the fraction

$$\frac{x+2}{x}=\frac{x}{x+2}$$

$$\frac{\cancel{x}(x+2)}{1}\cdot\frac{x+2}{\cancel{x}}=\frac{x\cancel{(x+2)}}{1}\cdot\frac{x}{\cancel{x+2}}$$

$$(x+2)(x+2)=x^2$$

$$x^2+4x+4=x^2$$

$$4x+4=0$$

$$4x=-4$$

$$x=-1$$

Then $x + 2 = -1 + 2 = 1$. Thus, the fraction is $\frac{-1}{1}$.

CHECK: 1 is two more than -1, and the reciprocal of $\frac{-1}{1}$ is $\frac{1}{-1}$, which is equal to the original fraction.

25. If someone can do a job in h hours, then he or she can complete $\frac{1}{h}$ of the job per hour. This is the rate at which the person works. When this rate is multiplied by the number of hours that the person works, we get the fractional part of the entire job that he or she accomplishes.

Chapter 9 Review Exercises

1. $\frac{x^2}{x^2+2}$ cannot be reduced.

3. $\frac{x^2+3x-4}{x^2-16}=\frac{\cancel{(x+4)}(x-1)}{(x-4)\cancel{(x+4)}}$

$$=\frac{x-1}{x-4}$$

5. $\dfrac{a^2+8a+16}{a^2+6a+8} = \dfrac{(a+4)\cancel{(a+4)}}{(a+2)\cancel{(a+4)}}$

$= \dfrac{a+4}{a+2}$

7. $\dfrac{3z^2-12}{3z^2+9z+6} = \dfrac{3(z^2-4)}{3(z^2+3z+2)}$

$= \dfrac{\cancel{3}(z-2)\cancel{(z+2)}}{\cancel{3}(z+1)\cancel{(z+2)}}$

$= \dfrac{z-2}{z+1}$

9. $\dfrac{x}{x+2} + \dfrac{x}{2}$

$= \dfrac{x(2)}{(x+2)(2)} + \dfrac{x(x+2)}{2(x+2)}$

$= \dfrac{2x}{2(x+2)} + \dfrac{x^2+2x}{2(x+2)}$

$= \dfrac{2x+x^2+2x}{2(x+2)} = \dfrac{x^2+4x}{2(x+2)}$

$= \dfrac{x(x+4)}{2(x+2)}$

11. $\dfrac{3}{2x+4} + \dfrac{6}{x^2+2x}$

$= \dfrac{3}{2(x+2)} + \dfrac{6}{x(x+2)}$

$= \dfrac{3(x)}{2(x+2)(x)} + \dfrac{6(2)}{x(x+2)(2)}$

$= \dfrac{3x}{2x(x+2)} + \dfrac{12}{2x(x+2)}$

$= \dfrac{3x+12}{2x(x+2)} = \dfrac{3(x+4)}{2x(x+2)}$

13. $\dfrac{x^2-5x-6}{2x-12} \div \dfrac{x^2+2x+1}{8x^2}$

$= \dfrac{x^2-5x-6}{2x-12} \cdot \dfrac{8x^2}{x^2+2x+1}$

$= \dfrac{\cancel{(x-6)}\cancel{(x+1)}}{\cancel{2}\cancel{(x-6)}_1} \cdot \dfrac{{}^4\cancel{8}x^2}{\cancel{(x+1)}(x+1)}$

$= \dfrac{4x^2}{x+1}$

15. $\dfrac{5}{z^2+z-6}-\dfrac{3}{z^2+3z}$

$=\dfrac{5}{(z+3)(z-2)}-\dfrac{3}{z(z+3)}$

$=\dfrac{5(z)}{(z+3)(z-2)(z)}$

$-\dfrac{3(z-2)}{z(z+3)(z-2)}$

$=\dfrac{5z}{z(z+3)(z-2)}$

$-\dfrac{3z-6}{z(z+3)(z-2)}$

$=\dfrac{5z-(3z-6)}{z(z+3)(z-2)}$

$=\dfrac{5z-3z+6}{z(z+3)(z-2)}$

$=\dfrac{2z+6}{z(z+3)(z-2)}$

$=\dfrac{2\cancel{(z+3)}}{z\cancel{(z+3)}(z-2)}=\dfrac{2}{z(z-2)}$

17. $2+\dfrac{3}{x+2}-\dfrac{1}{x}$

$=\dfrac{2}{1}+\dfrac{3}{x+2}-\dfrac{1}{x}$

$=\dfrac{2(x(x+2))}{1(x(x+2))}+\dfrac{3(x)}{(x+2)(x)}$

$-\dfrac{1(x+2)}{x(x+2)}$

$=\dfrac{2x^2+4x}{x(x+2)}+\dfrac{3x}{x(x+2)}-\dfrac{x+2}{x(x+2)}$

$=\dfrac{2x^2+4x+3x-(x+2)}{x(x+2)}$

$=\dfrac{2x^2+4x+3x-x-2}{x(x+2)}$

$=\dfrac{2x^2+6x-2}{x(x+2)}=\dfrac{2(x^2+3x-1)}{x(x+2)}$

19. $\dfrac{5}{x}-\dfrac{2}{3x}=\dfrac{13}{6}$

$6x\left(\dfrac{5}{x}-\dfrac{2}{3x}\right)=6x\left(\dfrac{13}{6}\right)$

$\dfrac{6\cancel{x}}{1}\cdot\dfrac{5}{\cancel{x}_1}-\dfrac{\cancel{6}^{2}\cancel{x}}{1}\cdot\dfrac{2}{\cancel{3}\cancel{x}_1}$

$=\dfrac{\cancel{6}x}{1}\cdot\dfrac{13}{\cancel{6}_1}$

$$30 - 4 = 13x$$
$$26 = 13x$$
$$2 = x$$

21.
$$\frac{y+2}{y} + \frac{4}{y+2} = \frac{y}{y+2}$$
$$y(y+2)\left(\frac{y+2}{y} + \frac{4}{y+2}\right) = y(y+2)\left(\frac{y}{y+2}\right)$$
$$\frac{\cancel{y}(y+2)}{1} \cdot \frac{y+2}{\cancel{y}_1} + \frac{y(\cancel{y+2})}{1} \cdot \frac{4}{\cancel{y+2}_1} = \frac{y(\cancel{y+2})}{1} \cdot \frac{y}{\cancel{y+2}_1}$$
$$(y+2)(y+2) + 4y = y^2$$
$$y^2 + 4y + 4 + 4y = y^2$$
$$y^2 + 8y + 4 = y^2$$
$$8y + 4 = 0$$
$$8y = -4$$
$$y = \frac{-4}{8} = -\frac{1}{2}$$

23.
$$\frac{x+2}{x-3} + \frac{4}{3} = \frac{5}{x-3}$$
$$3(x-3)\left(\frac{x+2}{x-3} + \frac{4}{3}\right) = 3(x-3)\left(\frac{5}{x-3}\right)$$
$$\frac{3(\cancel{x-3})}{1} \cdot \frac{x+2}{\cancel{x-3}_1} + \frac{\cancel{3}(x-3)}{1} \cdot \frac{4}{\cancel{3}_1} = \frac{3(\cancel{x-3})}{1} \cdot \frac{5}{\cancel{x-3}_1}$$
$$3(x+2) + 4(x-3) = 15$$
$$3x + 6 + 4x - 12 = 15$$
$$7x - 6 = 15$$
$$7x = 21$$
$$x = 3$$

This value does not check in the original equation, since it produces fractions with denominators of 0. Thus, the equation has no solution.

25.

$$\begin{array}{rcl}
3x - 4y + 7y & = & 8x - 7y + 3 \\
-8x & & -8x \\
\hline
-5x - 4y + 7 & = & -7y + 3 \\
+4y & & +4y \\
\hline
-5x \quad + 7 & = & -3y + 3 \\
-7 & & -7 \\
\hline
-5x & = & -3y - 4 \\
\dfrac{\cancel{-5}x}{\cancel{-5}} & = & \dfrac{-3y-4}{-5} \\
x & = & \dfrac{-3y-4}{-5} \\
& \text{or} & \dfrac{3y+4}{5}
\end{array}$$

27. Let x = # of black marbles in the bag

x + 80 = # of red marbles in the bag

$$\frac{x+80}{x} = \frac{7}{5}$$

$$\frac{5\cancel{x}}{1} \cdot \frac{x+80}{\cancel{x}_1} = \frac{\cancel{5}x}{1} \cdot \frac{7}{\cancel{5}_1}$$

$$5(x+80) = 7x$$

$$5x + 400 = 7x$$

$$400 = 2x$$

$$200 = x$$

Then 280 = x + 80. Thus, there are 200 black marbles and 280 red marbles in the bag, or 480 marbles altogether.

CHECK: 280 is 80 more than 200, and $\frac{280}{200} = \frac{7}{5}$.

29. Let x = # of hours John needs to do the job alone

$$4\left(\frac{1}{x}\right) + 4\left(\frac{1}{6}\right) = 1$$

$$\frac{4}{x} + \frac{4}{6} = 1$$

$$6x\left(\frac{4}{x} + \frac{4}{6}\right) = 6x \cdot 1$$

$$\frac{6\cancel{x}}{1} \cdot \frac{4}{\cancel{x}_1} + \frac{\cancel{6}x}{1} \cdot \frac{4}{\cancel{6}_1} = 6x$$

$$24 + 4x = 6x$$

$$24 = 2x$$

$$12 = x$$

Thus, John needs 12 hours to do the job alone.

CHECK: In 4 hours, Susan does $4x\left(\frac{1}{6}\right) = \frac{4}{6} = \frac{2}{3}$ of the job; in 4 hours, John does $4\left(\frac{1}{12}\right) = \frac{4}{12} = \frac{1}{3}$ of the job.

$\frac{2}{3} + \frac{1}{3} = 1$ (the entire job)

Chapter 9 Practice Test

1. $\dfrac{4x^2}{x^2-4x}=\dfrac{4\cancel{x}^{\,2}}{\cancel{x}(x-4)}=\dfrac{4x}{x-4}$

3. $\dfrac{3}{x+3}+\dfrac{2}{x+2}$

$$=\frac{3(x+2)}{(x+3)(x+2)}+\frac{2(x+3)}{(x+2)(x+3)}$$

$$=\frac{3x+6}{(x+3)(x+2)}+\frac{2x+6}{(x+3)(x+2)}$$

$$=\frac{3x+6+2x+6}{(x+3)(x+2)}+\frac{5x+12}{(x+3)(x+2)}$$

5. $\dfrac{5}{2x}-\dfrac{10}{x^2+4x}=\dfrac{5}{2x}-\dfrac{10}{x(x+4)}$

$$=\frac{5(x+4)}{2x(x+4)}-\frac{10(2)}{x(x+4)(2)}$$

$$=\frac{5x+20}{2x(x+4)}-\frac{20}{2x(x+4)}$$

$$=\frac{5x+20-20}{2x(x+4)}=\frac{5\cancel{x}}{2\cancel{x}(x+4)}$$

$$=\frac{5}{2(x+4)}$$

7. $\dfrac{2x-5}{x^2-3x}-\dfrac{3x+7}{x^2-3x}+\dfrac{6x-3}{x^2-3x}$

$$=\frac{2x-5-(3x+7)+6x-3}{x^2-3x}$$

$$=\frac{2x-5-3x-7+6x-3}{x^2-3x}$$

$$=\frac{5x-15}{x^2-3x}=\frac{5\cancel{(x-3)}}{x\cancel{(x-3)}}=\frac{5}{x}$$

9.

$$at+b=\frac{3t}{2}+7$$

$$2(at+b)=2\left(\frac{3t}{2}+7\right)$$

$$2at+2b=3t+14$$

$$\underline{\quad -3t \qquad -3t\quad}$$

$$2at-3t+2b=14$$

$$\underline{\quad -2b \qquad -2b\quad}$$

$$2at-3t=14-2b$$

$$(2a-3)t=14-2b$$

$$t=\frac{14-2b}{2a-3}$$

1. Let t = # of hours needed to go from A to B

14 – t = # of hours needed to go from B to A

$$45t = 60(14 - t)$$
$$45t = 840 - 60t$$
$$105t = 840$$
$$t = 8$$

Thus, the distance between the towns is 45(8) = 360 kilometers.

CHECK: Time going $= \frac{360}{45} = 8$ hours. Time returning $= \frac{360}{60} = 6$ hours. $8 + 6 \stackrel{\checkmark}{=} 14$ hours.

CUMULATIVE REVIEW CHAPTERS 7–9

1. $(x+8)(x-5)$
$= x^2 - 5x + 8x - 40$
$= x^2 + 3x - 40$

3. $(a+b+c)(a+b-c)$
$= a(a+b+c)$
$+ b(a+b-c)$
$+ c(a+b-c)$
$= a^2 + ab - ac$
$+ ab + b^2 - bc$
$+ ac + bc - c^2$
$= a^2 + 2ab + b^2 - c^2$

5. $(5a-3c)(4a+3c)$
$= 20a^2 + 15ac - 12ac - 9c^2$
$= 20a^2 + 3ac - 9c^2$

7. $(x+3)(x-12) + (x+6)^2$
$= (x+3)(x-12) + (x+6)(x+6)$
$= x^2 - 12x + 3x - 36 + x^2 + 6x + 6x + 36$
$= 2x^2 + 3x$

9. $(a-3)(2a+3)(2a-3)$
$= (a-3)(4a^2 - 6a + 6a - 9)$
$= (a-3)(4a^2 - 9)$
$= 4a^3 - 9a - 12a^2 + 27$
$= 4a^3 - 12a^2 - 9a + 27$

11. $(x^2 - xy + 3y^2)$
$+ (5x^2 - 8y^2)$
$+ (y^2 - 6x^2)$
$= x^2 + 5x^2 - 6x^2 - xy$
$+ 3y^2 - 8y^2 + y^2$
$= -xy - 4y^2$

13. (a) degree of $5x^4 - 3x^2 + 6x - 1$ is 4.

(b) coefficient of second degree term is –3.

15. $\frac{x^2+8x}{2x} = \frac{x^2}{2x} + \frac{8x}{2x}$

$= \frac{x}{2} + 4 = \frac{x+8}{2}$

17.
$$\begin{array}{r} y-1 \\ y-2\,\overline{)\,y^2-3y+4} \\ \underline{-(y^2-2y)} \\ -y+4 \\ \underline{-(-y+2)} \\ 2 \end{array}$$

19.
$$\begin{array}{r} 6x^2+8x\;+9 \\ 3x-4\,\overline{)\,18x^3+0x^2-5x-28} \\ \underline{-(18x^3-24x^2)} \\ 24x^2-5x \\ \underline{-(24x^2-32x)} \\ 27x-28 \\ \underline{-(27x-36)} \\ 8 \end{array}$$

21. $5^0 + 2^{-3} + 2^{-4}$

$= 1 + \frac{1}{2^3} + \frac{1}{2^4}$

$= 1 + \frac{1}{8} + \frac{1}{16}$

$= \frac{16}{16} + \frac{2}{16} + \frac{1}{16}$

$= \frac{16+2+1}{16} = \frac{19}{16}$

23. $\frac{(x^2)^3}{x^2x^3} = \frac{x^6}{x^5} = x^{6-5} = x$

25. $\frac{(2x^3)^4}{4(x^5)^3} = \frac{2^4(x^3)^4}{4(x^5)^3} = \frac{16x^{12}}{4x^{15}}$

$= \frac{16}{4} \cdot \frac{x^{12}}{x^{15}} = \frac{4}{1} \cdot x^{-3}$

$= \frac{4}{1} \cdot \frac{1}{x^3} = \frac{4}{x^3}$

27. $\dfrac{(3a^{-3}t^{2})^{-3}}{(a^{-1}t^{-2})^{2}}$

$= \dfrac{3^{-3}(a^{-3})^{-3}(t^{2})^{-3}}{(a^{-1})^{2}(t^{-2})^{2}}$

$= \dfrac{3^{-3}a^{9}t^{-6}}{a^{-2}t^{-4}}$

$= \dfrac{1}{3^{3}} \cdot \dfrac{a^{9}}{a^{-2}} \cdot \dfrac{t^{-6}}{t^{-4}}$

$= \dfrac{1}{27}a^{11}t^{-2}$

$= \dfrac{1}{27} \cdot \dfrac{a^{11}}{1} \cdot \dfrac{1}{t^{2}}$

$= \dfrac{a^{11}}{27t^{2}}$

29. $.000439 = 4.39 \times 10^{-4}$

31. $\dfrac{(4 \times 10^{-3})(5 \times 10^{4})}{2 \times 10^{-3}}$

$= \dfrac{(4)(5)}{2} \times \dfrac{10^{-3}10^{4}}{10^{-3}}$

$= 10 \times \dfrac{10^{1}}{10^{-3}}$

$= 10 \times 10^{4} = 1 \times 10^{5}$

33. $x^{2} + 6x + 5 = (x + 1)(x + 5)$

35. $x^{2} - 5x + 6 = (x - 2)(x - 3)$

37. $6x^{3}y - 12xy^{2} - 9x^{2}y$

$= 3xy(2x^{2} - 4y - 3x)$

39. $u^{2} - 49 = (u - 7)(u + 7)$

41. $2r^{2} + r - 15$

$= (2r - 5)(r + 3)$

43. $5t^{2} + 10t + 15$

$= 5(t^{2} + 2t + 3)$

45. $6x^{2} - 17xy + 12y^{2}$

$= (3x - 4y)(2x - 3y)$

47. $x^{2} + 16x = x(x + 16)$

49. $x^2 + ax + xy + ay$

$= (x^2 + ax) + (xy + ay)$

$= x(x + a) + y(x + a)$

$= (x + a)(x + y)$

51. $x^2 - 4x - ax + 4a$

$= (x^2 - 4x) + (-ax + 4a)$

$= x(x - 4) - a(x - 4)$

$= (x - 4)(x - a)$

53. $\dfrac{x^2 - 4x}{x^2 - 16} = \dfrac{x(x-4)}{(x+4)(x-4)} = \dfrac{x}{x+4}$

55. $\dfrac{3}{4x} + \dfrac{5}{x+4}$

$= \dfrac{3(x+4)}{4x(x+4)} + \dfrac{5(4x)}{(x+4)(4x)}$

$= \dfrac{3x+12}{4x(x+4)} + \dfrac{20x}{4x(x+4)}$

$= \dfrac{3x+12+20x}{4x(x+4)} = \dfrac{23x+12}{4x(x+4)}$

57. $\dfrac{x^2 - 5x}{10x} \cdot \dfrac{x^2}{x^2 - 25} = \dfrac{x(x-5)}{10x} \cdot \dfrac{x^2}{(x-5)(x+5)} = \dfrac{x^2}{10(x+5)}$

59. $\dfrac{6}{x^2 + 2x} - \dfrac{4}{x^2 - 2x} = \dfrac{6}{x(x+2)} - \dfrac{4}{x(x-2)}$

$= \dfrac{6(x-2)}{x(x+2)(x-2)} - \dfrac{4(x+2)}{x(x-2)(x+2)}$

$= \dfrac{6x-12}{x(x+2)(x-2)} - \dfrac{4x+8}{x(x+2)(x-2)}$

$= \dfrac{6x-12-(4x+8)}{x(x+2)(x-2)}$

$= \dfrac{6x-12-4x-8}{x(x+2)(x-2)}$

$= \dfrac{2x-20}{x(x+2)(x-2)}$

$= \dfrac{2(x-10)}{x(x+2)(x-2)}$

61. $$\frac{\frac{a}{2}-\frac{8}{a}}{\frac{a^2-8a+16}{4}} = \frac{4a\left(\frac{a}{2}-\frac{8}{a}\right)}{4a\left(\frac{a^2-8a+16}{4}\right)}$$

$$= \frac{\frac{4a}{1}\left(\frac{a}{2}\right)-\frac{4a}{1}\left(\frac{8}{a}\right)}{\frac{4a}{1}\left(\frac{a^2-8a+16}{4}\right)}$$

$$= \frac{2a^2-32}{a(a^2-8a+16)} = \frac{2(a^2-16)}{a(a^2-8a+16)}$$

$$= \frac{2(a-4)(a+4)}{a(a-4)(a-4)} = \frac{2(a+4)}{a(a-4)}$$

63. $$\frac{2z+9}{4z+12}-\frac{5z+8}{4z+12}+\frac{3z+1}{4z+12}$$

$$= \frac{2z+9-(5z+8)+3z+1}{4z+12}$$

$$= \frac{2z+9-5z-8+3z+1}{4z+12}$$

$$= \frac{2}{4z+12}$$

$$= \frac{2}{4(z+3)}$$

$$= \frac{1}{2(z+3)}$$

65. $$\frac{4}{3a+6}-\frac{3}{2}=\frac{5}{6a+12}$$

$$\frac{4}{3(a+2)}-\frac{3}{2}=\frac{5}{6(a+2)}$$

$$6(a+2)\left(\frac{4}{3(a+2)}-\frac{3}{2}\right)=6(a+2)\left(\frac{5}{6(a+2)}\right)$$

$$\frac{6(a+2)}{1}\left(\frac{4}{3(a+2)}\right)-\frac{6(a+2)}{1}\left(\frac{3}{2}\right)=\frac{6(a+2)}{1}\left(\frac{5}{6(a+2)}\right)$$

$$8-9(a+2)=5$$

$$8-9a-18=5$$

$$-9a-10=5$$

$$-9a=15$$

$$a=-\frac{5}{3}$$

67.
$$\frac{3y}{5} - a = 4y + 3a - 6$$
$$5\left(\frac{3y}{5} - a\right) = 5(4y + 3a - 6)$$
$$3y - 5a = 20y + 15a - 30$$
$$\underline{-20y \qquad -20y}$$
$$-17y - 5a = 15a - 30$$
$$\underline{+5a \qquad +5a}$$
$$-17y = 20a - 30$$
$$y = \frac{20a - 30}{-17} = \frac{30 - 20a}{17}$$

69.
$$\frac{x+8}{8} - \frac{x+6}{6} = \frac{x+3}{3} - \frac{x+4}{4}$$
$$24\left(\frac{x+8}{8} - \frac{x+6}{6}\right) = 24\left(\frac{x+3}{3} - \frac{x+4}{4}\right)$$
$$\overset{3}{\cancel{24}}\left(\frac{x+8}{\cancel{8}}\right) - \overset{4}{\cancel{24}}\left(\frac{x+6}{\cancel{6}}\right) = \overset{8}{\cancel{24}}\left(\frac{x+3}{\cancel{3}}\right) - \overset{6}{\cancel{24}}\left(\frac{x+4}{\cancel{4}}\right)$$
$$3(x+8) - 4(x+6) = 8(x+3) - 6(x+4)$$
$$3x + 24 - 4x - 24 = 8x + 24 - 6x - 24$$
$$-x = 2x$$
$$0 = 3x$$
$$0 = x$$

71.
$$\frac{3c+1}{3c-2} = \frac{3c}{3c-1}$$
$$\frac{\cancel{(3c-2)}(3c-1)}{1}\left(\frac{3c+1}{\cancel{3c-2}}\right) = \frac{(3c-2)\cancel{(3c-1)}}{1}\left(\frac{3c}{\cancel{3c-1}}\right)$$
$$(3c-1)(3c+1) = (3c-2)(3c)$$
$$9c^2 + 3c - 3c - 1 = 9c^2 - 6c$$
$$9c^2 - 1 = 9c^2 - 6c$$
$$\underline{-9c^2 \qquad -9c^2}$$
$$-1 = -6c$$
$$\frac{1}{6} = c$$

73. Let x = # of hours Pat needs working alone

$$5\left(\frac{1}{8}\right)+5\left(\frac{1}{x}\right)=1$$
$$\frac{5}{8}+\frac{5}{x}=1$$
$$8x\left(\frac{5}{8}+\frac{5}{x}\right)=8x(1)$$
$$\cancel{8}x\left(\frac{5}{\cancel{8}}\right)+8\cancel{x}\left(\frac{5}{\cancel{x}}\right)=8x$$
$$5x+40=8x$$
$$40=3x$$
$$\frac{40}{3}=x$$

So Pat needs $\frac{40}{3}$ or $13\frac{1}{3}$ hours to overhaul the engine working alone.

Cumulative Practice Test: Chapters 7–9

1. $(x-2y)(x^2-3xy-y^2)$
$$\begin{aligned}&=x(x^2-3xy-y^2)\\&\quad-2y(x^2-3xy-y^2)\\&=x^3-3x^2y-xy^2-2x^2y\\&\quad+6xy^2+2y^3\\&=x^3-3x^2y-2x^2y-xy^2\\&\quad+6xy^2+2y^3\\&=x^3-5x^2y+5xy^2+2y^3\end{aligned}$$

3. $2a(3a-5)+(a-6)(a-4)$
$$\begin{aligned}&=6a^2-10a+a^2-4a-6a+24\\&=7a^2-20a+24\end{aligned}$$

5. $$\frac{(2x^3)^{-4}}{(x^{-3})^3}=\frac{2^{-4}(x^3)^{-4}}{(x^{-3})^3}=\frac{x^{-12}}{2^4\cdot x^{-9}}=\frac{x^{-3}}{16}=\frac{1}{16x^3}$$

7.

$$\begin{array}{r} 4x^2 + 5x + 15 \\ x-2 \overline{)\,4x^3 - 3x^2 + 5x - 20} \\ \underline{-(4x^3 - 8x^2)} \\ 5x^2 + 5x \\ \underline{-(5x^2 - 10x)} \\ 15x - 20 \\ \underline{-(15x - 30)} \\ 10 \end{array}$$

9. (a) $.000916 = 9.16 \times 10^{-4}$

(b) $916,000 = 9.16 \times 10^{5}$

11. $x^3 - x^2 - x + 1 \; - (x^3 - 3x + x^2 - 5x)$

$= x^3 - x^2 - x + 1 - (x^3 + x^2 - 8x)$

$= x^3 - x^2 - x + 1 - x^3 - x^2 + 8x$

$= -2x^2 + 7x + 1$

13. $6a^2b^5 - 3ab^3 = 3ab^3(2ab^2 - 1)$

15. $6x^2 - 36x + 72 = 6\left(x^2 - 6x + 12\right)$

17. $a(a+5) - 7(a+5) = (a+5)(a-7)$

19. $2u^4 - 32 = 2(u^4 - 16)$

$= 2(u^2 + 4)(u^2 - 4)$

$= 2(u^2 + 4)(u + 2)(u - 2)$

21. $\dfrac{t^2 - t - 6}{t^2 + t - 6} = \dfrac{(t-3)(t+2)}{(t+3)(t-2)}$

(cannot be reduced)

23. $\dfrac{w^2 - 3w - 10}{4w^2 + 8w} \cdot \dfrac{w^2}{w^2 - 10w + 25}$

$= \dfrac{(w-5)(w+2)}{4w(w+2)} \cdot \dfrac{w^2}{(w-5)(w-5)}$

$= \dfrac{w}{4(w-5)}$

25. $\frac{u^2-9u}{u^2} \div (u^2-81)$

$= \frac{u^2-9u}{u^2} \cdot \frac{1}{u^2-81}$

$= \frac{u(u-9)}{u^2} \cdot \frac{1}{(u+9)(u-9)}$

$= \frac{1}{u(u+9)}$

27.

$$\frac{9}{4t-12} - \frac{2}{3} = \frac{11}{12t-36}$$

$$\frac{9}{4(t-3)} - \frac{2}{3} = \frac{11}{12(t-3)}$$

$$12(t-3)\left(\frac{9}{4(t-3)} - \frac{2}{3}\right) = 12(t-3)\left(\frac{11}{12(t-3)}\right)$$

$$\frac{12(t-3)}{1} \cdot \frac{9}{4(t-3)} - \frac{12(t-3)}{1} \cdot \frac{2}{3} = \frac{12(t-3)}{1} \cdot \frac{11}{12(t-3)}$$

$$27 - 8(t-3) = 11$$

$$27 - 8t + 24 = 11$$

$$-8t + 51 = 11$$

$$-8t = -40$$

$$t = 5$$

29.

$$\frac{10}{x+4} + \frac{3}{5} = \frac{6-x}{x+4}$$

$$5(x+4)\left(\frac{10}{x+4} + \frac{3}{5}\right) = 5(x+4)\left(\frac{6-x}{x+4}\right)$$

$$\frac{5(x+4)}{1} \cdot \frac{10}{x+4} + \frac{5(x+4)}{1} \cdot \frac{3}{5} = \frac{5(x+4)}{1} \cdot \frac{6-x}{x+4}$$

$$50 + 3(x+4) = 5(6-x)$$

$$50 + 3x + 12 = 30 - 5x$$

$$3x + 62 = 30 - 5x$$

$$8x + 62 = 30$$

$$8x = -32$$

$$x = -4$$

When this value of x is checked in the original equation, it produces fractions with zero denominators. Therefore, the equation has no solution.

31. Let x = Roni's speed (in kph)

x + 20 = Lamar's speed (in kph)

$$\frac{140}{x} = \frac{160}{x+20}$$

$$\frac{\cancel{x}(x+20)}{1}\left(\frac{140}{\cancel{x}}\right) = \frac{x\cancel{(x+20)}}{1}\left(\frac{160}{\cancel{x+20}}\right)$$

$$140(x+20) = 160x$$

$$140x + 2800 = 160x$$

$$2800 = 20x$$

$$140 = x$$

Since Roni drives 140 km at a speed of 140 kph, her driving time is 1 hour.

CHAPTER 10
RADICAL EXPRESSIONS

Exercises 10.1

1. $\sqrt{4} = 2$ because $2 \cdot 2 = 4$

3. $-\sqrt{4} = -2$

5. $\sqrt{-4}$ does not exist

11. $\sqrt{64} = 8$ because $8 \cdot 8 = 64$

15. $\sqrt{169} = 13$ because $13 \cdot 13 = 169$

19. $\sqrt{289} = 17$ because $17 \cdot 17 = 289$

23. $\sqrt{3}\sqrt{3} = 3$

27. $\left(\sqrt{11}\right)^2 = 11$

31. $\sqrt{x}\sqrt{x} = x$

35. $\left(\sqrt{7}\right)^5 = \underbrace{\sqrt{7} \cdot \sqrt{7}} \cdot \underbrace{\sqrt{7} \cdot \sqrt{7}} \cdot \sqrt{7}$
$= 7 \cdot 7 \cdot \sqrt{7}$
$= 49\sqrt{7}$

37. $\sqrt{25 - 9} = \sqrt{16} = 4$

39. $\sqrt{25} - \sqrt{9} = 5 - 3 = 2$

41. $\left(\sqrt{25} - \sqrt{9}\right)^2 = 2^2 = 4$

43. $\left(\sqrt{25 - 9}\right)^2 = \left(\sqrt{16}\right)^2 = 16$

45. 20.62

47. 25.24

49. 8.58

51. 1.45

53. .19

55. $\sqrt{17} = 4.12$ correct to 1 place
$= 4.123$ correct to 2 places
$= 4.1231$ correct to 3 places

57. $\sqrt{110} = 10.49$ correct to 1 place
$= 10.488$ correct to 2 places
$= 10.4881$ correct to 3 places

59. Since $20^2 = 400$, $30^2 = 900$, and 648 is between 400 and 900, $\sqrt{648}$ must be between 20 and 30. Consider the following table:

$0^2 = \underline{0}$ $\quad 5^2 = 2\underline{5}$
$1^2 = \underline{1}$ $\quad 6^2 = 36$
$2^2 = \underline{4}$ $\quad 7^2 = 4\underline{9}$
$3^2 = \underline{9}$ $\quad 8^2 = 6\underline{4}$
$4^2 = 1\underline{6}$ $\quad 9^2 = 8\underline{1}$

This implies that any perfect square must end in one of the digits underlined: 0, 1, 4, 9, 6, or 5. Therefore, any number that ends in either 2, 3, 7, or 8 cannot be a perfect square. So 648 cannot be a perfect square.

60. Using the same argument as in (59), $\sqrt{841}$ must be between 20 and 30. Since 841 ends in the digit 1, there are only two possibilities, if 841 is a perfect square: $(21)^2 = 841$ or $(29)^2 = 841$. Since 841 is closer to 900 than it is to 400, try $(29)^2$ first, since 29 is closer to 30 than it is to 20. Since 29 · 29 = 841, conclude that 841 <u>is</u> a perfect square, and that $\sqrt{841} = 29$.

61. It is incorrect to claim that $\sqrt{1+1} = \sqrt{1} + \sqrt{1}$. In fact, if a and b are positive, it is <u>never</u> true that $\sqrt{a+b} = \sqrt{a} + \sqrt{b}$. That is, the square root of the sum of two positive numbers is never equal to the sum of the square roots of those numbers.

62. For any non-negative number a, $\sqrt{a}\sqrt{a} = a$ tells us that $\sqrt{a}$ is the non-negative quantity whose square is equal to a.

63. $\dfrac{x^2 - 6x}{x^2} \cdot \dfrac{2x}{x^2 - 12x + 36}$

$= \dfrac{\cancel{x}\cancel{(x-6)}}{\cancel{x^2}} \cdot \dfrac{2\cancel{x}}{\cancel{(x-6)}(x-6)}$

$= \dfrac{2}{x-6}$

65. $\dfrac{3}{2x} + \dfrac{5}{x+2}$

$= \dfrac{3(x+2)}{2x(x+2)} + \dfrac{5(2x)}{2x(x+2)}$

$= \dfrac{3(x+2) + 5(2x)}{2x(x+2)}$

$= \dfrac{13x + 6}{2x(x+2)}$

67. Let x = number of regular bulbs

40 – x = number of long-life bulbs

$.75x + .89(40 - x) = 31.68$

$75x + 89(40 - x) = 3168$

$75x + 3560 - 89x = 3168$

$3560 - 14x = 3168$

$-14x = -392$

$x = 28$

Then 40 – x = 40 – 28 = 12, so 28 regular bulbs and 12 long-life bulbs are sold.

Exercises 10.2

3. $\sqrt{18} = \sqrt{9 \cdot 2} = \sqrt{9}\sqrt{2} = 3\sqrt{2}$

7. $\sqrt{50} = \sqrt{25 \cdot 2} = \sqrt{25}\sqrt{2} = 5\sqrt{2}$

11. $\sqrt{x^6} = x^3$, because $x^3x^3 = x^6$

13. $\sqrt{x^7} = \sqrt{x^6 \cdot x} = \sqrt{x^6}\sqrt{x} = x^3\sqrt{x}$

15. $\sqrt{16x^{16}} = \sqrt{16}\sqrt{x^{16}} = 4x^8$

19. $\sqrt{40x^8} = \sqrt{4x^8}\sqrt{10} = \sqrt{4}\sqrt{x^8}\sqrt{10}$

$= 2x^4\sqrt{10}$

23. $\sqrt{12x^5} = \sqrt{4x^4}\sqrt{3x} = \sqrt{4}\sqrt{x^4}\sqrt{3x}$
$= 2x^2\sqrt{3x}$

27. $\sqrt{x^6 + y^8}$ cannot be simplified

31. $\sqrt{28x^9y^6} = \sqrt{4x^8y^6}\sqrt{7x} = \sqrt{4}\sqrt{x^8}\sqrt{y^6}\sqrt{7x}$
$= 2x^4y^3\sqrt{7x}$

37. $\sqrt{48x^6y^8z^9} = \sqrt{16x^6y^8z^8}\sqrt{32}$
$= \sqrt{16}\sqrt{x^6}\sqrt{y^8}\sqrt{z^8}\sqrt{3z}$
$= 4x^3y^4z^4\sqrt{3z}$

39. $\sqrt{\frac{4}{9}} = \frac{\sqrt{4}}{\sqrt{9}} = \frac{2}{3}$

41. $\sqrt{\frac{7}{25}} = \frac{\sqrt{7}}{\sqrt{25}} = \frac{\sqrt{7}}{5}$

43. $\sqrt{\frac{5}{6}} = \frac{\sqrt{5}}{\sqrt{6}} = \frac{\sqrt{5}\cdot\sqrt{6}}{\sqrt{6}\cdot\sqrt{6}} = \frac{\sqrt{30}}{6}$

47. $\frac{1}{\sqrt{2}} = \frac{1\cdot\sqrt{2}}{\sqrt{2}\cdot\sqrt{2}} = \frac{\sqrt{2}}{2}$

49. $\frac{18}{\sqrt{10}} = \frac{18\cdot\sqrt{10}}{\sqrt{10}\cdot\sqrt{10}} = \frac{\overset{9}{\cancel{18}}\sqrt{10}}{\cancel{10}\,5} = \frac{9\sqrt{10}}{5}$

53. $\frac{15}{2\sqrt{7}} = \frac{15\cdot\sqrt{7}}{2\sqrt{7}\cdot\sqrt{7}} = \frac{15\sqrt{17}}{2\cdot 7} = \frac{15\sqrt{7}}{14}$

57. $\frac{8x}{\sqrt{2x}} = \frac{8x\sqrt{2x}}{\sqrt{2x}\sqrt{2x}} = \frac{\overset{4}{\cancel{8x}}\sqrt{2x}}{\cancel{2x}\,1} = 4\sqrt{2x}$

61. $\frac{x^2}{\sqrt{xy}} = \frac{x^2\sqrt{xy}}{\sqrt{xy}\sqrt{xy}} = \frac{\overset{x}{\cancel{x^2}}\sqrt{xy}}{\cancel{x}y} = \frac{x\sqrt{xy}}{y}$

63. $\frac{\sqrt{8}}{\sqrt{6}} = \frac{\sqrt{8}\cdot\sqrt{6}}{\sqrt{6}\cdot\sqrt{6}} = \frac{\sqrt{48}}{6} = \frac{\sqrt{16}\sqrt{3}}{6}$
$= \frac{\overset{2}{\cancel{4}}\sqrt{3}}{\cancel{6}\,3} = \frac{2\sqrt{3}}{3}$

65. When we square a real number other than 0 or 1, we get an answer that is different from the original number. So it is incorrect to say that $\frac{2}{\sqrt{5}} = \frac{2^2}{(\sqrt{5})^2}$. When we rationalize the denominator properly, we multiply $\frac{2}{\sqrt{5}}$ by $\frac{\sqrt{5}}{\sqrt{5}}$. This means that we multiply by 1, which does not change the value of the original number.

66. $\frac{1}{\sqrt{5}} = \frac{1\cdot\sqrt{5}}{\sqrt{5}\cdot\sqrt{5}} = \frac{\sqrt{5}}{5}$
$\frac{1}{\sqrt{5}} = \frac{1}{2.2360} = .4472$ correct to 3 places
$\frac{\sqrt{5}}{5} = \frac{2.2360}{5} = .4472$ correct to 3 places

It is easier to compute $\frac{\sqrt{5}}{5}$ than to compute $\frac{1}{\sqrt{5}}$, since $\frac{1}{\sqrt{5}}$ involves division by a decimal quantity, whereas $\frac{\sqrt{5}}{5}$ does not.

67. $\frac{\sqrt{3}}{\sqrt{7}} = \frac{1.7321}{2.6458} = .6457$ correct to 3 places
$\frac{\sqrt{3}}{\sqrt{7}} = \frac{\sqrt{3}\cdot\sqrt{7}}{\sqrt{7}\cdot\sqrt{7}} = \frac{\sqrt{21}}{7} = \frac{4.5828}{7} = .6547$ correct to 3 places

69. $$\begin{array}{r} 5x + 2y = 9x - 3y \\ +3y \qquad +3y \\ \hline 5x + 5y = 9x \\ -5x \qquad -5x \\ \hline 5y = 4x \\ y = \frac{4}{5}x \end{array}$$

71. $$\frac{u}{2} - \frac{v}{3} = u + v$$
$$6\left(\frac{u}{2} - \frac{v}{3}\right) = 6(u + v)$$
$$3u - 2v = 6u + 6v$$
$$-3u = 8v$$
$$u = \frac{-8}{3}v$$

73. Let x = number of representatives in favor of the bill

435 – x = number of representatives against the bill

$$\frac{x}{435 - x} = \frac{8}{7}$$
$$\frac{7(435 - x)}{1}\left(\frac{x}{435 - x}\right) = \frac{7(435 - x)}{1}\left(\frac{8}{7}\right)$$
$$7x = 8(435 - x)$$
$$7x = 3480 - 8x$$
$$15x = 3480$$
$$x = 232$$

So 232 representatives were in favor of the bill.

Exercises 10.3

3. $\sqrt{5} + 2\sqrt{5} + 3\sqrt{5} = (1 + 2 + 3)\sqrt{5}$
$= 6\sqrt{5}$

7. $4\sqrt{6} - \sqrt{6} = (4 - 1)\sqrt{6} = 3\sqrt{6}$

11. $3\sqrt{5} + 5\sqrt{3}$ cannot be simplified.

15. $\sqrt{5}\sqrt{5} = 5$

19. $\sqrt{5} + \sqrt{5} = 2\sqrt{5}$

21. $\sqrt{5} + 3\sqrt{7} - 4\sqrt{5} - 5\sqrt{7}$
$= \sqrt{5} - 4\sqrt{5} + 3\sqrt{7} - 5\sqrt{7}$
$= -3\sqrt{5} - 2\sqrt{7}$

27. $2\left(\sqrt{5} - \sqrt{3}\right) + 3\left(\sqrt{3} - \sqrt{5}\right)$
$= 2\sqrt{5} - 2\sqrt{3} + 3\sqrt{3} - 3\sqrt{5}$
$= 2\sqrt{5} - 3\sqrt{5} - 2\sqrt{3} + 3\sqrt{3}$
$= -\sqrt{5} + \sqrt{3}$

31. $6\left(\sqrt{m}-\sqrt{n}\right)-\left(3\sqrt{m}+6\sqrt{n}\right)$

$= 6\sqrt{m}-6\sqrt{n}-3\sqrt{m}-6\sqrt{n}$

$= 6\sqrt{m}-3\sqrt{m}-6\sqrt{n}-6\sqrt{n}$

$= 3\sqrt{m}-12\sqrt{n}$

33. $\sqrt{8}+\sqrt{18}=\sqrt{4}\sqrt{2}+\sqrt{9}\sqrt{2}$

$= 2\sqrt{2}+3\sqrt{2}=5\sqrt{2}$

35. $\sqrt{25}+\sqrt{24}=5+\sqrt{4}\sqrt{6}=5+2\sqrt{6}$

39. $4\sqrt{12}-\sqrt{75}=4\sqrt{4}\sqrt{3}=\sqrt{25}\sqrt{3}$

$= 4\left(2\sqrt{3}\right)-5\sqrt{3}$

$= 8\sqrt{3}-5\sqrt{3}=3\sqrt{3}$

43. $3\sqrt{72}-5\sqrt{32}=3\left(\sqrt{36}\sqrt{2}\right)-5\left(\sqrt{16}\sqrt{2}\right)$

$= 3\left(6\sqrt{2}\right)-5\left(4\sqrt{2}\right)$

$= 18\sqrt{2}-20\sqrt{2}$

$= -2\sqrt{2}$

45. $5\sqrt{36}+4\sqrt{30}=5\cdot 6+4\sqrt{30}$

$= 30+4\sqrt{30}$

49. $\sqrt{12w}+\sqrt{27w}=\sqrt{4}\sqrt{3w}+\sqrt{9}\sqrt{3w}$

$= 2\sqrt{3w}+3\sqrt{3w}=5\sqrt{3w}$

53. $\sqrt{20y^3}-\sqrt{45y^3}=\sqrt{4y^2}\sqrt{5y}-\sqrt{9y^2}\sqrt{5y}$

$= 2y\sqrt{5y}-3y\sqrt{5y}$

$= -y\sqrt{5y}$

55. $x\sqrt{28xy^3}+y\sqrt{63x^3y}$

$= x\sqrt{4y^2}\sqrt{7xy}+y\sqrt{9x^2}\sqrt{7xy}$

$= x(2y)\sqrt{7xy}+y(3x)\sqrt{7xy}$

$= 2xy\sqrt{7xy}+3xy\sqrt{7xy}$

$= 5xy\sqrt{7xy}$

57. $\dfrac{\sqrt{32x^3y^2}}{2xy}-\sqrt{8x}$

$= \dfrac{\sqrt{16x^2y^2}\sqrt{2x}}{2xy}-\sqrt{4}\sqrt{2x}$

$= \dfrac{4xy\sqrt{2x}}{2xy}-2\sqrt{2x}$

$= 2\sqrt{2x}-2\sqrt{2x}=0$

61. $\sqrt{27} + \dfrac{4}{\sqrt{3}} = \sqrt{9}\sqrt{3} + \dfrac{4 \cdot \sqrt{3}}{\sqrt{3} \cdot \sqrt{3}}$

$= 3\sqrt{3} + \dfrac{4}{3}\sqrt{3}$

$= \left(3 + \dfrac{4}{3}\right)\sqrt{3} = \dfrac{13}{3}\sqrt{3}$

65. $\sqrt{\dfrac{2}{7}} + \sqrt{\dfrac{7}{2}} = \dfrac{\sqrt{2}}{\sqrt{7}} + \dfrac{\sqrt{7}}{\sqrt{2}}$

$= \dfrac{\sqrt{2} \cdot \sqrt{7}}{\sqrt{7} \cdot \sqrt{7}} + \dfrac{\sqrt{7} \cdot \sqrt{2}}{\sqrt{2} \cdot \sqrt{2}}$

$= \dfrac{\sqrt{14}}{7} + \dfrac{\sqrt{14}}{2}$

$= \dfrac{1}{7}\sqrt{14} + \dfrac{1}{2}\sqrt{14}$

$= \left(\dfrac{1}{7} + \dfrac{1}{2}\right)\sqrt{14} = \dfrac{9}{14}\sqrt{14}$

67. $\sqrt{80} = \sqrt{16 \cdot 5} = \sqrt{16}\sqrt{5} = 4\sqrt{5}$

$\sqrt{80} = 8.9442719$

$4\sqrt{5} = 4(2.360679) = 8.9442716$

The two results are the same, to 6 places. These results should be equal, and only appear to differ because of rounding off.

68. $\sqrt{150} = \sqrt{25 \cdot 6} = \sqrt{25}\sqrt{6} = 5\sqrt{6}$

$\sqrt{150} = 12.247448$

$5\sqrt{6} = 5(2.4494897) = 12.247448$

69. $4x - 3y = 12$

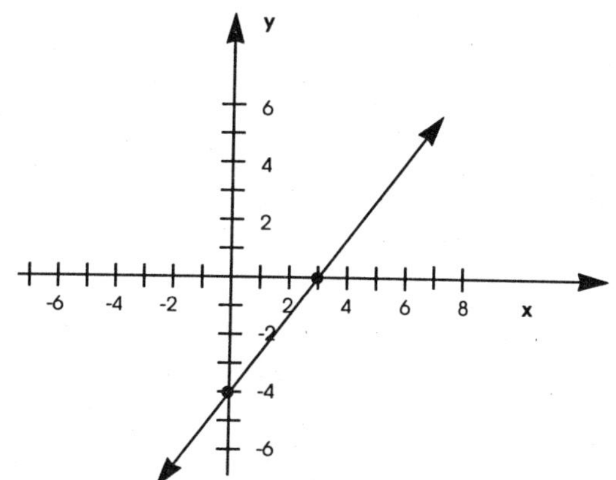

71. $x = 2$

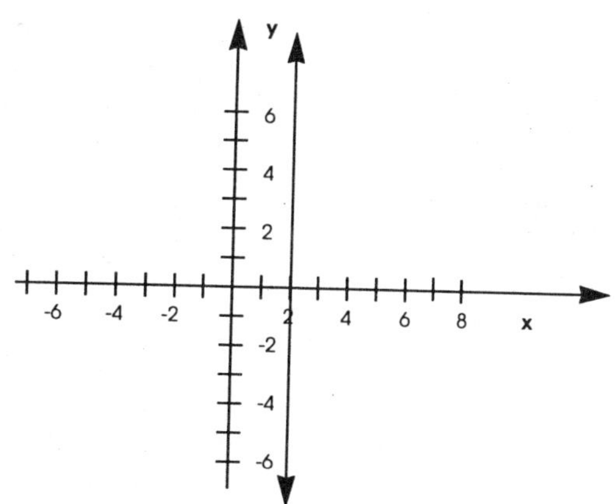

73. Let x = number of hours that associate worked

$14 - x$ = number of hours that law clerk worked

$$120x + 40(14 - x) = 1200$$
$$120x + 560 - 40x = 1200$$
$$80x + 560 = 1200$$
$$80x = 640$$
$$x = 8$$

Then $14 - x = 14 - 8 = 6$, so the associate worked 8 hours and the law clerk worked 6 hours.

Exercises 10.4

3. $\sqrt{3}\sqrt{5}\sqrt{13} = \sqrt{3 \cdot 5 \cdot 13} = \sqrt{195}$

7. $\sqrt{6}\sqrt{24} = \sqrt{6}\left(\sqrt{4}\sqrt{6}\right) = \sqrt{6}\left(2\sqrt{6}\right)$
$= 2(\sqrt{6}\sqrt{6}) = 2 \cdot 6 = 12$

11. $\sqrt{3}\left(\sqrt{5} + \sqrt{6}\right) = \sqrt{3}\sqrt{5} + \sqrt{3}\sqrt{6}$
$= \sqrt{15} + \sqrt{18}$
$= \sqrt{15} + \sqrt{9}\sqrt{2}$
$= \sqrt{15} + 3\sqrt{2}$

15. $\sqrt{3}\left(2\sqrt{3} - 3\sqrt{2}\right)$
$= \sqrt{3}\left(2\sqrt{3}\right) - \sqrt{3}\left(3\sqrt{2}\right)$
$= 2\left(\sqrt{3}\sqrt{3}\right) - 3\left(\sqrt{3}\sqrt{2}\right)$
$= 2 \cdot 3 - 3\sqrt{6} = 6 - 3\sqrt{6}$

19. $3\sqrt{2}\left(\sqrt{2} - 4\right) + \sqrt{2}\left(5 - \sqrt{2}\right)$
$3\sqrt{2}\left(\sqrt{2}\right) - 3\sqrt{2}(4) + \sqrt{2}(5)$
$- \sqrt{2}\sqrt{2}$
$= 3\left(\sqrt{2}\sqrt{2}\right) - 3 \cdot 4\sqrt{2} + 5\sqrt{2} - 2$
$= 3 \cdot 2 - 12\sqrt{2} + 5\sqrt{2} - 2$
$= 6 - 12\sqrt{2} + 5\sqrt{2} - 2 = 4 - 7\sqrt{2}$

1. $4\sqrt{x}\left(\sqrt{x}-\sqrt{2}\right)-\sqrt{x}\left(3\sqrt{x}-2\sqrt{2}\right)$

$$= 4\sqrt{x}\sqrt{x} - 4\sqrt{x}\sqrt{2} - 3\sqrt{x}\sqrt{x}$$
$$+ 2\sqrt{x}\sqrt{2}$$
$$= 4x - 4\sqrt{2x} - 3x + 2\sqrt{2x}$$
$$= x - 2\sqrt{2x}$$

25. $\left(\sqrt{x}+\sqrt{3}\right)^2$

$$= \left(\sqrt{x}+\sqrt{3}\right)\left(\sqrt{x}+\sqrt{3}\right)$$
$$= \sqrt{x}\sqrt{x} + \sqrt{x}\sqrt{3} + \sqrt{3}\sqrt{x} + \sqrt{3}\sqrt{3}$$
$$= x + \sqrt{3x} + \sqrt{3x} + 3$$
$$= x + 2\sqrt{3x} + 3$$

27. $\left(\sqrt{x}+\sqrt{3}\right)\left(\sqrt{x}-\sqrt{3}\right)$

$$= \sqrt{x}\sqrt{x} - \sqrt{x}\sqrt{3} + \sqrt{3}\sqrt{x}$$
$$- \sqrt{3}\sqrt{3}$$
$$= x - \sqrt{3x} + \sqrt{3x} - 3 = x - 3$$

29. $\left(3\sqrt{2}-2\sqrt{5}\right)^2$

$$= \left(3\sqrt{2}-2\sqrt{5}\right)\left(3\sqrt{2}-2\sqrt{5}\right)$$
$$= \left(3\sqrt{2}\right)\left(3\sqrt{2}\right) - \left(3\sqrt{2}\right)\left(2\sqrt{5}\right)$$
$$- \left(2\sqrt{5}\right)\left(3\sqrt{2}\right) + \left(2\sqrt{5}\right)\left(2\sqrt{5}\right)$$
$$= 3\cdot 3\sqrt{2}\sqrt{2} - 3\cdot 2\sqrt{2}\sqrt{5}$$
$$- 2\cdot 3\sqrt{5}\sqrt{2} + 2\cdot 2\sqrt{5}\sqrt{5}$$
$$= 9\cdot 2 - 6\sqrt{10} - 6\sqrt{10} + 4\cdot 5$$
$$= 18 - 6\sqrt{10} - 6\sqrt{10} + 20$$
$$= 38 - 12\sqrt{10}$$

35. $\left(\sqrt{28}-\sqrt{24}\right)\left(\sqrt{7}-\sqrt{6}\right)$

$$= \left(\sqrt{4}\sqrt{7} - \sqrt{4}\sqrt{6}\right)\left(\sqrt{7}-\sqrt{6}\right)$$
$$= \left(2\sqrt{7}-2\sqrt{6}\right)\left(\sqrt{7}-\sqrt{6}\right)$$
$$= 2\sqrt{7}\sqrt{7} - 2\sqrt{7}\sqrt{6} - 2\sqrt{6}\sqrt{7}$$
$$+ 2\sqrt{6}\sqrt{6}$$
$$= 2\cdot 7 - 2\sqrt{42} - 2\sqrt{42} + 2\cdot 6$$
$$= 14 - 2\sqrt{42} - 2\sqrt{42} + 12$$
$$= 26 - 4\sqrt{42}$$

39. $\left(\sqrt{x}+2\right)^2-\left(\sqrt{x+2}\right)^2$

$=\left(\sqrt{x}+2\right)\left(\sqrt{x}+2\right)-(x+2)$

$=\sqrt{x}\sqrt{x}+2\sqrt{x}+2\sqrt{x}$

$+4-(x+2)$

$=x+2\sqrt{x}+2\sqrt{x}+4-x-2$

$=4\sqrt{x}+2$

43. $\dfrac{\sqrt{54}}{\sqrt{3}}=\sqrt{\dfrac{54}{3}}=\sqrt{18}=\sqrt{9}\sqrt{2}=3\sqrt{2}$

47. $\dfrac{\sqrt{a^2b^5}}{\sqrt{ab^8}}=\sqrt{\dfrac{a^2b^5}{ab^8}}=\sqrt{\dfrac{a}{b^3}}=\dfrac{\sqrt{a}}{\sqrt{b^3}}$

$=\dfrac{\sqrt{a}}{\sqrt{b^2}\sqrt{b}}=\dfrac{\sqrt{a}}{b\sqrt{b}}$

$=\dfrac{\sqrt{a}\cdot\sqrt{b}}{b\sqrt{b}\cdot\sqrt{b}}=\dfrac{\sqrt{ab}}{b\cdot b}$

$=\dfrac{\sqrt{ab}}{b^2}$

49. $\dfrac{10}{4-\sqrt{11}}=\dfrac{10\left(4+\sqrt{11}\right)}{\left(4-\sqrt{11}\right)\left(4+\sqrt{11}\right)}$

$=\dfrac{10\left(4+\sqrt{11}\right)}{16-11}=\dfrac{10\left(4+\sqrt{11}\right)}{5}$

$=2\left(4+\sqrt{11}\right)$

53. $\dfrac{\sqrt{3}}{2+\sqrt{3}}=\dfrac{\sqrt{3}\left(2-\sqrt{3}\right)}{\left(2+\sqrt{3}\right)\left(2-\sqrt{3}\right)}$

$=\dfrac{2\sqrt{3}-\sqrt{3}\sqrt{3}}{4-3}$

$=\dfrac{2\sqrt{3}-3}{1}=2\sqrt{3}-3$

55. $\dfrac{\sqrt{5}+\sqrt{3}}{\sqrt{5}-\sqrt{3}}=\dfrac{\left(\sqrt{5}+\sqrt{3}\right)\left(\sqrt{5}+\sqrt{3}\right)}{\left(\sqrt{5}-\sqrt{3}\right)\left(\sqrt{5}+\sqrt{3}\right)}$

$=\dfrac{\sqrt{5}\sqrt{5}+\sqrt{5}\sqrt{3}+\sqrt{3}\sqrt{5}+\sqrt{3}\sqrt{3}}{5-3}$

$=\dfrac{5+\sqrt{15}+\sqrt{15}+3}{5-3}=\dfrac{8+2\sqrt{15}}{2}$

$=\dfrac{2\left(4+\sqrt{15}\right)}{2}=4+\sqrt{15}$

59. $\frac{6}{3-\sqrt{7}} - \frac{21}{\sqrt{7}}$

$= \frac{6(3+\sqrt{7})}{(3-\sqrt{7})(3+\sqrt{7})} - \frac{21\sqrt{7}}{\sqrt{7}\sqrt{7}}$

$= \frac{18+6\sqrt{7}}{9-7} - \frac{21\sqrt{7}}{7}$

$= \frac{\overset{3}{\cancel{6}}(3+\sqrt{7})}{\cancel{2}\,1} - \frac{\overset{3}{\cancel{21}}\sqrt{7}}{\cancel{7}\,1}$

$= 9 + 3\sqrt{7} - 3\sqrt{7} = 9$

61. $(3+\sqrt{5})^2 + \frac{8}{3-\sqrt{5}}$

$= (3+\sqrt{5})(3+\sqrt{5})$

$+ \frac{8(3+\sqrt{5})}{(3-\sqrt{5})(3+\sqrt{5})}$

$= 9 + 3\sqrt{5} + 3\sqrt{5} + 5$

$+ \frac{24+8\sqrt{5}}{9-5}$

$= 14 + 6\sqrt{5} + \frac{\overset{2}{\cancel{8}}(3+\sqrt{5})}{\cancel{4}\,1}$

$= 14 + 6\sqrt{5} + 6 + 2\sqrt{5}$

$= 20 + 8\sqrt{5}$

65. $\frac{12-\sqrt{20}}{10} = \frac{12-\sqrt{4}\sqrt{5}}{10}$

$= \frac{12-2\sqrt{5}}{10} = \frac{\cancel{2}(6-\sqrt{5})}{\cancel{10}\,5}$

$= \frac{6-\sqrt{5}}{5}$

67. $(2+\sqrt{10})^2 - 4(2+\sqrt{10}) - 6$

$= (2+\sqrt{10})(2+\sqrt{10}) - 4(2+\sqrt{10}) - 6$

$= 4 + 2\sqrt{10} + 2\sqrt{10} + 10 - 8 - 4\sqrt{10} - 6$

$= 0$

69. (a) The "2" in the numerator is under the square root and is thus $\sqrt{2}$. This cannot be canceled with the "2" in the denominator.

(b) The cancellation is not valid since 2 is not a common factor of the numerator. Remember that terms cannot be canceled.

71. $x^2 - 6x - 7 = (x-7)(x+1)$

73. $x^2 + 10x - 16$ is not factorable.

75. $w^3 - 8w^2 + 16w = (w^2 - 8w + 16)$
$= w(w - 4)(w - 4)$

77. $6u^2 - 5u - 14 = (6u + 7)(u - 2)$

79. Let x = price of a ticket sold at the door (in dollars)

x – 1.50 = price of a ticket sold in advance (in dollars)

$$
\begin{aligned}
200(x - 1.50) + 75x &= 1075 \\
200x - 300 + 75x &= 1075 \\
275x - 300 &= 1075 \\
275x &= 1375 \\
x &= 5
\end{aligned}
$$

So a ticket sold at the door costs $5.

Exercises 10.5

1.
$$
\begin{aligned}
\sqrt{x} &= 5 \\
\left(\sqrt{x}\right)^2 &= (5)^2 \\
x &= 25
\end{aligned}
$$

CHECK x = 25:
$$
\begin{aligned}
\sqrt{x} &= 5 \\
\sqrt{25} &\stackrel{?}{=} 5 \\
5 &\stackrel{\checkmark}{=} 5
\end{aligned}
$$

5.
$$
\begin{aligned}
\sqrt{u} &= 6 \\
\left(\sqrt{u}\right)^2 &= (6)^2 \\
u &= 36
\end{aligned}
$$
CHECK u = 36:
$$
\begin{aligned}
\sqrt{u} &= 6 \\
\sqrt{36} &\stackrel{?}{=} 6 \\
6 &\stackrel{\checkmark}{=} 6
\end{aligned}
$$

9.
$$
\begin{aligned}
\sqrt{x+3} &= 10 \\
\left(\sqrt{x+3}\right)^2 &= (10)^2 \\
x + 3 &= 100 \\
x &= 97
\end{aligned}
$$

CHECK x = 97:
$$
\begin{aligned}
\sqrt{x+3} &= 10 \\
\sqrt{97+3} &\stackrel{?}{=} 10 \\
\sqrt{100} &\stackrel{?}{=} 10 \\
10 &\stackrel{\checkmark}{=} 10
\end{aligned}
$$

11.
$$\begin{aligned} \sqrt{x} + 3 &= 10 \\ -3 \quad & -3 \\ \hline \sqrt{x} \quad &= 7 \\ \left(\sqrt{x}\right)^2 &= (7)^2 \\ x \quad &= 49 \end{aligned}$$

CHECK x = 49:

$$\begin{aligned} \sqrt{x} + 3 &= 10 \\ \sqrt{49} + 3 &\stackrel{?}{=} 10 \\ 7 + 3 &\stackrel{?}{=} 10 \\ 10 &\stackrel{\checkmark}{=} 10 \end{aligned}$$

17.
$$\begin{aligned} \sqrt{5 - 4x} &= 7 \\ \left(\sqrt{5 - 4x}\right)^2 &= (7)^2 \\ 5 - 4x &= 49 \\ -4x &= 44 \\ x &= -11 \end{aligned}$$

CHECK x = −11:

$$\begin{aligned} \sqrt{5 - 4x} &= 7 \\ \sqrt{5 - 4(-11)} &\stackrel{?}{=} 7 \\ \sqrt{5 + 44} &\stackrel{?}{=} 7 \\ \sqrt{49} &\stackrel{?}{=} 7 \\ 7 &\stackrel{\checkmark}{=} 7 \end{aligned}$$

21.
$$\begin{aligned} \sqrt{1 + x} + 8 &= 4 \\ -8 \quad & -8 \\ \hline \sqrt{1 + x} \quad &= -4 \\ \left(\sqrt{1 + x}\right)^2 &= (-4)^2 \\ 1 + x &= 16 \\ x &= 15 \end{aligned}$$

CHECK x = 15:

$$\begin{aligned} \sqrt{1 + x} + 8 &= 4 \\ \sqrt{1 + 15} + 8 &\stackrel{?}{=} 4 \\ \sqrt{16} + 8 &\stackrel{?}{=} 4 \\ 4 + 8 &\stackrel{?}{=} 4 \\ 12 &\neq 4 \end{aligned}$$

So the given equation has no solution.

27. $10 - \sqrt{2y} = 4$

$\underline{-10 \qquad -10}$

$-2\sqrt{y} = -6$

$\sqrt{y} = 3$

$\left(\sqrt{y}\right)^2 = (3)^2$

$y = 9$

CHECK y = 9:

$10 - 2\sqrt{y} = 4$

$10 - 2\sqrt{9} \stackrel{?}{=} 4$

$10 - 2(3) \stackrel{?}{=} 4$

$10 - 6 \stackrel{?}{=} 4$

$4 \stackrel{\checkmark}{=} 4$

29. $10 - \sqrt{2y} = 2$

$\underline{-10 \qquad -10}$

$-\sqrt{2y} = -8$

$\sqrt{2y} = 8$

$\left(\sqrt{2y}\right)^2 = (8)^2$

$2y = 64$

$y = 32$

CHECK y = 32:

$10 - \sqrt{2y} = 2$

$10 - \sqrt{2(32)} \stackrel{?}{=} 2$

$10 - \sqrt{64} \stackrel{?}{=} 2$

$10 - 8 \stackrel{?}{=} 2$

$2 \stackrel{\checkmark}{=} 2$

33. $6\sqrt{c} - 3 = 3\sqrt{c}$

$\underline{-6\sqrt{c} \qquad -6\sqrt{c}}$

$-3 = -3\sqrt{c}$

$1 = \sqrt{c}$

$(1)^2 = \left(\sqrt{c}\right)^2$

$1 = c$

CHECK c = 1:

$$6\sqrt{c} - 3 = 3\sqrt{c}$$
$$6\sqrt{1} - 3 \stackrel{?}{=} 3\sqrt{1}$$
$$6(1) - 3 \stackrel{?}{=} 3(1)$$
$$6 - 3 \stackrel{?}{=} 3$$
$$3 \stackrel{\checkmark}{=} 3$$

37.

$$\begin{array}{r} 6 + \sqrt{5x+1} = 12 \\ \underline{-6 \qquad\qquad -6} \\ \sqrt{5x+1} = 6 \end{array}$$
$$\left(\sqrt{5x+1}\right)^2 = (6)^2$$
$$5x + 1 = 36$$
$$5x = 35$$
$$x = 7$$

CHECK x = 7:

$$6 + \sqrt{5x+1} = 12$$
$$6 + \sqrt{5(7)+1} \stackrel{?}{=} 12$$
$$6 + \sqrt{35+1} \stackrel{?}{=} 12$$
$$6 + \sqrt{36} \stackrel{?}{=} 12$$
$$6 + 6 \stackrel{?}{=} 12$$
$$12 \stackrel{\checkmark}{=} 12$$

41.

$$\begin{array}{r} 2\sqrt{w} + 5 = 7\sqrt{w} - 10 \\ \underline{-2\sqrt{w} \qquad -2\sqrt{w}} \\ 5 = 5\sqrt{w} - 10 \\ \underline{+10 \qquad\quad +10} \\ 15 = 5\sqrt{w} \end{array}$$
$$3 = \sqrt{w}$$
$$(3)^2 = \left(\sqrt{w}\right)^2$$
$$9 = w$$

CHECK w = 9:

$$2\sqrt{w} + 5 = 7\sqrt{w} - 10$$
$$2\sqrt{9} + 5 \stackrel{?}{=} 7\sqrt{9} - 10$$
$$2(3) + 5 \stackrel{?}{=} 7(3) - 10$$
$$6 + 5 \stackrel{?}{=} 21 - 10$$
$$11 \stackrel{\checkmark}{=} 11$$

45.

$$\begin{aligned} 4\sqrt{7-8x} + 3 &= 15 \\ -3 \quad & \;\; -3 \\ \hline 4\sqrt{7-8x} &= 12 \\ \sqrt{7-8x} &= 3 \\ \left(\sqrt{7-8x}\right)^2 &= (3)^2 \\ 7-8x &= 9 \\ -8x &= 2 \\ x &= -\frac{1}{4} \end{aligned}$$

CHECK $x = -\frac{1}{4}$:

$$\begin{aligned} 4\sqrt{7-8x} + 3 &= 15 \\ 4\sqrt{7-8\left(-\frac{1}{4}\right)} + 3 &\stackrel{?}{=} 15 \\ 4\sqrt{7+2} + 3 &\stackrel{?}{=} 15 \\ 4\sqrt{9} + 3 &\stackrel{?}{=} 15 \\ 4(3) + 3 &\stackrel{?}{=} 15 \\ 12 + 3 &\stackrel{?}{=} 15 \\ 15 &\stackrel{\checkmark}{=} 15 \end{aligned}$$

49.

$$\begin{aligned} \sqrt{5x-1} &= \sqrt{3x+9} \\ \left(\sqrt{5x-1}\right)^2 &= \left(\sqrt{3x+9}\right)^2 \\ 5x \;\; -1 &= \;\; 3x+9 \\ -3x \qquad & \quad\; -3x \\ \hline 2x \;\; -1 &= 9 \\ 2x &= 10 \\ x &= 5 \end{aligned}$$

CHECK $x = 5$:

$$\begin{aligned} \sqrt{5x-1} &= \sqrt{3x+9} \\ \sqrt{5(5)-1} &\stackrel{?}{=} \sqrt{3(5)+9} \\ \sqrt{25-1} &\stackrel{?}{=} \sqrt{15+9} \\ \sqrt{24} &\stackrel{\checkmark}{=} \sqrt{24} \end{aligned}$$

51.

$$\begin{aligned} I &= \sqrt{\frac{W}{R}} \\ 12 &= \sqrt{\frac{W}{9}} \\ (12)^2 &= \left(\sqrt{\frac{W}{9}}\right)^2 \\ 144 &= \frac{W}{9} \\ 1296 &= W \end{aligned}$$

3. $d = \sqrt{8000a}$

$d = \sqrt{8000(5)}$

$d = \sqrt{40000}$

$d = 200$

55. The equations $x = 5$ and $x^2 = 25$ are not equivalent, since their solution sets are not the same. The solution set of the second equation includes –5, which is not in the solution set of the first equation. In general, squaring both sides of an equation does not yield an equivalent equation, since the "squared" equation might have a solution set that fails to satisfy the original equation.

57. $$\frac{9ab^2 - 15ab + 6a^2b}{12ab} = \frac{9ab^2}{12ab} - \frac{15ab}{12ab} + \frac{6a^2b}{12ab}$$
$$= \frac{3b}{4} - \frac{5}{4} + \frac{a}{2}$$
$$= \frac{3b}{4} - \frac{5}{4} + \frac{2a}{4}$$
$$= \frac{3b - 5 + 2a}{4}$$

59. $$\begin{array}{r} 4x^2 + 2x - 2 \\ 2x - 1 \overline{\smash{)}8x^3 + 0x^2 - 6x + 3} \\ \underline{-(8x^3 - 4x^2)} \\ 4x^2 - 6x + 3 \\ \underline{-(4x^2 - 2x)} \\ -4x + 3 \\ \underline{-(-4x + 2)} \\ 1 \end{array}$$

Chapter 10 Review Exercises

1. $\sqrt{49} = 7$

3. $\sqrt{-16}$ does not exist

5. $\sqrt{96} = \sqrt{16 \cdot 6} = \sqrt{16}\sqrt{6} = 4\sqrt{6}$

7. $\sqrt{9x^9} = \sqrt{9x^8}\sqrt{x} = \sqrt{9}\sqrt{x^8}\sqrt{x} = 3x^4\sqrt{x}$

9. $\sqrt{\frac{4}{9}} = \frac{\sqrt{4}}{\sqrt{9}} = \frac{2}{3}$

11. $\sqrt{\frac{3}{5}} = \frac{\sqrt{3}}{\sqrt{5}} = \frac{\sqrt{3}\sqrt{5}}{\sqrt{5}\sqrt{5}} = \frac{\sqrt{15}}{5}$

13. $8\sqrt{7} - 5\sqrt{7} - \sqrt{7} = (8 - 5 - 1)\sqrt{7}$
$= 2\sqrt{7}$

15. $\sqrt{45} - \sqrt{20} = \sqrt{9}\sqrt{5} - \sqrt{4}\sqrt{5}$
$= 3\sqrt{5} - 2\sqrt{5} = \sqrt{5}$

17. $\sqrt{75x} + \sqrt{12x} = \sqrt{25}\sqrt{3x} + \sqrt{4}\sqrt{3x}$
$= 5\sqrt{3x} + 2\sqrt{3x} = 7\sqrt{3x}$

19. $\dfrac{\sqrt{12x^3y^2}}{xy} + \sqrt{27x}$

$= \dfrac{\sqrt{4x^2y^2}\sqrt{3x}}{xy} + \sqrt{9}\sqrt{3x}$

$= \dfrac{2xy\sqrt{3x}}{xy} + 3\sqrt{3x}$

$= 2\sqrt{3x} + 3\sqrt{3x} = 5\sqrt{3x}$

21. $\sqrt{5}\left(3\sqrt{5} + \sqrt{2}\right) = 3\sqrt{5}\sqrt{5} + \sqrt{5}\sqrt{2}$

$= 3 \cdot 5 + \sqrt{10}$

$= 15 + \sqrt{10}$

23. $\left(3\sqrt{7} - 2\sqrt{3}\right)\left(2\sqrt{7} + 5\sqrt{3}\right)$

$= \left(3\sqrt{7}\right)\left(2\sqrt{7}\right) + \left(3\sqrt{7}\right)\left(5\sqrt{3}\right)$

$\quad - \left(2\sqrt{3}\right)\left(2\sqrt{7}\right) - \left(2\sqrt{3}\right)\left(5\sqrt{3}\right)$

$= 3 \cdot 2\sqrt{7}\sqrt{7} + 3 \cdot 5\sqrt{7}\sqrt{3}$

$\quad - 2 \cdot 2\sqrt{3}\sqrt{7} - 2 \cdot 5\sqrt{3}\sqrt{3}$

$= 6 \cdot 7 + 15\sqrt{21} - 4\sqrt{21} - 10 \cdot 3$

$= 42 + 15\sqrt{21} - 4\sqrt{21} - 30$

$= 12 + 11\sqrt{21}$

25. $\left(\sqrt{x} - 3\right)^2 = \left(\sqrt{x} - 3\right)\left(\sqrt{x} - 3\right)$

$= \sqrt{x}\sqrt{x} - 3\sqrt{x} - 3\sqrt{x} + 9$

$= x - 6\sqrt{x} + 9$

27. $\dfrac{7}{\sqrt{3}} = \dfrac{7\sqrt{3}}{\sqrt{3}\sqrt{3}} = \dfrac{7\sqrt{3}}{3}$

29. $\dfrac{x^2}{\sqrt{x}} = \dfrac{x^2\sqrt{x}}{\sqrt{x}\sqrt{x}} = \dfrac{x^2\sqrt{x}}{x} = x\sqrt{x}$

31. $\dfrac{18}{\sqrt{12}} = \dfrac{18}{\sqrt{4}\sqrt{3}} = \dfrac{18}{2\sqrt{3}} = \dfrac{9}{\sqrt{3}}$

$= \dfrac{9\sqrt{3}}{\sqrt{3}\sqrt{3}} = \dfrac{9\sqrt{3}}{3} = 3\sqrt{3}$

33. $\frac{14}{3-\sqrt{2}} = \frac{14(3+\sqrt{2})}{(3-\sqrt{2})(3+\sqrt{2})}$

$= \frac{14(3+\sqrt{2})}{9-2} = \frac{\overset{2}{\cancel{14}}(3+\sqrt{2})}{\cancel{7}\,1}$

$= 2(3+\sqrt{2})$

35. $\frac{2+\sqrt{5}}{6+\sqrt{5}} = \frac{(2+\sqrt{5})(6-\sqrt{5})}{(3-\sqrt{2})(6-\sqrt{5})}$

$= \frac{12-2\sqrt{5}+6\sqrt{5}-5}{36-5}$

$= \frac{7+4\sqrt{5}}{31}$

37. $\left(\sqrt{x+7}\right)^2 - \left(\sqrt{x}+\sqrt{7}\right)^2$

$= x+7-\left(\sqrt{x}+\sqrt{7}\right)\left(\sqrt{x}+\sqrt{7}\right)$

$= x+7-\left(x+\sqrt{x}\sqrt{7}+\sqrt{7}\sqrt{x}+7\right)$

$= x+7-\left(x+\sqrt{7x}+\sqrt{7x}+7\right)$

$= x+7-x-\sqrt{7x}-\sqrt{7x}-7$

$= -2\sqrt{7x}$

39. $\sqrt{x+5} = 7$

$\left(\sqrt{x+5}\right)^2 = (7)^2$

$x+5 = 49$

$x = 44$

CHECK $x = 44$:

$\sqrt{x+5} = 7$

$\sqrt{44+5} \overset{?}{=} 7$

$\sqrt{49} \overset{?}{=} 7$

$7 \overset{\checkmark}{=} 7$

41. $10-\sqrt{4u} = 2$

$\underline{-10 \qquad\quad -10}$

$-\sqrt{4u} = -8$

$\sqrt{4u} = 8$

$\left(\sqrt{4u}\right)^2 = (8)^2$

$4u = 64$

$u = 16$

CHECK u = 16:

$$10-\sqrt{4u}=2$$
$$10-\sqrt{4(16)}\stackrel{?}{=}2$$
$$10-\sqrt{64}\stackrel{?}{=}2$$
$$10-8\stackrel{?}{=}2$$
$$2\stackrel{\checkmark}{=}2$$

43.
$$\begin{aligned} 8-\sqrt{y-5} &= 11 \\ -8 \quad\quad &\quad -8 \\ \hline -\sqrt{y-5} &= 3 \\ \sqrt{y-5} &= -3 \\ \left(\sqrt{y-5}\right)^2 &= (-3)^2 \\ y-5 &= 9 \\ y &= 14 \end{aligned}$$

CHECK y = 14:

$$8-\sqrt{y-5}=11$$
$$8-\sqrt{14-5}\stackrel{?}{=}11$$
$$8-\sqrt{9}\stackrel{?}{=}11$$
$$8-3\stackrel{?}{=}11$$
$$5\neq 11$$

So the given equation has no solution.

Chapter 10 Practice Test

1. $\sqrt{25x^{16}y^6}=\sqrt{25}\sqrt{x^{16}}\sqrt{y^6}=5x^8y^3$

3. $\sqrt{50x^3}-x\sqrt{32x}$
$$=\sqrt{25x^2}\sqrt{2x}-x\left(\sqrt{16}\sqrt{2x}\right)$$
$$=5x\sqrt{2x}-x\left(4\sqrt{2x}\right)$$
$$=5x\sqrt{2x}-4x\sqrt{2x}=x\sqrt{2x}$$

5. $\sqrt{20x^8y^9}+3x^4y^4\sqrt{5y}$
$$=\sqrt{4x^8y^8}\sqrt{5y}+3x^4y^4\sqrt{5y}$$
$$=2x^4y^4\sqrt{5y}+3x^4y^4\sqrt{5y}$$
$$=5x^4y^4\sqrt{5y}$$

[illegible]. $\left(2\sqrt{x}-\sqrt{5}\right)\left(\sqrt{x}+3\sqrt{5}\right)$

$$= 2\sqrt{x}\sqrt{x}+\left(2\sqrt{x}\right)\left(3\sqrt{5}\right)-\sqrt{5}\sqrt{x}-3\sqrt{5}\sqrt{5}$$

$$= 2x+6\sqrt{5x}-\sqrt{5x}-3\cdot 5$$

$$= 2x+5\sqrt{5x}-15$$

9. $\left(\sqrt{x}-4\right)^2-\left(\sqrt{x-4}\right)^2$

$$= \left(\sqrt{x}-4\right)\left(\sqrt{x}-4\right)-(x-4)$$

$$= \sqrt{x}\sqrt{x}-4\sqrt{x}-4\sqrt{x}+16-x+4$$

$$= x-8\sqrt{x}+16-x+4$$

$$= -8\sqrt{x}+20$$

11. $x^2-2x=1$

$$\left(1-\sqrt{2}\right)^2-2\left(1-\sqrt{2}\right)\stackrel{?}{=}1$$

$$\left(1-\sqrt{2}\right)\left(1-\sqrt{2}\right)-2\left(1-\sqrt{2}\right)\stackrel{?}{=}1$$

$$1-\sqrt{2}-\sqrt{2}+2-2+2\sqrt{2}\stackrel{?}{=}1$$

$$1\stackrel{\checkmark}{=}1$$

So $1-\sqrt{2}$ is a solution to the equation.

CHAPTER 11
QUADRATIC EQUATIONS

Exercises 11.1

1. $(x-2)(x+3)=0$

   $x-2=0$ or $x+3=0$

   $x=2$ or $x=-3$

3. $(x-2)(x+3)=6$

   $x^2+x-6=6$

   $x^2+x-12=0$

   $(x+4)(x-3)=0$

   $x+4=0$ or $x-3=0$

   $x=-4$ or $x=3$

   CHECK

   $x=-4$:

   $(x-2)(x+3)=6$

   $(-4-2)(-4+3)\stackrel{?}{=}6$

   $(-6)(-1)\stackrel{?}{=}6$

   $6\stackrel{\checkmark}{=}6$

   $x=3$:

   $(x-2)(x+3)=6$

   $(3-2)(3+3)\stackrel{?}{=}6$

   $(1)(6)\stackrel{?}{=}6$

   $6\stackrel{\checkmark}{=}6$

9. $x^2-x-6=0$

   $(x-3)(x+2)=0$

   $x-3=0$ or $x+2=0$

   $x=3$ or $x=-2$

   CHECK

   $x=3$:

   $x^2-x-6=0$

   $(3)^2-(3)-6\stackrel{?}{=}0$

   $9-3-6\stackrel{?}{=}0$

   $0\stackrel{\checkmark}{=}0$

$x = -2$:

$$x^2 - x - 6 = 0$$
$$(-2)^2 - (-2) - 6 \stackrel{?}{=} 0$$
$$4 + 2 - 6 \stackrel{?}{=} 0$$
$$0 \stackrel{\surd}{=} 0$$

11. $$x^2 - 3x = 10$$
$$x^2 - 3x - 10 = 0$$
$$(x - 5)(x + 2) = 0$$
$$x - 5 = 0 \text{ or } x + 2 = 0$$
$$x = 5 \text{ or } x = -2$$

CHECK

$x = 5$:

$$x^2 - 3x = 10$$
$$(5)^2 - 3(5) \stackrel{?}{=} 10$$
$$25 - 15 \stackrel{?}{=} 10$$
$$10 \stackrel{\surd}{=} 10$$

$x = -2$:

$$x^2 - 3x = 10$$
$$(-2)^2 - 3(-2) \stackrel{?}{=} 10$$
$$4 + 6 \stackrel{?}{=} 10$$
$$10 \stackrel{\surd}{=} 10$$

15. $$-m^2 = 8 - 9m$$
$$-m^2 + 9m - 8 = 0$$
$$m^2 - 9m + 8 = 0$$
$$(m - 1)(m - 8) = 0$$
$$m - 1 = 0 \text{ or } m - 8 = 0$$
$$m = 1 \text{ or } m = 8$$

CHECK

$m = 1$:

$$-m^2 = 8 - 9m$$
$$-(1)^2 \stackrel{?}{=} 8 - 9(1)$$
$$-1 \stackrel{?}{=} 8 - 9$$
$$-1 \stackrel{\surd}{=} -1$$

$m = 8$:

$$-m^2 = 8 - 9m$$
$$-(8)^2 \stackrel{?}{=} 8 - 9(8)$$
$$-64 \stackrel{?}{=} 8 - 72$$
$$-64 \stackrel{\surd}{=} -64$$

17. $$\begin{aligned} p^2 + 3p &= p(p+4) \\ p^2 + 3p &= p^2 + 4p \\ 3p &= 4p \\ 0 &= p \end{aligned}$$

CHECK

$p = 0$:

$$\begin{aligned} p^2 + 3p &= p(p+4) \\ (0)^2 + 3(0) &\overset{?}{=} 0(0+4) \\ 0 + 0 &\overset{?}{=} 0(4) \\ 0 &\overset{\checkmark}{=} 0 \end{aligned}$$

21. $$\begin{aligned} 5w^2 &= 8w \\ 5w^2 - 8w &= 0 \\ w(5w-8) &= 0 \end{aligned}$$

$$\begin{aligned} w = 0 \quad &\text{or} \quad 5w - 8 = 0 \\ w = 0 \quad &\text{or} \quad 5w = 8 \\ w = 0 \quad &\text{or} \quad w = \frac{8}{5} \end{aligned}$$

CHECK

$w = 0$:

$$\begin{aligned} 5w^2 &= 8w \\ 5(0)^2 &\overset{?}{=} 8(0) \\ 5(0) &\overset{?}{=} 8(0) \\ 0 &\overset{\checkmark}{=} 0 \end{aligned}$$

$w = \frac{8}{5}$:

$$\begin{aligned} 5w^2 &= 8w \\ 5\left(\frac{8}{5}\right)^2 &\overset{?}{=} 8\left(\frac{8}{5}\right) \\ 5\left(\frac{64}{25}\right) &\overset{?}{=} \frac{64}{5} \\ \frac{64}{5} &\overset{\checkmark}{=} \frac{64}{5} \end{aligned}$$

27. $$\begin{aligned} 2a(a+3) &= 20 \\ 2a^2 + 6a &= 20 \\ 2a^2 + 6a - 20 &= 0 \\ 2(a^2 + 3a - 10) &= 0 \\ a^2 + 3a - 10 &= 0 \\ (a+5)(a-2) &= 0 \end{aligned}$$

$a + 5 = 0$ or $a - 2 = 0$

$a = -5$ or $a = 2$

CHECK

$a = -5$:

$$2a(a + 3) = 20$$
$$2(-5)(-5 + 3) \stackrel{?}{=} 20$$
$$2(-5)(-2) \stackrel{?}{=} 20$$
$$20 \stackrel{\checkmark}{=} 20$$

$a = 2$:

$$2a(a + 3) = 20$$
$$2(2)(2 + 3) \stackrel{?}{=} 20$$
$$2(2)(5) \stackrel{?}{=} 20$$
$$20 \stackrel{\checkmark}{=} 20$$

31. $a^2 - 4a - 2 = 2a^2 - 9a - 16$

$$-4a - 2 = a^2 - 9a - 16$$
$$-2 = a^2 - 5a - 16$$
$$0 = a^2 - 5a - 14$$
$$0 = (a - 7)(a + 2)$$

$0 = a - 7$ or $0 = a + 2$

$7 = a$ or $-2 = a$

CHECK

$a = 7$:

$$a^2 - 4a - 2 = 2a^2 - 9a - 16$$
$$(7)^2 - 4(7) - 2 \stackrel{?}{=} 2(7)^2 - 9(7) - 16$$
$$49 - 28 - 2 \stackrel{?}{=} 98 - 63 - 16$$
$$19 \stackrel{\checkmark}{=} 19$$

$a = -2$:

$$a^2 - 4a - 2 = 2a^2 - 9a - 16$$
$$(-2)^2 - 4(-2) - 2 \stackrel{?}{=} 2(-2)^2 - 9(-2) - 16$$
$$4 + 8 - 2 \stackrel{?}{=} 8 + 18 - 16$$
$$10 \stackrel{\checkmark}{=} 10$$

35. $(x + 3)^2 = 3x^2 - 10$

$$x^2 + 6x + 9 = 3x^2 - 10$$
$$6x + 9 = 2x^2 - 10$$
$$9 = 2x^2 - 6x - 10$$
$$0 = 2x^2 - 6x - 19$$

The quadratic expression $2x^2 - 6x - 19$ cannot be factored.

39. $(x+2)^2 = 25$

$$(x+2)(x+2) = 25$$
$$x^2 + 4x + 4 = 25$$
$$x^2 + 4x - 21 = 0$$
$$(x+7)(x-3) = 0$$

$x + 7 = 0$ or $x - 3 = 0$

$x = -7$ or $x = 3$

CHECK

$x = -7$:

$$(x+2)^2 = 25$$
$$(-7+2)^2 \stackrel{?}{=} 25$$
$$(-5)^2 \stackrel{?}{=} 25$$
$$25 \stackrel{\checkmark}{=} 25$$

$x = 3$:

$$(x+2)^2 = 25$$
$$(3+2)^2 \stackrel{?}{=} 25$$
$$5^2 \stackrel{?}{=} 25$$
$$25 \stackrel{\checkmark}{=} 25$$

43. $(2x-4)(x+1) = (x-3)(x-2)$

$$2x^2 - 2x - 4 = x^2 - 5x + 6$$
$$x^2 - 2x - 4 = -5x + 6$$
$$x^2 + 3x - 4 = 6$$
$$x^2 + 3x - 10 = 0$$
$$(x+5)(x-2) = 0$$

$x + 5 = 0$ or $x - 2 = 0$

$x = -5$ or $x = 2$

CHECK

$x = -5$:

$$(2x-4)(x+1) = (x-3)(x-2)$$
$$(2(-5)-4)(-5+1) \stackrel{?}{=} (-5-3)(-5-2)$$
$$(-14)(-4) \stackrel{?}{=} (-8)(-7)$$
$$56 \stackrel{\checkmark}{=} 56$$

$x = 2$:

$$(2x-4)(x+1) = (x-3)(x-2)$$
$$(2(2)-4)(2+1) \stackrel{?}{=} (2-3)(2-2)$$
$$(0)(3) \stackrel{?}{=} (-1)(0)$$
$$0 \stackrel{\checkmark}{=} 0$$

45.
$$x+\frac{1}{x}=2$$
$$x\left(x+\frac{1}{x}\right)=x(2)$$
$$x\cdot x+\frac{x}{1}\cdot\frac{1}{x}=2x$$
$$x^2+1=2x$$
$$x^2-2x+1=0$$
$(x-1)(x-1)=0$

$x-1=0$ or $x-1=0$

$x=1$ or $x=1$

So $x=1$

CHECK

$x=1$:
$$x+\frac{1}{x}=2$$
$$(1)+\frac{1}{(1)}\stackrel{?}{=}2$$
$$1+1\stackrel{?}{=}2$$
$$2\stackrel{\checkmark}{=}2$$

49.
$$\frac{2x}{x+2}+1=x$$
$$(x+2)\left(\frac{2x}{x+2}+1\right)=(x+2)x$$
$$\frac{x+2}{1}\cdot\frac{2x}{x+2}+(x+2)\cdot 1=(x+2)x$$
$$2x+x+2=x^2+2x$$
$$3x+2=x^2+2x$$
$$2=x^2-x$$
$$0=x^2-x-2$$
$$0=(x-2)(x+1)$$
$0=x-2$ or $0=x+1$

$2=x$ or $-1=x$

CHECK

$x = 2$:

$$\frac{2x}{x+2} + 1 = x$$

$$\frac{2(2)}{(2)+2} + 1 \stackrel{?}{=} 2$$

$$\frac{4}{4} + 1 \stackrel{?}{=} 2$$

$$1 + 1 \stackrel{?}{=} 2$$

$$2 \stackrel{\checkmark}{=} 2$$

$x = -1$:

$$\frac{2x}{x+2} + 1 = x$$

$$\frac{2(-1)}{(-1)+2} + 1 \stackrel{?}{=} -1$$

$$\frac{-2}{1} + 1 \stackrel{?}{=} -1$$

$$-2 + 1 \stackrel{?}{=} -1$$

$$-1 \stackrel{\checkmark}{=} -1$$

51. $$a - \frac{5a}{a+1} = \frac{5}{a+1}$$

$$(a+1)\left(a - \frac{5a}{a+1}\right) = (a+1)\left(\frac{5}{a+1}\right)$$

$$(a+1)a - \frac{\cancel{a+1}}{1} \cdot \frac{5a}{\cancel{a+1}} = \frac{\cancel{a+1}}{1} \cdot \frac{5}{\cancel{a+1}}$$

$$(a+1)a - 5a = 5$$

$$a^2 + a - 5a = 5$$

$$a^2 - 4a = 5$$

$$a^2 - 4a - 5 = 0$$

$$(a-5)(a+1) = 0$$

$a - 5 = 0$ or $a + 1 = 0$

$a = 5$ or $a = -1$

CHECK

$a = 5$:

$$a - \frac{5a}{a+1} = \frac{5}{a+1}$$

$$5 - \frac{5(5)}{5+1} \stackrel{?}{=} \frac{5}{5+1}$$

$$5 - \frac{25}{6} \stackrel{?}{=} \frac{5}{6}$$

$$\frac{30}{6} - \frac{25}{6} \stackrel{?}{=} \frac{5}{6}$$

$$\frac{5}{6} \stackrel{\checkmark}{=} \frac{5}{6}$$

$a = -1$:

$$a - \frac{5a}{a+1} = \frac{5}{a+1}$$

$$-1 - \frac{5(-1)}{-1+1} \stackrel{?}{=} \frac{5}{-1+1}$$

$$-1 + \frac{5}{0} \stackrel{?}{=} \frac{5}{0}$$

$$-1 + \frac{5}{0} \neq \frac{5}{0}$$

since we cannot divide by 0. Therefore, we only have one solution: $a = 5$.

53.

$$\frac{3}{x-2} + \frac{7}{x+2} = \frac{x+1}{x-2}$$

$$(x-2)(x+2)\left(\frac{3}{x-2} + \frac{7}{x+2}\right) = (x-2)(x+2)\left(\frac{x+1}{x-2}\right)$$

$$\frac{(x-2)(x+2)}{1} \cdot \frac{3}{x-2} + \frac{(x-2)(x+2)}{1} \cdot \frac{7}{x+2} = \frac{(x-2)(x+2)}{1} \cdot \frac{x+1}{x-2}$$

$$3(x+2) + 7(x-2) = (x+2)(x+1)$$

$$3x + 6 + 7x - 14 = x^2 + 3x + 2$$

$$10x - 8 = x^2 + 3x + 2$$

$$-8 = x^2 - 7x + 2$$

$$0 = x^2 - 7x + 10$$

$$0 = (x-2)(x-5)$$

$0 = x - 2$ or $0 = x - 5$

$2 = x$ or $5 = x$

CHECK

$x = 2$:

$$\frac{3}{x-2} + \frac{7}{x+2} = \frac{x+1}{x-2}$$

$$\frac{3}{2-2} + \frac{7}{2+2} \stackrel{?}{=} \frac{2+1}{2-2}$$

$$\frac{3}{0} + \frac{7}{4} \stackrel{?}{=} \frac{3}{0}$$

$$\frac{3}{0} + \frac{7}{4} \neq \frac{3}{0}$$

since we cannot divide by 0.

$x = 5$:

$$\frac{3}{x-2} + \frac{7}{x+2} = \frac{x+1}{x-2}$$

$$\frac{3}{5-2} + \frac{7}{5+2} \stackrel{?}{=} \frac{5+1}{5-2}$$

$$\frac{3}{3} + \frac{7}{7} \stackrel{?}{=} \frac{6}{3}$$

$$1 + 1 \stackrel{?}{=} 2$$

$$2 \stackrel{\checkmark}{=} 2$$

Therefore, we have only one solution: $x = 5$.

55. (a) We cannot set each of the factors equal to 7. The zero-product rule requires that the product of the factors be equal to 0.

(b) $x = 3$ is not a possible solution. The first factor on the left side of the equation can never be equal to zero. We can either ignore its presence or divide both sides of the equation by it. (This logic is valid for constant factors of a zero product, but <u>not</u> for variable factors.)

56. Since the square root of any real number must be non-negative, x^2 must be at least zero. Therefore, $x^2 + 4$ must be at least 4, which means that $x^2 + 4$ cannot ever equal 0. Thus, the equation $x^2 + 4 = 0$ cannot have any solution in the real number system.

57. $(x-5)^2 = (x-5)(x-5)$

$$= x^2 - 5x - 5x + 25$$

$$= x^2 - 10x + 25$$

59. $(x+5)(x-5)$

$$= x^2 - 5x + 5x - 25$$

$$= x^2 - 25$$

61. $(x+p)^2 = (x+p)(x+p)$

$$= x^2 + px + px + p^2$$

$$= x^2 + 2px + p^2$$

63. Let x = number of hours assistant works

x – 1 = number of hours handyman works

$$11x + 19(x - 1) = 146$$
$$11x + 19x - 19 = 146$$
$$30x - 19 = 146$$
$$30x = 165$$
$$x = 5\frac{1}{2}$$

Since the assistant began at 10:00 a.m. and worked for $5\frac{1}{2}$ hours, the job was finished at 3:30 p.m.

Exercises 11.2

3. $b^2 - 16 = 0$

$$b^2 = 16$$
$$b = \pm\sqrt{16} = \pm 4$$

5. $9b^2 - 16 = 0$

$$9b^2 = 16$$
$$b^2 = \frac{16}{9}$$
$$b = \pm\sqrt{\frac{16}{9}} = \pm\frac{\sqrt{16}}{\sqrt{9}} = \pm\frac{4}{3}$$

7. $b^2 + 16 = 0$

$$b^2 = -16$$
$$b = \pm\sqrt{-16}$$

Since $\sqrt{-16}$ is not a real number, the equation has no real solutions.

11. $36x^2 - 15 = 0$

$$36x^2 = 15$$
$$x^2 = \frac{15}{36}$$
$$x = \pm\sqrt{\frac{15}{36}} = \pm\frac{\sqrt{15}}{\sqrt{36}}$$
$$= \pm\frac{\sqrt{15}}{6}$$

13. $3b^2 = 11$

$$b^2 = \frac{11}{3}$$

$$b = \pm\sqrt{\frac{11}{3}} = \pm\frac{\sqrt{11}}{\sqrt{3}} = \pm\frac{\sqrt{11}\sqrt{3}}{\sqrt{3}\sqrt{3}}$$

$$= \pm\frac{\sqrt{33}}{3}$$

17. $9a^2 = 20$

$$a^2 = \frac{20}{9}$$

$$a = \pm\sqrt{\frac{20}{9}} = \pm\frac{\sqrt{20}}{\sqrt{9}}$$

$$= \pm\frac{\sqrt{20}}{3}$$

$$= \pm\frac{\sqrt{4}\sqrt{5}}{3} = \pm\frac{2\sqrt{5}}{3}$$

21. $7y^2 - 4 = 5y^2 + 6$

$$2y^2 - 4 = 6$$

$$2y^2 = 10$$

$$y^2 = 5$$

$$y = \pm\sqrt{5}$$

23. $5a^2 - 3a + 4 = 2a^2 - 3a + 13$

$$3a^2 - 3a + 4 = -3a + 13$$

$$3a^2 + 4 = 13$$

$$3a^2 = 9$$

$$a^2 = 3$$

$$a = \pm\sqrt{3}$$

27. $(y - 3)(y + 4) = y$

$$y^2 + y - 12 = y$$

$$y^2 - 12 = 0$$

$$y^2 = 12$$

$$y = \pm\sqrt{12}$$

$$= \pm\sqrt{4}\sqrt{3} = \pm 2\sqrt{3}$$

31. $(x+2)^2 = 4(x+7)$

$$x^2 + 4x + 4 = 4x + 28$$
$$x^2 + 4 = 28$$
$$x^2 = 24$$
$$x = \pm\sqrt{24}$$
$$= \pm\sqrt{4}\sqrt{6} = \pm 2\sqrt{6}$$

33. $(t-2)^2 = 9$

$$t - 2 = \pm\sqrt{9}$$
$$t - 2 = \pm 3$$
$$t - 2 = 3 \text{ or } t - 2 = -3$$
$$t = 5 \text{ or } t = -1$$

35. $(a+5)^2 = 7$

$$a + 5 = \pm\sqrt{7}$$
$$a = -5 \pm \sqrt{7}$$

39. $\left(m - \frac{2}{3}\right)^2 = \frac{4}{9}$

$$m - \frac{2}{3} = \pm\sqrt{\frac{4}{9}}$$
$$m - \frac{2}{3} = \pm\frac{\sqrt{4}}{\sqrt{9}}$$
$$m - \frac{2}{3} = \pm\frac{2}{3}$$
$$m - \frac{2}{3} = \frac{2}{3} \text{ or } m - \frac{2}{3} = -\frac{2}{3}$$
$$m = \frac{4}{3} \text{ or } m = 0$$

41. $\left(x + \frac{2}{5}\right)^2 = \frac{3}{25}$

$$x + \frac{2}{5} = \pm\sqrt{\frac{3}{25}}$$
$$x + \frac{2}{5} = \pm\frac{\sqrt{3}}{\sqrt{25}}$$
$$x + \frac{2}{5} = \pm\frac{\sqrt{3}}{5}$$
$$x = -\frac{2}{5} \pm \frac{\sqrt{3}}{5} = \frac{-2 \pm \sqrt{3}}{5}$$

45. $2x^2 + 7x - 5 = 3x^2 + 9x - 4$

$$7x - 5 = x^2 + 9x - 4$$
$$-5 = x^2 + 2x - 4$$
$$0 = x^2 + 2x + 1$$
$$0 = (x + 1)(x + 1)$$
$$0 = x + 1 \text{ or } 0 = x + 1$$
$$-1 = x \text{ or } -1 = x$$
So $x = -1$

47. $(y - 2)(y + 3) = y + 10$

$$y^2 + y - 6 = y + 10$$
$$y^2 - 6 = 10$$
$$y^2 = 16$$
$$y = \pm\sqrt{16} = \pm 4$$

49. $(y + 2)(y + 3) = (2y - 7)(y + 4)$

$$y^2 + y - 6 = 2y^2 + y - 28$$
$$y - 6 = y^2 + y - 28$$
$$22 = y^2$$
$$\pm\sqrt{22} = y$$

53. $4(x + 1) = \frac{9}{x + 1}$

$$(x + 1)(4(x + 1)) = \frac{\overset{1}{\cancel{x + 1}}}{1} \cdot \frac{9}{\underset{1}{\cancel{x + 1}}}$$
$$4(x + 1)^2 = 9$$
$$(x + 1)^2 = \frac{9}{4}$$
$$x + 1 = \pm\sqrt{\frac{9}{4}}$$
$$x + 1 = \pm\frac{\sqrt{9}}{\sqrt{4}}$$
$$x + 1 = \pm\frac{3}{2}$$
$$x + 1 = \frac{3}{2} \text{ or } x + 1 = -\frac{3}{2}$$
$$x = \frac{1}{2} \text{ or } x = -\frac{5}{2}$$

55. $x = \pm 2.65$

57. $k = \pm 2.58$

59. $x = \pm 2.22$

61. $a = 1.52,\ a = -1.22$

63. $y = .23,\ y = -1.19$

65. Step 1: We can add the same quantity, 7, to both sides of an equation to obtain an equivalent equation.

Step 2: We can add the same quantity, 4, to both sides of an equation to obtain an equivalent equation.

Step 3: $7 + 4 = 11$

Step 4: $x^2 + 4x + 4$ factors as the square of the sum $x + 2$.

Step 5: $u^2 = d$ implies that $u = \pm\sqrt{d}$.

Step 6: We can subtract the same quantity, 2, from both sides of an equation to obtain an equivalent equation.

66. $(2+\sqrt{3})^2 - 4(2+\sqrt{3}) + 1$
$= (2+\sqrt{3})(2+\sqrt{3}) - 4(2+\sqrt{3}) + 1$
$= 4 + 2\sqrt{3} + 2\sqrt{3} + 3 - 8 - 4\sqrt{3} + 1$
$= 0$

So $2+\sqrt{3}$ is a solution to the equation $x^2 - 4x + 1 = 0$.

$(2-\sqrt{3})^2 - 4(2-\sqrt{3}) + 1$
$= (2-\sqrt{3})(2-\sqrt{3}) - 4(2-\sqrt{3}) + 1$
$= 4 - 2\sqrt{3} - 2\sqrt{3} + 3 - 8 + 4\sqrt{3} + 1$
$= 0$

So $2-\sqrt{3}$ is also a solution to the equation $x^2 - 4x + 1 = 0$. (This is no coincidence. In fact, whenever $a+\sqrt{b}$ satisfies an equation, it must be true that $a-\sqrt{b}$ also satisfies the equation.)

67. $\begin{cases} 2x + y = 6 \xrightarrow{\text{mutiply by 2}} \\ 3x - 2y = 23 \xrightarrow{\text{as is}} \end{cases}$

$4x + 2y = 12$
$3x - 2y = 23$

Add: $7x = 35$
$x = 5$

$2x + y = 6$
$2(5) + y = 6$
$10 + y = 6$
$y = -4$

Solution: $(5, -4)$

CHECK:

$3x - 2y = 23$
$3(5) - 2(-4) \stackrel{?}{=} 23$
$15 + 8 \stackrel{?}{=} 23$
$23 \stackrel{\checkmark}{=} 23$

69. $\begin{cases} 6x - 4y = 9 \xrightarrow{\text{as is}} \\ 3x - 2y = 2 \xrightarrow{\text{multiply by } -2} \end{cases}$

$$\begin{array}{r} 6x - 4y = 9 \\ \underline{-6x + 4y = -4} \\ \text{Add: } 0 = 5 \end{array}$$

a contradiction. So the system has no solution.

71. Let x = number of hours needed if they work together.

$$\frac{x}{4} + \frac{x}{3} = 1$$

$$12\left(\frac{x}{4} + \frac{x}{3}\right) = 12(1)$$

$$3x + 4x = 12$$

$$7x = 12$$

$$x = \frac{12}{7} \text{ or } 1\frac{5}{7}$$

To prepare the meal together, Lorraine and Renaldo need $1\frac{5}{7}$ hours.

Exercises 11.3

1.
$$\begin{array}{l} x^2 + 8x + 6 = 0 \\ x^2 + 8x \quad = -6 \\ \underline{\qquad +16 \quad +16} \\ x^2 + 8x + 16 = 10 \\ (x + 4)^2 = 10 \\ x + 4 = \pm\sqrt{10} \\ x = -4 \pm \sqrt{10} \end{array}$$

$$\left[\left(\frac{8}{2}\right) = 4,\ 4^2 = 16\right]$$

5.
$$\begin{array}{l} x^2 - 10x \quad = 15 \\ \underline{\qquad +25 \quad +25} \\ x^2 - 10x + 25 = 40 \\ (x - 5)^2 = 40 \\ x - 5 = \pm\sqrt{40} \\ x - 5 = \pm 2\sqrt{10} \\ x = 5 \pm 2\sqrt{10} \end{array}$$

$$\left[\left(\frac{-10}{2}\right) = -5,\ (-5)^2 = 25\right]$$

$$a^2 - 8a - 20 = 0$$
$$a^2 - 8a \quad = 20$$
$$\underline{\quad +16 \quad +16}$$
$$a^2 - 8a + 16 = 36$$
$$(a-4)^2 = 36$$
$$a - 4 = \pm\sqrt{36}$$
$$a - 4 = \pm 6$$
$$a - 4 = 6 \text{ or } a - 4 = -6$$
$$a = 10 \text{ or } a = -2$$
$$\left[\left(\frac{-8}{2}\right) = -4,\ (-4)^2 = 16\right]$$

11. $$2z^2 - 12z + 4 = 0$$
$$2z^2 - 12z \quad = -4$$
$$z^2 - 6z \quad = -2$$
$$\underline{\quad +9 \quad +9}$$
$$z^2 - 6z + 9 = 7$$
$$(z-3)^2 = 7$$
$$z - 3 = \pm\sqrt{7}$$
$$z = 3 \pm \sqrt{7}$$
$$\left[\left(\frac{-6}{2}\right) = -3,\ (-3)^2 = 9\right]$$

15. $$u^2 + 5u - 2 = 0$$
$$u^2 + 5u \quad = 2$$
$$\underline{\quad +\frac{25}{4} \quad +\frac{25}{4}}$$
$$u^2 + 5u + \frac{25}{4} = 2 + \frac{25}{4}$$
$$\left(u + \frac{5}{2}\right)^2 = \frac{8}{4} + \frac{25}{4}$$
$$\left(u + \frac{5}{2}\right)^2 = \frac{33}{4}$$
$$u + \frac{5}{2} = \pm\sqrt{\frac{33}{4}} = \pm\frac{\sqrt{33}}{2}$$
$$u = -\frac{5}{2} \pm \frac{\sqrt{33}}{2}$$
$$u = \frac{-5 \pm \sqrt{33}}{2}$$

19. $$\begin{aligned} w^2 - 3w &= 2w^2 - 7w + 2 \\ -w^2 - 3w &= -7w + 2 \\ -w^2 + 4w &= 2 \\ w^2 - 4w &= -2 \\ +4 & \quad +4 \\ \hline w^2 - 4w + 4 &= 2 \\ (w-2)^2 &= 2 \\ w - 2 &= \pm\sqrt{2} \\ w &= 2 \pm \sqrt{2} \end{aligned}$$

$$\left[\left(-\frac{4}{2}\right) = -2,\ (-2)^2 = 4\right]$$

23. $$\begin{aligned} (x-3)(x+2) &= 9x - 1 \\ x^2 - x - 6 &= 9x - 1 \\ x^2 - 10x - 6 &= -1 \\ x^2 - 10x &= 5 \\ +25 & \quad +25 \\ \hline x^2 - 10x + 25 &= 30 \\ (x-5)^2 &= 30 \\ x - 5 &= \pm\sqrt{30} \\ x &= 5 \pm \sqrt{30} \end{aligned}$$

$$\left[\left(\frac{-10}{2}\right) = -5,\ (-5)^2 = 25\right]$$

5.

$$
\begin{aligned}
(a-2)(a+1) &= 6 \\
a^2 - a - 2 &= 6 \\
a^2 - a &= 8 \\
+\frac{1}{4} \quad & \quad +\frac{1}{4} \\
\hline
a^2 - a + \frac{1}{4} &= 8 + \frac{1}{4} \\
\left(a - \frac{1}{2}\right)^2 &= \frac{32}{4} + \frac{1}{4} \\
\left(a - \frac{1}{2}\right)^2 &= \frac{33}{4} \\
a - \frac{1}{2} &= \pm\sqrt{\frac{33}{4}} = \pm\frac{\sqrt{33}}{2} \\
a &= \frac{1}{2} \pm \frac{\sqrt{33}}{2} \\
a &= \frac{1 \pm \sqrt{33}}{2}
\end{aligned}
$$

$$\left[\left(\frac{-1}{2}\right)^2 = \frac{1}{4}\right]$$

29.

$$
\begin{aligned}
2x^2 + 3 &= 6x \\
2x^2 - 6x + 3 &= 0 \\
2x^2 - 6x &= -3 \\
x^2 - 3x &= -\frac{3}{2} \\
+\frac{9}{4} \quad & \quad +\frac{9}{4} \\
\hline
x^2 - 3x + \frac{9}{4} &= -\frac{3}{2} + \frac{9}{4} \\
\left(x - \frac{3}{2}\right)^2 &= -\frac{6}{4} + \frac{9}{4} \\
\left(x - \frac{3}{2}\right)^2 &= \frac{3}{4} \\
x - \frac{3}{2} &= \pm\sqrt{\frac{3}{4}} = \pm\frac{\sqrt{3}}{2} \\
x &= \frac{3}{2} \pm \frac{\sqrt{3}}{2} \\
x &= \frac{3 \pm \sqrt{3}}{2}
\end{aligned}
$$

$$\left[\left(-\frac{3}{2}\right)^2 = \frac{9}{4}\right]$$

33. $4z^2 + 20z + 19 = 0$

$$4z^2 + 20z = -19$$

$$z^2 + 5z = -\frac{19}{4}$$

$$+\frac{25}{4} \qquad +\frac{25}{4}$$

$$z^2 + 5z + \frac{25}{4} = \frac{-19}{4} + \frac{25}{4}$$

$$\left(z + \frac{5}{2}\right)^2 = \frac{6}{4}$$

$$z + \frac{5}{2} = \pm\sqrt{\frac{6}{4}} = \pm\frac{\sqrt{6}}{2}$$

$$z = -\frac{5}{2} \pm \frac{\sqrt{6}}{2}$$

$$z = \frac{-5 \pm \sqrt{6}}{2}$$

$$\left[\left(\frac{5}{2}\right)^2 = \frac{25}{4}\right]$$

35. $(x + 3)^2 = 6$

$$x + 3 = \pm\sqrt{6}$$

$$x = -3 \pm \sqrt{6}$$

37. $(x + 3)^2 = 6x$

$$x^2 + 6x + 9 = 6x$$

$$x^2 + 9 = 0$$

$$x^2 = -9$$

This has no real solution.

39. $x^2 + 8x - 9 = 0$

$(x + 9)(x - 1) = 0$

$x + 9 = 0$ or $x - 1 = 0$

$x = -9$ or $x = 1$

41. $3x^2 + 4 = 8x$

$$3x^2 - 8x + 4 = 0$$

$(3x - 2)(x - 2) = 0$

$3x - 2 = 0$ or $x - 2 = 0$

$x = \frac{2}{3}$ or $x = 2$

43. The constant of a perfect square is the square of one-half of the coefficient of the middle term.

$$\frac{x}{x-1}=\frac{2}{x-2}$$

$$\frac{(\cancel{x-1})(x-2)}{1}\cdot\frac{x}{\cancel{x-1}}=\frac{(x-1)(\cancel{x-2})}{1}\cdot\frac{2}{\cancel{x-2}}$$

$$x(x-2)=2(x-1)$$

$$x^2-2x=2x-2$$

$$x^2-4x=-2$$

$$\underline{\qquad +4 \quad +4 \qquad}$$

$$x^2-4x+4=2$$

$$(x-2)^2=2$$

$$x-2=\pm\sqrt{2}$$

$$x=2\pm\sqrt{2}$$

CHECK $x=2+\sqrt{2}$:

$$\frac{x}{x-1}=\frac{2}{x-2}$$

$$\frac{2+\sqrt{2}}{2+\sqrt{2}-1}\stackrel{?}{=}\frac{2}{2+\sqrt{2}-2}$$

$$\frac{2+\sqrt{2}}{1+\sqrt{2}}\stackrel{?}{=}\frac{2}{\sqrt{2}}$$

$$\frac{2+\sqrt{2}}{1+\sqrt{2}}=\frac{(2+\sqrt{2})(1-\sqrt{2})}{(1+\sqrt{2})(1-\sqrt{2})}$$

$$=\frac{2-2\sqrt{2}+\sqrt{2}-2}{1-2}=\frac{-\sqrt{2}}{-1}$$

$$=\sqrt{2}$$

$$\frac{2}{\sqrt{2}}=\frac{2\sqrt{2}}{\sqrt{2}\sqrt{2}}=\frac{2\sqrt{2}}{2}=\sqrt{2}$$

so $\frac{2+\sqrt{2}}{1+\sqrt{2}}\stackrel{\checkmark}{=}\frac{2}{\sqrt{2}}$

CHECK $x = 2 - \sqrt{2}$:

$$\frac{x}{x-1} = \frac{2}{x-2}$$

$$\frac{2-\sqrt{2}}{2-\sqrt{2}-1} \stackrel{?}{=} \frac{2}{2-\sqrt{2}-2}$$

$$\frac{2-\sqrt{2}}{1-\sqrt{2}} \stackrel{?}{=} \frac{2}{-\sqrt{2}}$$

$$\frac{2-\sqrt{2}}{1-\sqrt{2}} = \frac{(2-\sqrt{2})(1+\sqrt{2})}{(1-\sqrt{2})(1+\sqrt{2})}$$

$$= \frac{2+2\sqrt{2}-\sqrt{2}-2}{1-2} = \frac{\sqrt{2}}{-1} = -\sqrt{2}$$

$$\frac{2}{-\sqrt{2}} = \frac{2\sqrt{2}}{-\sqrt{2}\sqrt{2}} = \frac{2\sqrt{2}}{-2} = -\sqrt{2}$$

so $\frac{2-\sqrt{2}}{1-\sqrt{2}} \stackrel{\checkmark}{=} \frac{2}{-\sqrt{2}}$

45. $\sqrt{48} - \sqrt{75} = \sqrt{16 \cdot 3} - \sqrt{25 \cdot 3}$

$= \sqrt{16}\sqrt{3} - \sqrt{25}\sqrt{3}$

$= 4\sqrt{3} - 5\sqrt{3}$

$= -\sqrt{3}$

47. $\sqrt{\frac{2}{7}} = \frac{\sqrt{2}}{\sqrt{7}} = \frac{\sqrt{2}\sqrt{7}}{\sqrt{7}\sqrt{7}} = \frac{\sqrt{14}}{7}$

49. $\frac{3}{3-\sqrt{2}} = \frac{3(3+\sqrt{2})}{(3-\sqrt{2})(3+\sqrt{2})}$

$= \frac{3(3+\sqrt{2})}{9-2} = \frac{3(3+\sqrt{2})}{7}$

51. $\frac{6+4\sqrt{3}}{2} = \frac{6}{2} + \frac{4\sqrt{3}}{2} = 3 + 2\sqrt{3}$

53. Let x = number of hours Yvonne needs to finish the puzzle.

$$\frac{3}{10} + \frac{x}{8} = 1$$

$$40\left(\frac{3}{10} + \frac{x}{8}\right) = 40(1)$$

$$12 + 5x = 40$$

$$5x = 28$$

$$x = \frac{28}{5} \text{ or } 5\frac{3}{5}$$

It takes Yvonne $5\frac{3}{5}$ hours to finish the puzzle.

Exercises 11.4

1. $x^2 + 3x - 5 = 0$

   $a = 1, \ b = 3, \ c = -5$

5. $2u^2 = 8u$

   $2u^2 - 8u = 0$

   $a = 2, \ b = -8, \ c = 0$

9. $x^2 + 3x - 5 = 0$

   $a = 1, \ b = 3, \ c = -5$

   $$x = \frac{-b \pm \sqrt{b^2 - 4ac}}{2a}$$

   $$x = \frac{-3 \pm \sqrt{(3)^2 - 4(1)(-5)}}{2(1)}$$

   $$x = \frac{-3 \pm \sqrt{9 + 20}}{2}$$

   $$x = \frac{-3 \pm \sqrt{29}}{2}$$

13. $u^2 - 2u + 3 = 0$

   $a = 1, \ b = -2, \ c = 3$

   $$u = \frac{-b \pm \sqrt{b^2 - 4ac}}{2a}$$

   $$u = \frac{-(-2) \pm \sqrt{(-2)^2 - 4(1)(3)}}{2(1)}$$

   $$u = \frac{2 \pm \sqrt{4 - 12}}{2}$$

   $$u = \frac{2 \pm \sqrt{-8}}{2}$$

   No real solutions, since the answer involves the square root of a negative number.

17. $2x^2 - 3x - 1 = 0$

$a = 2,\ b = -3,\ c = -1$

$$x = \frac{-b \pm \sqrt{b^2 - 4ac}}{2a}$$

$$x = \frac{-(-3) \pm \sqrt{(-3)^2 - 4(2)(-1)}}{2(2)}$$

$$x = \frac{3 \pm \sqrt{9 + 8}}{4}$$

$$x = \frac{3 \pm \sqrt{17}}{4}$$

19. $5x^2 - x = 2$

$5x^2 - x - 2 = 0$

$a = 5,\ b = -1,\ c = -2$

$$x = \frac{-b \pm \sqrt{b^2 - 4ac}}{2a}$$

$$x = \frac{-(-1) \pm \sqrt{(-1)^2 - 4(5)(-2)}}{2(5)}$$

$$x = \frac{1 \pm \sqrt{1 + 40}}{10}$$

$$x = \frac{1 \pm \sqrt{41}}{10}$$

21. $t^2 - 3t + 4 = 2t^2 + 4t - 3$

$-3t + 4 = t^2 + 4t - 3$

$4 = t^2 + 7t - 3$

$0 = t^2 + 7t - 7$

$a = 1,\ b = 7,\ c = -7$

$$t = \frac{-b \pm \sqrt{b^2 - 4ac}}{2a}$$

$$t = \frac{-7 \pm \sqrt{(7)^2 - 4(1)(-7)}}{2(1)}$$

$$t = \frac{-7 \pm \sqrt{49 + 28}}{2}$$

$$t = \frac{-7 \pm \sqrt{77}}{2}$$

23. $(5w + 2)(w - 1) = 3w + 1$

$5w^2 - 3w - 2 = 3w + 1$

$5w^2 - 6w - 2 = 1$

$5w^2 - 6w - 3 = 0$

$a = 5, \; b = -6, \; c = -3$

$$w = \frac{-b \pm \sqrt{b^2 - 4ac}}{2a}$$

$$w = \frac{-(-6) \pm \sqrt{(-6)^2 - 4(5)(-3)}}{2(5)}$$

$$w = \frac{6 \pm \sqrt{36 + 60}}{10}$$

$$w = \frac{6 \pm \sqrt{96}}{10}$$

$$w = \frac{6 \pm 4\sqrt{6}}{10} = \frac{2(3 \pm 2\sqrt{6})}{10}$$

$$w = \frac{3 \pm 2\sqrt{6}}{5}$$

27. $3x^2 - 5x + 7 = 2x(x - 5) + 9x + 5$

$3x^2 - 5x + 7 = 2x^2 - 10x + 9x + 5$

$3x^2 - 5x + 7 = 2x^2 - x + 5$

$x^2 - 5x + 7 = -x + 5$

$x^2 - 4x + 7 = 5$

$x^2 - 4x + 2 = 0$

$a = 1, \; b = -4, \; c = 2$

$$x = \frac{-b \pm \sqrt{b^2 - 4ac}}{2a}$$

$$x = \frac{-(-4) \pm \sqrt{(-4)^2 - 4(1)(2)}}{2(1)}$$

$$x = \frac{4 \pm \sqrt{16 - 8}}{2}$$

$$x = \frac{4 \pm \sqrt{8}}{2}$$

$$x = \frac{4 \pm 2\sqrt{2}}{2} = \frac{2(2 \pm \sqrt{2})}{2}$$

$= 2 \pm \sqrt{2}$

31. $x^2(x-1)=(x-1)^3$

$x^3-x^2=x^3-3x^2+3x-1$

$-x^2=-3x^2+3x-1$

$0=-2x^2+3x-1$

$0=2x^2-3x+1$

$0=(2x-1)(x-1)$

$0=2x-1$ or $0=x-1$

$1=2x$ or $1=x$

$\frac{1}{2}=x$

33. $4x=9x^2$

$0=9x^2-4x$

$0=x(9x-4)$

$0=x$ or $0=9x-4$

$0=x$ or $4=9x$

$0=x$ or $\frac{4}{9}=x$

37. $\frac{w}{2}=\frac{3}{w+2}$

$\frac{\cancel{2}(w+2)}{1}\cdot\frac{w}{\cancel{2}}=\frac{2\cancel{(w+2)}}{1}\cdot\frac{3}{\cancel{w+2}}$

$w(w+2)=6$

$w^2+2w=6$

$w^2+2w-6=0$

$a=1,\ b=2,\ c=-6$

$w=\frac{-b\pm\sqrt{b^2-4ac}}{2(1)}$

$w=\frac{-2\pm\sqrt{2^2-4(1)(-6)}}{2(1)}$

$w=\frac{-2\pm\sqrt{4+24}}{2}$

$w=\frac{-2\pm\sqrt{28}}{2}$

$w=\frac{-2\pm2\sqrt{7}}{2}=\frac{\cancel{2}(-1\pm\sqrt{7})}{\cancel{2}}$

$=-1\pm\sqrt{7}$

41. $x=1.62,\ x=-.62$

43. $t=8.22,\ t=-1.22$

45. $w=-1.72,\ w=-4.06$

47. $x=3.06,\ x=-1.17$

9. $2x^2 + 7x + 4 = 0$

$a = 2, b = 7, c = 4$

$$x = \frac{-b \pm \sqrt{b^2 - 4ac}}{2a}$$

$$x = \frac{-7 \pm \sqrt{7^2 - 4(2)(4)}}{2(2)}$$

$$x = \frac{-7 \pm \sqrt{49 - 32}}{4}$$

$$x = \frac{-7 \pm \sqrt{17}}{4}$$

Using the quadratic formula is easier than the method of completing the square.

50. The factoring method, when it works, is usually the easiest method to use. However, it does not always work. Completing the square and the quadratic formula work for any quadratic equation. Generally, completing the square is the most complicated of these two methods.

51. (a) Since $b = -3$ and $c = -1$,

$$x = \frac{-(-3) \pm \sqrt{9 - 4(-1)}}{2}$$

$$= \frac{3 \pm \sqrt{9 + 4}}{2} = \frac{3 \pm \sqrt{13}}{2}$$

(b) The minus sign under the square root should be a plus sign.

(c) The "5" should be divided by 2 as well. That is,

$$x = \frac{5 \pm \sqrt{25 - 12}}{2} = \frac{5 \pm \sqrt{13}}{2}$$

(d) This is correct up until the last step. Then

$$\frac{6 \pm 4\sqrt{3}}{2} = \frac{\cancel{2}(3 \pm 2\sqrt{3})}{\cancel{2}}$$

$= 3 \pm 2\sqrt{3}$, not $6 \pm 2\sqrt{3}$.

53. $3\sqrt{2}(\sqrt{2} + \sqrt{7}) + \sqrt{7}(5\sqrt{2} - \sqrt{7})$

$= 3\sqrt{2}\sqrt{2} + 3\sqrt{2}\sqrt{7}$

$\quad + 5\sqrt{2}\sqrt{7} - \sqrt{7}\sqrt{7}$

$= 3(2) + 3\sqrt{14} + 5\sqrt{14} - 7$

$= 6 + 8\sqrt{14} - 7$

$= 8\sqrt{14} - 1$

55. $(2\sqrt{x}-\sqrt{5})(3\sqrt{x}-4\sqrt{5})$

$$= (2\sqrt{x})(3\sqrt{x}) - (2\sqrt{x})(4\sqrt{5})$$
$$- (\sqrt{5})(3\sqrt{x}) + (\sqrt{5})(4\sqrt{5})$$
$$= 6x - 8\sqrt{5x} - 3\sqrt{5x} + 20$$
$$= 6x - 11\sqrt{5x} + 20$$

57. Let p = price of a pastrami sandwich

t = price of a tuna sandwich

$$5p + 6t = 42.00 \xrightarrow{\text{multiply by } 3}$$
$$4p + 9t = 47.25 \xrightarrow{\text{multiply by } -2}$$
$$15p + 18t = 126.00$$
$$\underline{-8p - 18t = \ 94.50}$$
$$\text{Add: } 7p = 31.50$$
$$p = 4.50$$
$$5p + 6t = 42.00$$
$$5(4.50) + 6t = 42.00$$
$$22.50 + 6t = 42.00$$
$$6t = 19.50$$
$$t = 3.25$$

So a pastrami sandwich costs \$4.50 and a tuna sandwich costs \$3.25.

Exercises 11.5

1. $x^2 + 6x + 5 = 0$

$$(x + 1)(x + 5) = 0$$
$$x + 1 = 0 \text{ or } x + 5 = 0$$
$$x = -1 \text{ or } x = -5$$

$$x^2 + 6x \quad = -5$$
$$\underline{\qquad +9 \quad +9}$$
$$x^2 + 6x + 9 = 4$$
$$(x + 3)^2 = 4$$
$$x + 3 = \pm\sqrt{4} = \pm 2$$
$$x + 3 = 2 \text{ or } x + 3 = -2$$
$$x = -1 \text{ or } x = -5$$

6. $2r^2 + 1 = 3r$

$2r^2 - 3r + 1 = 0$

$(2r - 1)(r - 1) = 0$

$2r - 1 = 0 \text{ or } r - 1 = 0$

$2r = 1 \text{ or } r = 1$

$r = \frac{1}{2} \text{ or } r = 1$

$a = 2,\ b = -3,\ c = 1$

$r = \frac{-b \pm \sqrt{b^2 - 4ac}}{2a}$

$r = \frac{-(-3) \pm \sqrt{(-3)^2 - 4(2)(1)}}{2(2)}$

$r = \frac{3 \pm \sqrt{9 - 8}}{6}$

$r = \frac{3 \pm \sqrt{1}}{4} = \frac{3 \pm 1}{4}$

$r = \frac{3 + 1}{4} = \frac{4}{4} = 1$

$\text{or } r = \frac{3 - 1}{4} + \frac{2}{4} + \frac{1}{2}$

7. $w^2 = 4w + 5$

$w^2 - 4w = 5$

$w^2 - 4w - 5 = 0$

$(w - 5)(w + 1) = 0$

$w - 5 = 0 \text{ or } w + 1 = 0$

$w = 5 \text{ or } w = -1$

$w^2 = 4w + 5$

$w^2 - 4w = 5$

$\underline{\quad +4 \quad +4}$

$w^2 - 4w + 4 = 9$

$(w - 2)^2 = 9$

$w - 2 = \pm\sqrt{9} = \pm 3$

$w - 2 = 3 \text{ or } w - 2 = -3$

$w = 5 \text{ or } w = -1$

13. $4x^2 = 16x - 28$

$$4x^2 - 16x = -28$$

$$x^2 - 4x = -7$$

$$+4 \quad +4$$

$$x^2 - 4x + 4 = -3$$

$$(x-2)^2 = -3$$

$$x - 2 = \pm\sqrt{-3}$$

Thus, there are no real solutions, since square roots of negative numbers are not real.

$$4x^2 = 16x - 28$$

$$4x^2 - 16x = -28$$

$$4x^2 - 16x + 28 = 0$$

$$x^2 - 4x + 7 = 0$$

$$a = 1, \quad b = -4, \quad c = 7$$

$$x = \frac{-b \pm \sqrt{b^2 - 4ac}}{2a}$$

$$x = \frac{-(-4) \pm \sqrt{(-4)^2 - 4(1)(7)}}{2(1)}$$

$$x = \frac{4 \pm \sqrt{16 - 28}}{2}$$

$$x = \frac{4 \pm \sqrt{-12}}{2},$$

which leads to the same conclusion for the same reason.

15. $(x-1)^2 = 5$

$x - 1 = \pm\sqrt{5}$

$x = 1 \pm \sqrt{5}$

$(x-1)^2 = 5$

$x^2 - 2x + 1 = 5$

$x^2 - 2x - 4 = 0$

$a = 1,\ \ b = -2,\ \ c = -4$

$$x = \frac{-b \pm \sqrt{b^2 - 4ac}}{2a}$$

$$x = \frac{-(-2) \pm \sqrt{(-2)^2 - 4(1)(-4)}}{2(1)}$$

$$x = \frac{2 \pm \sqrt{4 + 16}}{2}$$

$$x = \frac{2 \pm \sqrt{20}}{2}$$

$$x = \frac{2 \pm 2\sqrt{5}}{2}$$

$$x = \frac{\cancel{2}(1 \pm \sqrt{5})}{\cancel{2}}$$

$x = 1 \pm \sqrt{5}$

17. $(x-1)^2 = 5x$

$x^2 - 2x + 1 = 5x$

$x^2 - 7x + 1 = 0$

$a = 1,\ \ b = -7,\ \ c = 1$

$$x = \frac{-b \pm \sqrt{b^2 - 4ac}}{2a}$$

$$x = \frac{-(-7) \pm \sqrt{(-7)^2 - 4(1)(1)}}{2(1)}$$

$$x = \frac{7 \pm \sqrt{49 - 4}}{2}$$

$$x = \frac{7 \pm \sqrt{45}}{2}$$

$$x = \frac{7 \pm 3\sqrt{5}}{2}$$

$$(x-1)^2 = 5x$$
$$x^2 - 2x + 1 = 5x$$
$$x^2 - 7x + 1 = 0$$
$$x^2 - 7x \quad = -1$$
$$\underline{\quad + \frac{49}{4} \qquad + \frac{49}{4}}$$
$$x^2 - 7x + \frac{49}{4} = -1 + \frac{49}{4}$$
$$\left(x - \frac{7}{2}\right)^2 = \frac{-4}{4} + \frac{49}{4}$$
$$\left(x - \frac{7}{2}\right)^2 = \frac{45}{4}$$
$$x - \frac{7}{2} = \pm\sqrt{\frac{45}{4}} = \pm\frac{\sqrt{45}}{2}$$
$$x - \frac{7}{2} = \pm\frac{3\sqrt{5}}{2}$$
$$x = \frac{7}{2} \pm \frac{3\sqrt{5}}{2}$$
$$x = \frac{7 \pm 3\sqrt{5}}{2}$$

21. $y^2 - 4y + 10 = 5(y+2)$

$$y^2 - 4y + 10 = 5y + 10$$
$$y^2 - 9y + 10 = 10$$
$$y^2 - 9y = 0$$
$$y(y-9) = 0$$
$$y = 0 \text{ or } y - 9 = 0$$
$$y = 0 \text{ or } y = 9$$

$y^2 - 9y = 0$

$a = 1,\ b = -9,\ c = 0$

$$y = \frac{-b \pm \sqrt{b^2 - 4ac}}{2a}$$

$$y = \frac{-(-9) \pm \sqrt{(-9)^2 - 4(1)(0)}}{2(1)}$$

$$y = \frac{9 \pm \sqrt{81 - 0}}{2}$$

$$y = \frac{9 \pm \sqrt{81}}{2}$$

$$y = \frac{9 \pm 9}{2}$$

$$y = \frac{9 + 9}{2} = \frac{18}{2} = 9 \text{ or}$$

$$y = \frac{9 - 9}{2} = \frac{0}{2} = 0$$

23. $(t + 4)(t - 8) = 13$

$t^2 - 4t - 32 = 13$

$t^2 - 4t - 45 = 0$

$(t - 9)(t + 5) = 0$

$t - 9 = 0$ or $t + 5 = 0$

$t = 9$ or $t = -5$

$(t + 4)(t - 8) = 13$

$t^2 - 4t - 32 = 13$

$t^2 - 4t \quad = 45$

$\quad +4 \quad +4$

$t^2 - 4t + 4 = 49$

$(t - 2)^2 = 49$

$t - 2 = \pm\sqrt{49} = \pm 7$

$t - 2 = 7$ or $t - 2 = -7$

$t = 9$ or $t = -5$

27. $z^2 - 3z \quad = 3z - 9$

$z^2 - 6z \quad = -9$

$z^2 - 6z + 9 = 0$

$(z - 3)^2 = 0$

$z - 3 = 0$

$z = 3$

$$z^2 - 3z = 3z - 9$$

$$z^2 - 6z = -9$$

$$z^2 - 6z + 9 = 0$$

$a = 1,\ b = -6,\ c = 9$

$$z = \frac{-b \pm \sqrt{b^2 - 4ac}}{2a}$$

$$z = \frac{-(-6) \pm \sqrt{(-6)^2 - 4(1)(9)}}{2(1)}$$

$$z = \frac{6 \pm \sqrt{36 - 36}}{2}$$

$$z = \frac{6 \pm \sqrt{0}}{2}$$

$$z = \frac{6 \pm 0}{2} = \frac{6}{2} = 3$$

33. $$x^2 + 1 = \frac{5}{2}x$$

$$2(x^2 + 1) = 2\left(\frac{5}{2}x\right)$$

$$2x^2 + 2 = \frac{\not{2}}{1} \cdot \frac{5}{\not{2}}x$$

$$2x^2 + 2 = 5x$$

$$2x^2 - 5x + 2 = 0$$

$$(2x - 1)(x - 2) = 0$$

$2x - 1 = 0$ or $x - 2 = 0$

$2x = 1$ or $x = 2$

$x = \frac{1}{2}$ or $x = 2$

$$2x^2 - 5x + 2 = 0$$

$a = 2,\ b = -5,\ c = 2$

$$x = \frac{-b \pm \sqrt{b^2 - 4ac}}{2a}$$

$$x = \frac{-(-5) \pm \sqrt{(-5)^2 - 4(2)(2)}}{2(2)}$$

$$x = \frac{5 \pm \sqrt{25 - 16}}{4}$$

$$x = \frac{5 \pm \sqrt{9}}{4} = \frac{5 \pm 3}{4}$$

$x = \frac{5 + 3}{4} = \frac{8}{4} = 2$ or

$$x = \frac{5 - 3}{4} = \frac{2}{4} = \frac{1}{2}$$

35. $\dfrac{x}{x+1} = \dfrac{4}{x+4}$

$$\frac{(x+1)(x+4)}{1} \cdot \frac{x}{x+1} = \frac{(x+1)(x+4)}{1} \cdot \frac{4}{x+4}$$

$$x(x+4) = 4(x+1)$$

$$x^2 + 4x = 4x + 4$$

$$x^2 = 4$$

$$x = \pm\sqrt{4}$$

$$x = \pm 2$$

$x^2 = 4$

$x^2 - 4 = 0$

$(x-2)(x+2) = 0$

$x - 2 = 0$ or $x + 2 = 0$

$x = 2$ or $x = -2$

39. $$\frac{3x}{x+1} + \frac{2}{x-1} = 4$$

$$(x+1)(x-1)\left(\frac{3x}{x+1} + \frac{2}{x-1}\right) = (x+1)(x-1)4$$

$$\frac{(x+1)(x-1)}{1} \cdot \frac{3x}{x+1} + \frac{(x+1)(x-1)}{1} \cdot \frac{2}{x-1} = 4(x+1)(x-1)$$

$$3x(x-1) + 2(x+1) = 4(x+1)(x-1)$$

$$3x^2 - 3x + 2x + 2 = 4(x^2 - 1)$$

$$3x^2 - x + 2 = 4x^2 - 4$$

$$-x + 2 = x^2 - 4$$

$$2 = x^2 + x - 4$$

$$0 = x^2 + x - 6$$

$$0 = (x+3)(x-2)$$

$0 = x + 3$ or $0 = x - 2$

$-3 = x$ or $2 = x$

$x^2 + x - 6 = 0$

$a = 1,\ b = 1,\ c = -6$

$$x = \frac{-b \pm \sqrt{b^2 - 4ac}}{2a}$$

$$x = \frac{-1 \pm \sqrt{1^2 - 4(1)(-6)}}{2(1)}$$

$$x = \frac{-1 \pm \sqrt{1 + 24}}{2}$$

$$x = \frac{-1 \pm \sqrt{25}}{2}$$

$$x = \frac{-1 \pm 5}{2}$$

$$x = \frac{-1 + 5}{2} = \frac{4}{2} = 2 \text{ or } x = \frac{-1 - 5}{2} = \frac{-6}{2} = -3$$

41. $2x^2 + 3x = 20$

Method 1—factoring:

$2x^2 + 3x - 20 = 0$

$(2x - 5)(x + 4) = 0$

$2x - 5 = 0$ or $x + 4 = 0$

$2x = 5$ or $x = -4$

$x = \frac{5}{2}$ or $x = -4$

Method 2—completing the square:

$$2x^2 + 3x \quad = 20$$

$$x^2 + \frac{3}{2}x \quad = 10$$

$$\begin{array}{r}
\quad + \frac{9}{16} \qquad + \frac{9}{16} \\
\hline
x^2 + \frac{3}{2}x + \frac{9}{16} = 10 + \frac{9}{16}
\end{array}$$

$$\left(x + \frac{3}{4}\right)^2 = \frac{160}{16} + \frac{9}{16}$$

$$\left(x + \frac{3}{4}\right)^2 = \frac{169}{16}$$

$$x + \frac{3}{4} = \pm\sqrt{\frac{169}{16}}$$

$$= \pm\frac{\sqrt{169}}{\sqrt{16}}$$

$$x + \frac{3}{4} = \pm\frac{13}{4}$$

$$x = -\frac{3}{4} \pm \frac{13}{4}$$

$$x = \frac{-3 \pm 13}{4}$$

$$x = \frac{-3 + 13}{4} = \frac{10}{4} = \frac{5}{2}$$

$$\text{or } x = \frac{-3 - 13}{4} = \frac{-16}{4} = -4$$

Method 3—quadratic formula:

$$2x^2 + 3x - 20 = 0$$

$$a = 2, \; b = 3, \; c = -20$$

$$x = \frac{-b \pm \sqrt{b^2 - 4ac}}{2a}$$

$$x = \frac{-3 \pm \sqrt{3^2 - 4(2)(-20)}}{2(2)}$$

$$x = \frac{-3 \pm \sqrt{9 + 160}}{4}$$

$$x = \frac{-3 \pm \sqrt{169}}{4} = \frac{-3 \pm 13}{4}$$

$$x = \frac{-3 + 13}{4} = \frac{10}{4} = \frac{5}{2} \text{ or } x = \frac{-3 - 13}{4} = \frac{-16}{4} = -4$$

The easiest of these methods is the first, while the second one appears to be the most difficult.

42. $3x^2 - 5x - 1 = 0$

$a = 3,\ b = -5,\ c = -1$

$$x = \frac{-b \pm \sqrt{b^2 - 4ac}}{2a}$$

$$x = \frac{-(-5) \pm \sqrt{(-5)^2 - 4(3)(-1)}}{2(3)}$$

$$x = \frac{5 \pm \sqrt{25 + 12}}{6} = \frac{5 \pm \sqrt{37}}{6}$$

CHECK $x = \dfrac{5 + \sqrt{37}}{6}$:

$$3x^2 - 5x - 1 = 0$$

$$3\left(\frac{5 + \sqrt{37}}{6}\right)^2 - 5\left(\frac{5 + \sqrt{37}}{6}\right) - 1 \stackrel{?}{=} 0$$

$$3\left(\frac{5 + \sqrt{37}}{6}\right)\left(\frac{5 + \sqrt{37}}{6}\right) - 5\left(\frac{5 + \sqrt{37}}{6}\right) - 1 \stackrel{?}{=} 0$$

$$\frac{3}{1}\left(\frac{25 + 10\sqrt{37} + 37}{36\ 12}\right) - \frac{5}{1}\left(\frac{5 + \sqrt{37}}{6}\right) - 1 \stackrel{?}{=} 0$$

$$\frac{62 + 10\sqrt{37}}{12} - \frac{25 + 5\sqrt{37}}{6} - 1 \stackrel{?}{=} 0$$

$$\frac{2(31 + 5\sqrt{37})}{12\ 6} - \frac{25 + 5\sqrt{37}}{6} - 1 \stackrel{?}{=} 0$$

$$\frac{31 + 5\sqrt{37} - (25 + 5\sqrt{37})}{6} - 1 \stackrel{?}{=} 0$$

$$\frac{6}{6} - 1 \stackrel{?}{=} 0$$

$$1 - 1 \stackrel{?}{=} 0$$

$$0 \stackrel{\checkmark}{=} 0$$

$$3x^2 - 5x - 1 = 0$$

$$\underline{\quad +1 \qquad +1}$$

$$3x^2 - 5x = 1$$

$$x^2 - \frac{5}{3}x = \frac{1}{3}$$

$$\underline{+\frac{25}{36} \qquad +\frac{25}{36}}$$

$$x^2 - \frac{5}{3}x + \frac{25}{36} = \frac{1}{3} + \frac{25}{36}$$

$$\left(x - \frac{5}{6}\right)^2 = \frac{12}{36} + \frac{25}{36}$$

$$\left(x - \frac{5}{6}\right)^2 = \frac{37}{36}$$

$$x - \frac{5}{6} = \pm\sqrt{\frac{37}{36}}$$

$$= \pm\frac{\sqrt{37}}{\sqrt{36}}$$

$$x - \frac{5}{6} = \pm\frac{\sqrt{37}}{6}$$

$$x = \frac{5}{6} \pm \frac{\sqrt{37}}{6}$$

$$= \frac{5 \pm \sqrt{37}}{6}$$

The second method of checking was by far the easier in this example. The first check was more complicated than the actual solution. If the first check happened to fail, we really would not know whether an error was made in the solution to the problem or in the check of that solution. Further, the first check only verified that $x = \dfrac{5+\sqrt{37}}{6}$ is a valid solution. We would still have to examine $x = \dfrac{5-\sqrt{37}}{6}$ and verify that it satisfied the equation as well.

43. (a) $m = \dfrac{y_2 - y_1}{x_2 - x_1} = \dfrac{-4-3}{3-(-1)} = \dfrac{-7}{4}$

(b) $m = 0$, since the points have the same y-coordinate.

(c) m is undefined, since the points have the same x-coordinate.

45. $m = \frac{y_2 - y_1}{x_2 - x_1} = \frac{-1-(-5)}{1-(-2)} = \frac{-1+5}{1+2}$

$= \frac{4}{3}$

$(x_1, y_1) = (-2, -5)$

$y - y_1 = m(x - x_1)$

$y - (-5) = \frac{4}{3}(x - (-2))$

$y + 5 = \frac{4}{3}(x + 2)$ or $y = \frac{4}{3}x - \frac{7}{3}$

Exercises 11.6

1. Let x = one of the numbers

20 – x = the other number

$x(20 - x) = 96$

$20x - x^2 = 96$

$x^2 - 20x + 96 = 0$

$(x - 8)(x - 12) = 0$

$x - 8 = 0$ or $x - 12 = 0$

$x = 8$ or $x = 12$

If x = 8, then 20 – x = 20 – 8 = 12. If x = 12, then 20 – x = 20 – 12 = 8. In both cases, we conclude that the numbers are 8 and 12.

CHECK:

$8 + 12 \stackrel{\checkmark}{=} 20$

$8(12) \stackrel{\checkmark}{=} 96$

5. Let W = width of the rectangle

2W + 3 = length of the rectangle

$W(2W + 3) = 90$

$2W^2 + 3W = 90$

$2W^2 + 3W - 90 = 0$

$(2W + 15)(W - 6) = 0$

$2W + 15 = 0$ or $W - 6 = 0$

$W = -\frac{15}{2}$ or $W = 6$

↑

Reject this, since width cannot be negative.

So W = 6. Then 2W + 3 = 2(6) + 3 = 15. Thus the rectangle has a width of 6 meters and a length of 15 meters.

CHECK: 15 is 3 more than twice 6, and 15(6) = 90.

9. Let d = length of the diagonal. From the Pythagorean Theorem, we get

$$8^2 + 8^2 = d^2$$
$$64 + 64 = d^2$$
$$128 = d^2$$
$$\pm\sqrt{128} = d$$
$$\pm 8\sqrt{2} = d$$

Reject $d = -8\sqrt{2}$, since we cannot have a negative length. So $d = 8\sqrt{2}$. Thus the length of the diagonal is $8\sqrt{2}$ inches.

CHECK:

$$8^2 + 8^2 \stackrel{?}{=} (8\sqrt{2})^2$$
$$64 + 64 \stackrel{?}{=} 64 \cdot 2$$
$$128 \stackrel{\checkmark}{=} 128$$

13. Let x = length of the shortest side

x + 1 = length of the middle side

x + 2 = length of the longest side

Using the Pythagorean Theorem, it follows that

$$x^2 + (x+1)^2 = (x+2)^2$$
$$x^2 + x^2 + 2x + 1 = x^2 + 4x + 4$$
$$2x^2 + 2x + 1 = x^2 + 4x + 4$$
$$x^2 - 2x - 3 = 0$$
$$(x-3)(x+1) = 0$$
$$x - 3 = 0 \text{ or } x + 1 + 0$$
$$x = 3 \text{ or } x = -1$$

Reject x = –1, since the side of a triangle cannot have negative length. So x = 3. Then

x + 1 = 3 + 1 = 4 and x + 2 = 3 + 2 = 5. Thus the sides of the triangle have lengths of 3, 4, and 5.

CHECK: 3, 4, and 5 are three consecutive integers and $3^2 + 4^2 = 9 + 16 = 25 = 5^2$.

15. Let n = numerator of the original fraction

n + 1 = denominator of the original fraction

$$\frac{n+3}{n+1} = \frac{n}{n+1} + 1$$

$$(n+1)\left(\frac{n+3}{n+1}\right) = (n+1)\left(\frac{n}{n+1} + 1\right)$$

$$\frac{\cancel{n+1}}{1} \cdot \frac{n+3}{\cancel{n+1}} = \frac{\cancel{n+1}}{1} \cdot \frac{n}{\cancel{n+1}} + (n+1) \cdot 1$$

$$n + 3 = n + n + 1$$
$$n + 3 = 2n + 1$$
$$3 = n + 1$$
$$2 = n$$

Then $n + 1 = 2 + 1 = 3$. Thus, the original fraction is $\frac{2}{3}$.

CHECK: $\frac{2+3}{3} = \frac{5}{3}$, which is one more than $\frac{2}{3}$.

17. Let s = # of seats in each row

$s - 8$ = # of rows of seats

$$s(s-8) = 768$$
$$s^2 - 8s = 768$$
$$s^2 - 8s - 768 = 0$$
$$(s-32)(s+24) = 0$$
$$s - 32 = 0 \text{ or } s + 24 = 0$$
$$s = 32 \text{ or } s = -24$$

Reject $s = -24$, since the number of seats in a row cannot be negative. So $s = 32$. Then

$s - 8 = 32 - 8 = 24$. Thus, there are 24 rows of seats in the concert hall, and each row has 32 seats in it.

CHECK: 24 is eight less than 32, and $24 \cdot 32 = 768$.

19. Let x = speed of the motorist for the first part of the trip.

$x - 20$ = speed of the motorist for second part of the trip

$$\frac{120}{x} + \frac{30}{x-20} = 2$$
$$x(x-20)\left(\frac{120}{x} + \frac{30}{x-20}\right) = x(x-20)\cdot 2$$
$$\frac{\cancel{x}(x-20)}{1}\cdot\frac{120}{\cancel{x}_1} + \frac{x\cancel{(x-20)}}{1}\cdot\frac{30}{\cancel{x-20}_1} = 2x(x-20)$$
$$120(x-20) + 30x = 2x(x-20)$$
$$120x - 2400 + 30x = 2x^2 - 40x$$
$$150x - 2400 = 2x^2 - 40x$$
$$0 = 2x^2 - 190 + 2400$$
$$0 = x^2 - 95x + 1200$$
$$0 = (x-15)(x-80)$$
$$0 = x - 15 \text{ or } 0 = x - 80$$
$$15 = x \text{ or } 80 = x$$

We reject $x = 15$, since this would mean that $x - 20 = 15 - 20 = -5$, and a negative speed is impossible. So $x = 80$. Thus the motorist's speed for the first part of the trip is 80 kph.

CHECK: At 80 kph, the motorist covers 120 kilometers in $\frac{120}{80} = \frac{3}{2}$ hours. At (80 – 20) = 60 kph, the motorist covers the remaining 30 kilometers in $\frac{30}{60} = \frac{1}{2}$ hour. Then $\frac{3}{2} + \frac{1}{2} = \frac{4}{2} = 2$, as required.

23. Let x = distance from P to R (in meters)

By the Pythagorean Theorem,

$$\begin{aligned} x^2 &= (8.6)^2 + (4.9)^2 \\ &= 73.96 + 24.01 \\ &= 97.97 \end{aligned}$$

So $x = 9.9$

The distance between P and R is 9.9 meters.

27. (a) $h = 1000 - 16t^2$

$$\begin{aligned} 700 &= 1000 - 16t^2 \\ -300 &= -16t^2 \\ \frac{75}{4} &= t^2 \\ 18.75 &= t^2 \\ 4.3 &= t \end{aligned}$$

The object takes 4.3 seconds to reach a height of 700 feet.

(b) $h = 1000 - 16t^2$

$$\begin{aligned} 0 &= 1000 - 16t^2 \\ 16t^2 &= 1000 \\ t^2 &= \frac{125}{2} \\ t^2 &= 62.5 \\ t &= 7.9 \end{aligned}$$

The object hits the ground 7.9 seconds after it is dropped.

31. (a) $\frac{1}{\ell} = \frac{\ell}{1+\ell}$

$$\frac{\ell(1+\ell)}{1} \cdot \frac{1}{\ell} = \frac{\ell(1+\ell)}{1} \cdot \frac{\ell}{1+\ell}$$

$$1 + \ell = \ell^2$$

$$0 = \ell^2 - \ell - 1$$

$$a = 1, \quad b = -1, \quad c = -1$$

$$\ell = \frac{-b \pm \sqrt{b^2 - 4ac}}{2a}$$

$$\ell = \frac{-(-1) \pm \sqrt{(-1)^2 - 4(1)(-1)}}{2(1)}$$

$$\ell = \frac{1 \pm \sqrt{1+4}}{2} = \frac{1 \pm \sqrt{5}}{2}$$

so $\ell = \frac{1+\sqrt{5}}{2}$ or $\ell = \frac{1-\sqrt{5}}{2}$

Since $\sqrt{5}$ is greater than 1, $\frac{1-\sqrt{5}}{2}$ is negative, and thus must be rejected. So $\ell = \frac{1+\sqrt{5}}{2}$ inches.

(b)

$$\frac{w}{\ell} = \frac{\ell}{w+\ell}$$

$$\frac{\ell(w+\ell)}{1} \cdot \frac{w}{\ell} = \frac{\ell(w+\ell)}{1} \cdot \frac{\ell}{w+\ell}$$

$$w(w+\ell) = \ell^2$$

$$w^2 + w\ell = \ell^2$$

$$0 = \ell^2 - w\ell - w^2$$

$$a = 1, \quad b = -w, \quad c = -w^2$$

$$\ell = \frac{-b \pm \sqrt{b^2 - 4ac}}{2a}$$

$$\ell = \frac{-(-w) \pm \sqrt{(-w)^2 - 4(1)(-w^2)}}{2(1)}$$

$$\ell = \frac{w \pm \sqrt{w^2 + 4w^2}}{2}$$

$$\ell = \frac{w \pm \sqrt{5w^2}}{2}$$

$$\ell = \frac{w \pm \sqrt{5}w}{2} = \frac{(1 \pm \sqrt{5})w}{2}$$

As in part (a), we reject $\frac{(1-\sqrt{5})w}{2}$ because it is negative. Thus, $\ell = \frac{(1+\sqrt{5})w}{2}$.

(If w = 1, we get the golden ratio of part (a).)

32. $\frac{1}{x-1} = \frac{x}{1}$

$\frac{x-1}{1} \cdot \frac{1}{x-1} = \frac{x-1}{1} \cdot \frac{x}{1}$

$1 = x(x-1)$

$1 = x^2 - x$

$0 = x^2 - x - 1$

Proceed with the quadratic formula to find that $x = \frac{1+\sqrt{5}}{2}$.

(This quadratic equation is the same as the one encountered in part (a) of exercise (31), except for the letter used to represent the variable.)

33. Let x = length of the shortest side

$x + 2$ = length of the middle side

$x + 4$ = length of the longest side

Use the Pythagorean Theorem to find that

$x^2 + (x+2)^2 = (x+4)^2$

Then

$x^2 + x^2 + 4x + 4 = x^2 + 8x + 16$

$2x^2 + 4x + 4 = x^2 + 8x + 16$

$x^2 - 4x - 12 = 0$

$(x-6)(x+2) = 0$

$x - 6 = 0$ or $x + 2 = 0$

$x = 6$ or $x = -2$

Reject $x = 6$, since it is not odd. Reject $x = -2$, since we cannot have a negative length. Since both possible solutions to the problem have been rejected, we conclude that it is impossible to find three consecutive odd integers that are the sides of a right triangle.

Exercises 11.7

3. $y = x^2 + 2$

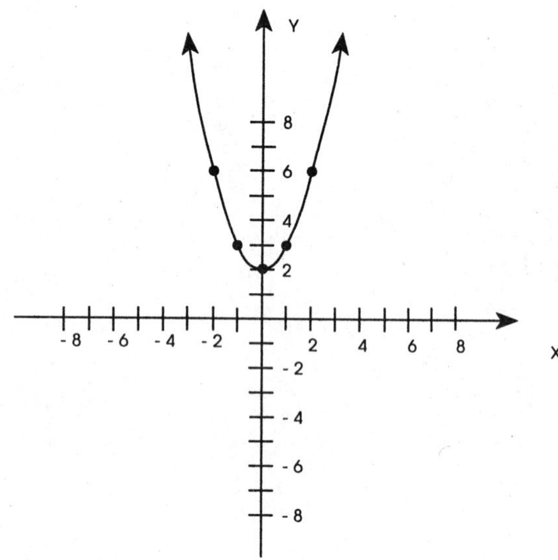

There are no x-intercepts.

7. $y = -x^2 + 1$

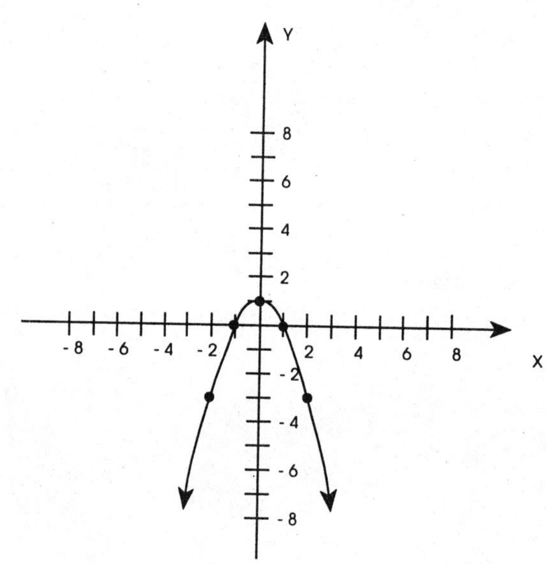

$$0 = -x^2 + 1$$
$$x^2 = 1$$
$$x = \pm 1$$

The x-intercepts are 1 and –1.

11. $y = 6x - 2x^2$

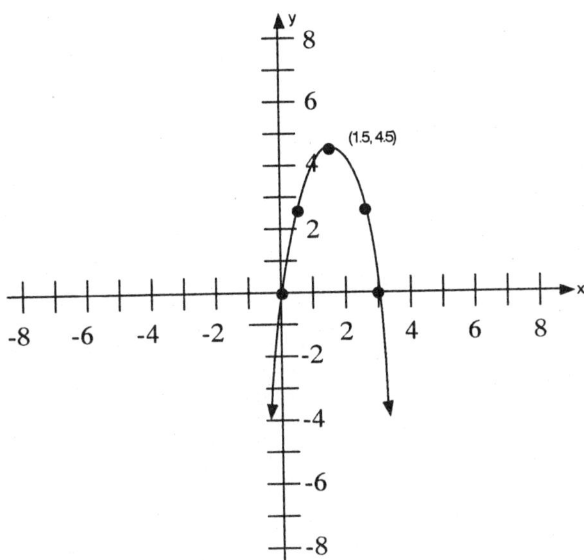

$0 = 6x - 2x^2$

$0 = 2x(3 - x)$

$0 = 2x$ or $0 = 3 - x$

$0 = x$ or $3 = x$

The x-intercepts are 0 and 3.

17. $y = x^2 - 4x + 3$

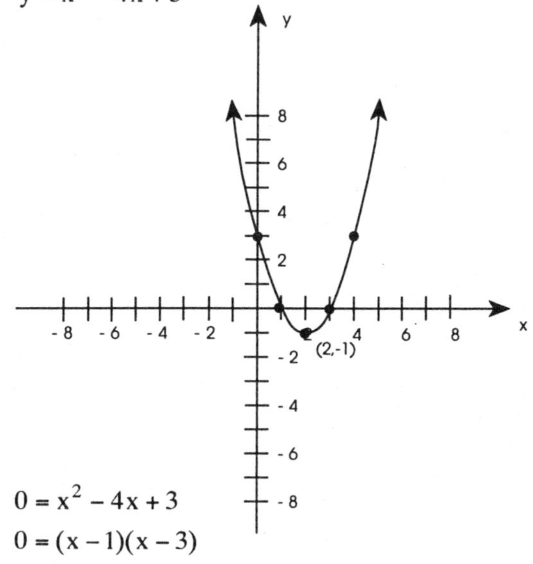

$0 = x^2 - 4x + 3$

$0 = (x - 1)(x - 3)$

$0 = x - 1$ or $0 = x - 3$

$1 = x$ or $3 = x$

The x-intercepts are 1 and 3.

23. $y = x^2 + 4x + 4$

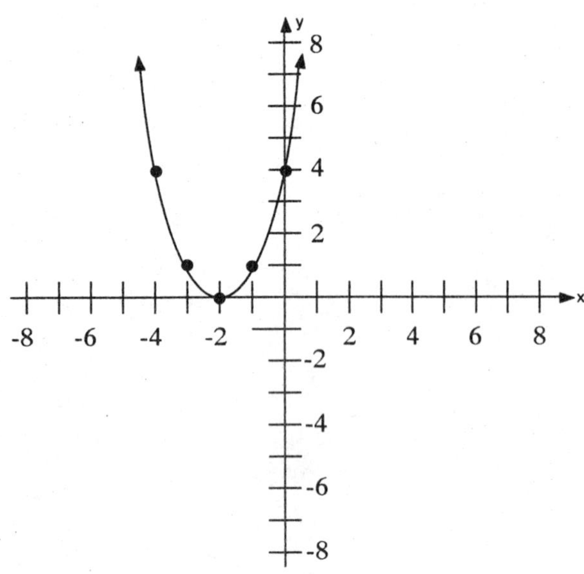

$0 = x^2 + 4x + 4$

$0 = (x + 2)(x + 2)$

$0 = x + 2$ or $0 = x + 2$

$-2 = x$ or $-2 = x$

The x-intercept is –2.

29. $y = -x^2 + x + 6$

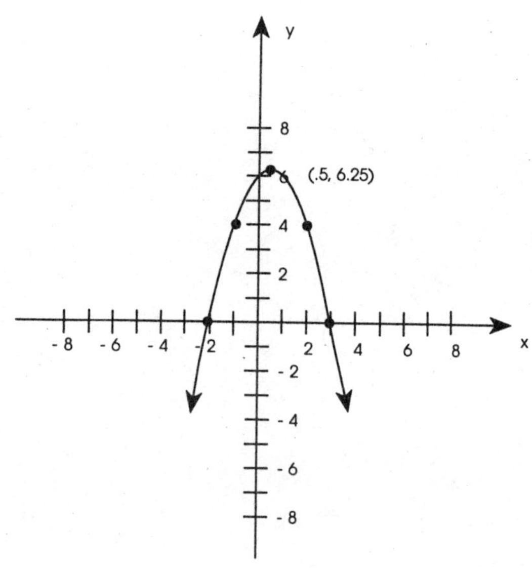

$0 = -x^2 + x + 6$

$0 = x^2 - x - 6$

$0 = (x - 3)(x + 2)$

$0 = x - 3$ or $0 = x + 2$

$3 = x$ or $-2 = x$

The x-intercepts are 3 and –2.

33. $y = -x^2 - 8x - 5$

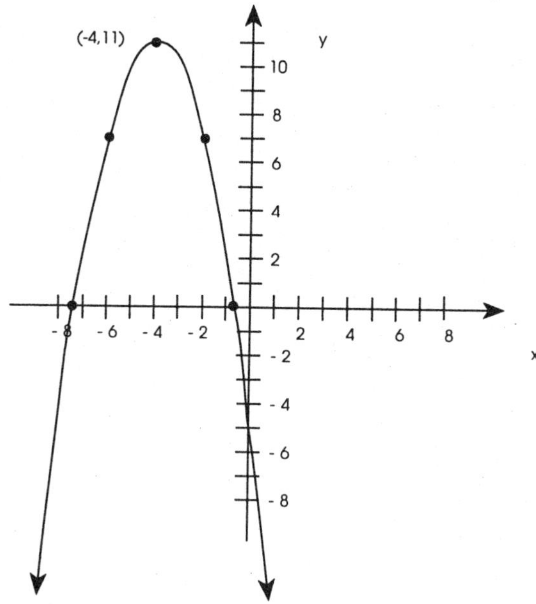

$0 = -x^2 - 8x - 5$

$0 = x^2 + 8x + 5$

$a = 1,\ b = 8,\ c = 5$

$$x = \frac{-b \pm \sqrt{b^2 - 4ac}}{2a}$$

$$= \frac{-8 \pm \sqrt{64 - 20}}{2}$$

$$= \frac{-8 \pm \sqrt{44}}{2}$$

$$= \frac{-8 \pm 2\sqrt{11}}{2}$$

$$= -4 \pm \sqrt{11}$$

The x-intercepts are $-4 + \sqrt{11} \approx -.7$ and $-4 - \sqrt{11} \approx -7.3$

39. $y = 4x^2 - 8x - 5$

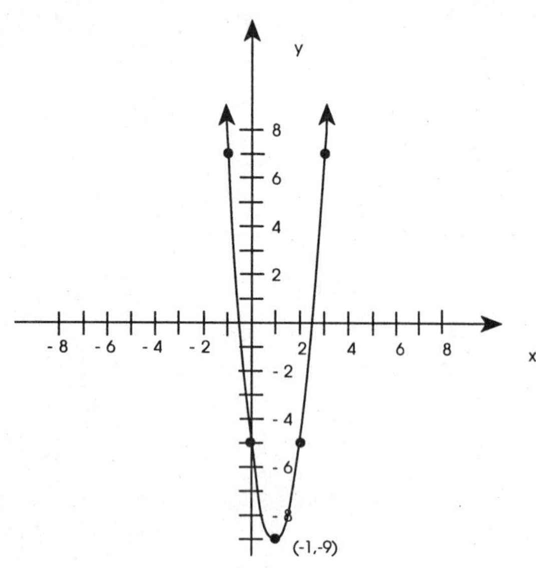

$$0 = 4x^2 - 8x - 5$$
$$0 = (2x + 1)(2x - 5)$$
$$0 = 2x + 1 \text{ or } 0 = 2x - 5$$
$$-1 = 2x \text{ or } 5 = 2x$$
$$-\frac{1}{2} = x \text{ or } \frac{5}{2} = x$$

The x-intercepts are $-\frac{1}{2}$ and $\frac{5}{2}$.

43.

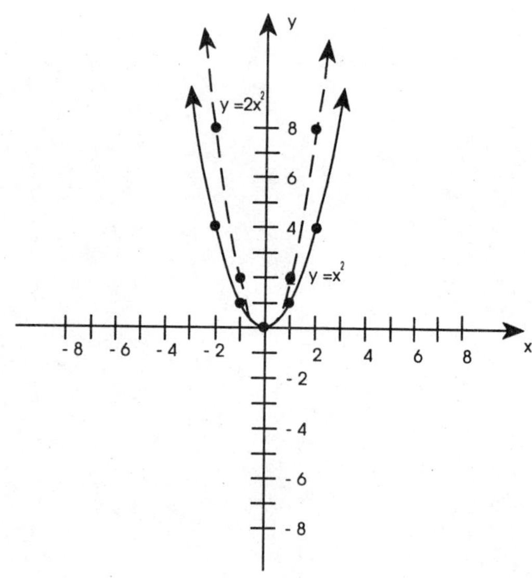

The coefficient of 2 narrows the graph of $y = x^2$.

44.

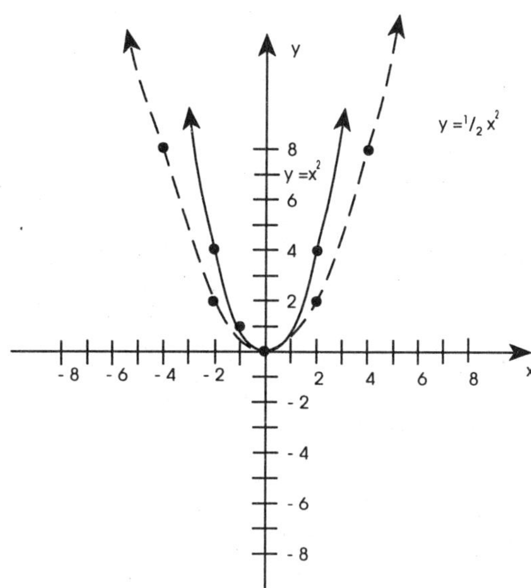

The coefficient of $\frac{1}{2}$ widens the graph of $y = x^2$.

45.

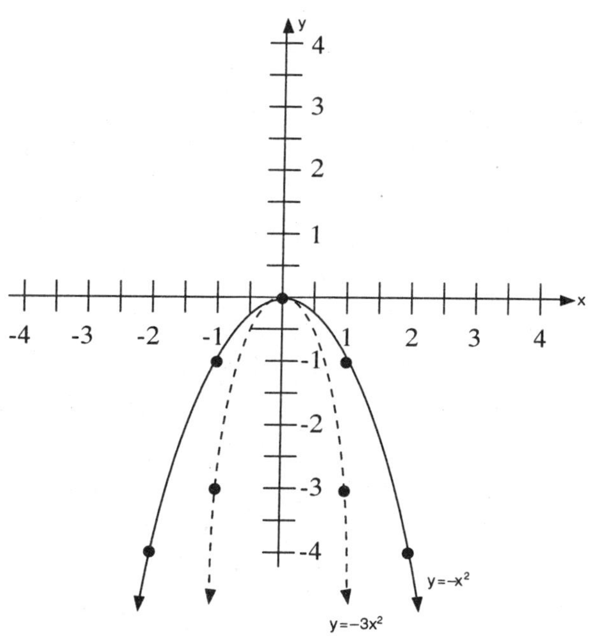

The coefficient of –3 narrows the graph of $y = x^2$ and then "flips" the resulting graph over the x-axis.

46.

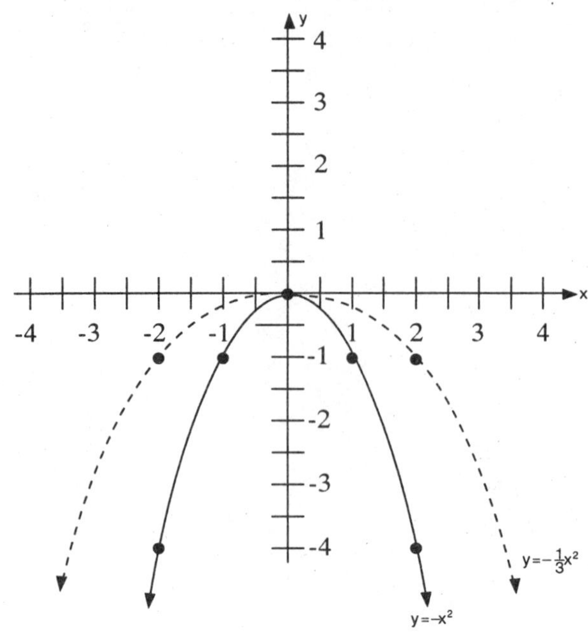

The coefficient of $-\frac{1}{3}$ widens the graph of $y = x^2$ and then "flips" the resulting graph over the x-axis.

Chapter 11 Review Exercises

1. $x^2 - 7x - 6 = 0$

$a = 1,\ b = -7,\ c = -6$

$$x = \frac{-b \pm \sqrt{b^2 - 4ac}}{2a}$$

$$x = \frac{-(-7) \pm \sqrt{(-7)^2 - 4(1)(-6)}}{2(1)}$$

$$x = \frac{7 \pm \sqrt{49 + 24}}{2} = \frac{7 \pm \sqrt{73}}{2}$$

3. $x^2 + 5 = 4x$

$$x^2 - 4x + 5 = 0$$

$$x^2 - 4x = -5$$

$$\underline{\quad +4 \quad +4}$$

$$x^2 - 4x + 4 = -1$$

$$(x - 2)^2 = -1$$

$$x - 2 = \pm\sqrt{-1}$$

Therefore, there are no real solutions.

$(u-6)^2 = 13$

$u - 6 = \pm\sqrt{13}$

$u = 6 \pm \sqrt{13}$

7. $2y^2 + 7y = 15$

$2y^2 + 7y - 15 = 0$

$(2y-3)(y+5) = 0$

$2y - 3 = 0$ or $y + 5 = 0$

$y = \frac{3}{2}$ or $y = -5$

9. $18x^2 - 24x + 6 = 0$

$3x^2 - 4x + 1 = 0$

$(3x-1)(x-1) = 0$

$3x - 1 = 0$ or $x - 1 = 0$

$x = \frac{1}{3}$ or $x = 1$

11. $(x-6)^2 = (x+3)(x-5)$

$x^2 - 12x + 36 = x^2 - 2x - 15$

$-12x + 36 = -2x - 15$

$36 = 10x - 15$

$51 = 10x$

$\frac{51}{10} = x$

13. $u^2 + 1 = \frac{13u}{6}$

$6(u^2+1) = \frac{\cancel{6}}{1} \cdot \frac{13u}{\cancel{6}_1}$

$6u^2 + 6 = 13u$

$6u^2 - 13u + 6 = 0$

$(3u-2)(2u-3) = 0$

$3u - 2 = 0$ or $2u - 3 = 0$

$u = \frac{2}{3}$ or $u = \frac{3}{2}$

15. $3x(x-2) = (x-3)^2$

$$3x^2 - 6x = x^2 - 6x + 9$$

$$2x^2 - 6x = -6x + 9$$

$$2x^2 = 9$$

$$x^2 = \frac{9}{2}$$

$$x = \pm\sqrt{\frac{9}{2}} = \pm\frac{\sqrt{9}}{\sqrt{2}} = \pm\frac{3}{\sqrt{2}} = \pm\frac{3\sqrt{2}}{\sqrt{2}\sqrt{2}}$$

$$x = \pm\frac{3\sqrt{2}}{2}$$

17. $\frac{x+3}{x+6} = \frac{x+2}{x+4}$

$$\frac{\cancel{(x+6)}(x+4)}{1}\cdot\frac{x+3}{\cancel{x+6}\ 1} = \frac{(x+6)\cancel{(x+4)}}{1}\cdot\frac{x+2}{\cancel{x+4}\ 1}$$

$$(x+4)(x+3) = (x+6)(x+2)$$

$$x^2 + 7x + 12 = x^2 + 8x + 12$$

$$7x + 12 = 8x + 12$$

$$7x = 8x$$

$$0 = x$$

19. $x^2 - 7x + 3 = 0$

$$x^2 - 7x = -3$$

$$+\frac{49}{4} \qquad +\frac{49}{4}$$

$$x^2 - 7x + \frac{49}{4} = -3 + \frac{49}{4}$$

$$\left(x - \frac{7}{2}\right)^2 = -\frac{12}{4} + \frac{49}{4}$$

$$\left(x - \frac{7}{2}\right)^2 = \frac{37}{4}$$

$$x - \frac{7}{2} = \pm\sqrt{\frac{37}{4}} = \pm\frac{\sqrt{37}}{\sqrt{4}}$$

$$= \pm\frac{\sqrt{37}}{2}$$

$$x = \frac{7}{2} \pm \frac{\sqrt{37}}{2} = \frac{7 \pm \sqrt{37}}{2}$$

21. $y = x^2 - 6x$

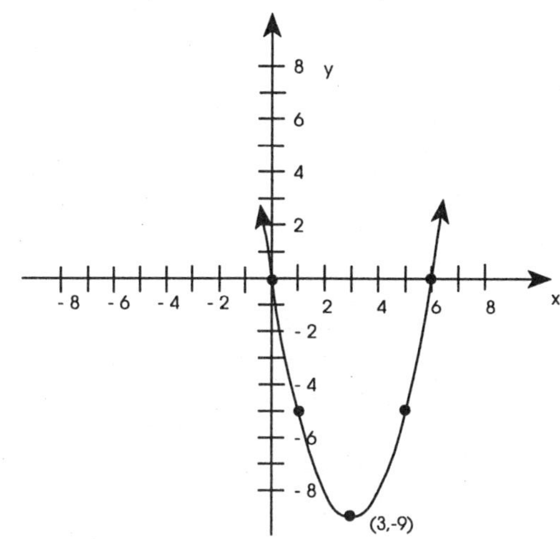

$0 = x^2 - 6x$

$0 = x(x - 6)$

$0 = x$ or $0 = x - 6$

$0 = x$ or $6 = x$

The x-intercepts are 0 and 6.

23. $y = -x^2 - 2x + 8$

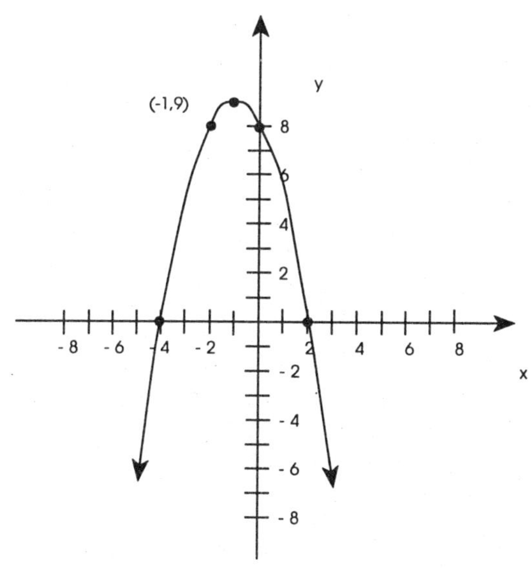

$0 = -x^2 - 2x + 8$

$0 = x^2 + 2x - 8$

$0 = (x + 4)(x - 2)$

$0 = x + 4$ or $0 = x - 2$

$-4 = x$ or $2 = x$

The x-intercepts are –4 and 2.

25. $y = 2x^2 - 4x + 3$

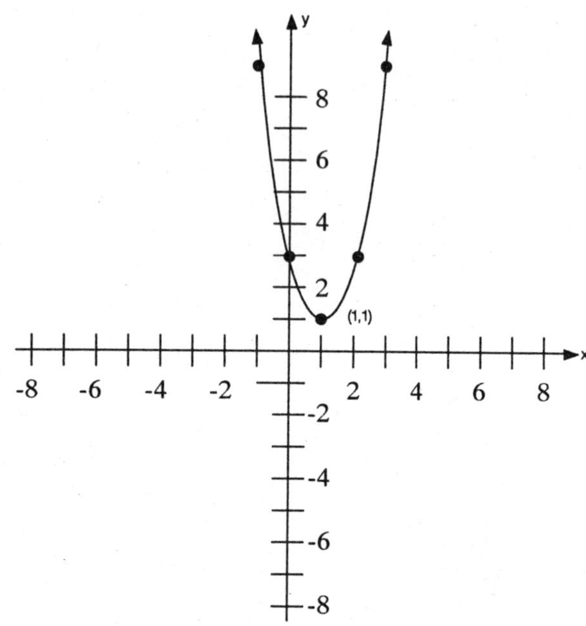

There are no x-intercepts.

Chapter 11 Practice Test

1. $(x + 5)(x - 2) = 18$

$x^2 + 3x - 10 = 18$

$x^2 + 3x - 28 = 0$

$(x + 7)(x - 4) = 0$

$x + 7 = 0$ or $x - 4 = 0$

$x = -7$ or $x = 4$

3. $x^2 - x - 14 = 2x(x - 3)$

$x^2 - x - 14 = 2x^2 - 6x$

$0 = x^2 - 5x + 14$

$a = 1, \; b = -5, \; c = 14$

$$x = \frac{-b \pm \sqrt{b^2 - 4ac}}{2a}$$

$$x = \frac{-(-5) \pm \sqrt{(-5)^2 - 4(1)(14)}}{2(1)}$$

$$x = \frac{5 \pm \sqrt{25 - 56}}{2}$$

$$x = \frac{5 \pm \sqrt{-31}}{2}$$

Since a negative number appears under the square root, there are no real solutions.

5. $(x + 5)^2 = 10x$

$x^2 + 10x + 25 = 10x$

$x^2 + 25 = 0$

$x^2 = -25$

$x = \pm\sqrt{-25}$

Since square roots of negative numbers are not real, the equation has no real solution.

7. $5x^2 + 15 = 30x$

$5x^2 - 30x + 15 = 0$

$5(x^2 - 6x + 3) = 0$

$x^2 - 6x + 3 = 0$

$a = 1, \; b = -6, \; c = 3$

$$x = \frac{-b \pm \sqrt{b^2 - 4ac}}{2a}$$

$$x = \frac{-(-6) \pm \sqrt{(-6)^2 - 4(1)(3)}}{2(1)}$$

$$x = \frac{6 \pm \sqrt{36 - 12}}{2}$$

$$x = \frac{6 \pm \sqrt{24}}{2}$$

$$x = \frac{6 \pm 2\sqrt{6}}{2}$$

$$x = \frac{\cancel{2}(3 \pm \sqrt{6})}{\cancel{2}} = 3 \pm \sqrt{6}$$

9.

$$3x^2 - 12x = 7$$

$$x^2 - 4x = \frac{7}{3}$$

$$\underline{\quad +4 \qquad +4 \quad}$$

$$x^2 - 4x + 4 = \frac{7}{3} + 4$$

$$(x-2)^2 = \frac{19}{3}$$

$$x - 2 = \pm\sqrt{\frac{19}{3}}$$

$$x - 2 = \pm\frac{\sqrt{19}}{\sqrt{3}}$$

$$x - 2 = \pm\frac{\sqrt{19}\sqrt{3}}{\sqrt{3}\sqrt{3}}$$

$$x - 2 = \pm\frac{\sqrt{57}}{3}$$

$$x = 2 \pm \frac{\sqrt{57}}{3}$$

$$x = \frac{6 \pm \sqrt{57}}{3}$$

$$3x^2 - 12x = 7$$

$$3x^2 - 12x - 7 = 0$$

$a = 3, \ b = -12, \ c = -7$

$$x = \frac{-b \pm \sqrt{b^2 - 4ac}}{2a}$$

$$x = \frac{-(-12) \pm \sqrt{(-12)^2 - 4(3)(-7)}}{2(3)}$$

$$x = \frac{12 \pm \sqrt{144 + 84}}{6}$$

$$x = \frac{12 \pm \sqrt{228}}{6}$$

$$x = \frac{12 \pm 2\sqrt{57}}{6}$$

$$x = \frac{2(6 \pm \sqrt{57})}{6}$$

$$x = \frac{6 \pm \sqrt{57}}{3}$$

1. $y = 6x - x^2$

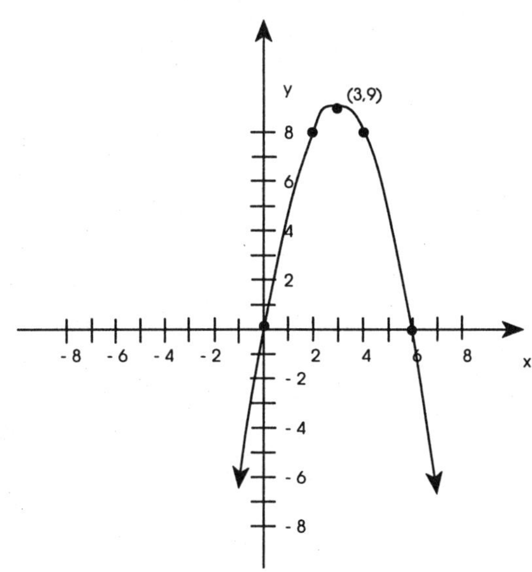

$0 = 6x - x^2$

$0 = x(6 - x)$

$0 = x$ or $0 = 6 - x$

$0 = x$ or $6 = x$

The x-intercepts are 0 and 6.

CUMULATIVE REVIEW CHAPTERS 10–11

1. $\sqrt{36x^{16}y^{12}} = \sqrt{36}\sqrt{x^{16}}\sqrt{y^{12}} = 6x^8y^6$

3. $\frac{7}{\sqrt{6}} = \frac{7\sqrt{6}}{\sqrt{6}\sqrt{6}} = \frac{7\sqrt{6}}{6}$

5. $\frac{20}{3-\sqrt{5}} = \frac{20\left(3+\sqrt{5}\right)}{(3-\sqrt{5})\left(3+\sqrt{5}\right)}$

$= \frac{20\left(3+\sqrt{5}\right)}{9-5} = \frac{\overset{5}{\cancel{20}}\left(3+\sqrt{5}\right)}{\cancel{4}\,1}$

$= 5\left(3+\sqrt{5}\right)$

7. $\sqrt{120} = \sqrt{4}\sqrt{30} = 2\sqrt{30}$

9. $\sqrt{45t} - \sqrt{20t}$

$= \sqrt{9\cdot 5t} - \sqrt{4\cdot 5t}$

$= \sqrt{9}\sqrt{5t} - \sqrt{4}\sqrt{5t}$

$= 3\sqrt{5t} - 2\sqrt{5t}$

$= \sqrt{5t}$

11. $\sqrt{\frac{3}{7}} + \sqrt{21} = \frac{\sqrt{3}}{\sqrt{7}} + \sqrt{21}$

$= \frac{\sqrt{3}\sqrt{7}}{\sqrt{7}\sqrt{7}} + \sqrt{21}$

$= \frac{\sqrt{21}}{7} + \sqrt{21}$

$= \left(\frac{1}{7}+1\right)\sqrt{21}$

$= \frac{8}{7}\sqrt{21}$

13. $\sqrt{27x^6y^5} - xy\sqrt{12x^4y^3}$

$= \sqrt{9x^6y^4\cdot 3y} - xy\sqrt{4x^4y^2\cdot 3y}$

$= \sqrt{9x^6y^4}\sqrt{3y} - xy\sqrt{4x^4y^2}\sqrt{3y}$

$= 3x^3y^2\sqrt{3y} - xy\left(2x^2y\right)\sqrt{3y}$

$= 3x^3y^2\sqrt{3y} - 2x^3y^2\sqrt{3y}$

$= x^3y^2\sqrt{3y}$

15. $\left(3\sqrt{2}-4\sqrt{5}\right)\left(4\sqrt{2}-2\sqrt{5}\right)$

$$= \left(3\sqrt{2}\right)\left(4\sqrt{2}\right)-\left(3\sqrt{2}\right)\left(2\sqrt{5}\right)-\left(4\sqrt{5}\right)\left(4\sqrt{2}\right)+\left(4\sqrt{5}\right)\left(2\sqrt{5}\right)$$

$$= 3\cdot 4\sqrt{2}\sqrt{2}-3\cdot 2\sqrt{2}\sqrt{5}-4\cdot 4\sqrt{5}\sqrt{2}+4\cdot 2\sqrt{5}\sqrt{5}$$

$$= 12\cdot 2-6\sqrt{10}-16\sqrt{10}+8\cdot 5$$

$$= 24-6\sqrt{10}-16\sqrt{10}+40$$

$$= 64-22\sqrt{10}$$

17. $\dfrac{15}{\sqrt{7}-\sqrt{2}}-\dfrac{10}{\sqrt{2}}$

$$= \frac{15\left(\sqrt{7}+\sqrt{2}\right)}{\left(\sqrt{7}-\sqrt{2}\right)\left(\sqrt{7}+\sqrt{2}\right)}-\frac{10\sqrt{2}}{\sqrt{2}\sqrt{2}}$$

$$= \frac{15\left(\sqrt{7}+\sqrt{2}\right)}{7-2}-\frac{10\sqrt{2}}{2}$$

$$= \frac{15\left(\sqrt{7}+\sqrt{2}\right)}{5}-5\sqrt{2}$$

$$= 3\left(\sqrt{7}+\sqrt{2}\right)-5\sqrt{2}$$

$$= 3\sqrt{7}+3\sqrt{2}-5\sqrt{2}=3\sqrt{7}-2\sqrt{2}$$

19. $x^2-3x=10$

$$x^2-3x-10=0$$

$$(x-5)(x+2)=0$$

$x-5=0$ or $x+2=0$

$x=5$ or $x=-2$

21. $\dfrac{x}{x-2}=\dfrac{x+1}{x-3}$

$$\frac{(x-2)(x-3)}{1}\left(\frac{x}{x-2}\right)=\frac{(x-2)(x-3)}{1}\left(\frac{x+1}{x-3}\right)$$

$$(x-3)x=(x-2)(x+1)$$

$$x^2-3x=x^2+x-2x-2$$

$$x^2-3x=x^2-x-2$$

$$-3x=-x-2$$

$$-2x=-2$$

$$x=1$$

23. $3x^2-6x-2=x^2-x-4$

$$2x^2-6x-2=-x-4$$

$$2x^2-5x-2=-4$$

$$2x^2-5x+2=0$$

$$(2x-1)(x-2)=0$$

$2x-1=0$ or $x-2=0$

$2x = 1$ or $x = 2$

$x = \frac{1}{2}$ or $x = 2$

25. $(x+3)(x-4) = 8$

$x^2 - 4x + 3x - 12 = 8$

$x^2 - x - 12 = 8$

$x^2 - x - 20 = 0$

$(x-5)(x+4) = 0$

$x - 5 = 0$ or $x + 4 = 0$

$x = 5$ or $x = -4$

27. $(2x-3)(x-4) = (x-2)(2x-1)$

$2x^2 - 8x - 3x + 12 = 2x^2 - x - 4x + 2$

$2x^2 - 11x + 12 = 2x^2 - 5x + 2$

$-11x + 12 = -5x + 2$

$12 = 6x + 2$

$10 = 6x$

$\frac{5}{3} = x$

29. $(s-2)^2 = 10$

$s - 2 = \pm\sqrt{10}$

$s = 2 \pm \sqrt{10}$

31. $\sqrt{x} - 3 = 4$

$\sqrt{x} = 7$

$\left(\sqrt{x}\right)^2 = 7^2$

$x = 49$

33. $3\sqrt{t+4} - 1 = 7$

$3\sqrt{t+4} = 8$

$\sqrt{t+4} = \frac{8}{3}$

$\left(\sqrt{t+4}\right)^2 = \left(\frac{8}{3}\right)^2$

$t + 4 = \frac{64}{9}$

$t = \frac{28}{9}$

35. $2x^2 - 10x + 6 = 0$

$$2x^2 - 10x = -6$$

$$x^2 - 5x = -3$$

$$x^2 - 5x + \frac{25}{4} = -3 + \frac{25}{4}$$

$$\left(x - \frac{5}{2}\right)^2 = \frac{13}{4}$$

$$x - \frac{5}{2} = \pm\sqrt{\frac{13}{4}}$$

$$x - \frac{5}{2} = \pm\frac{\sqrt{13}}{2}$$

$$x = \frac{5}{2} \pm \frac{\sqrt{13}}{2}$$

$$x = \frac{5 \pm \sqrt{13}}{2}$$

37. $y = (2x - 3)^2$

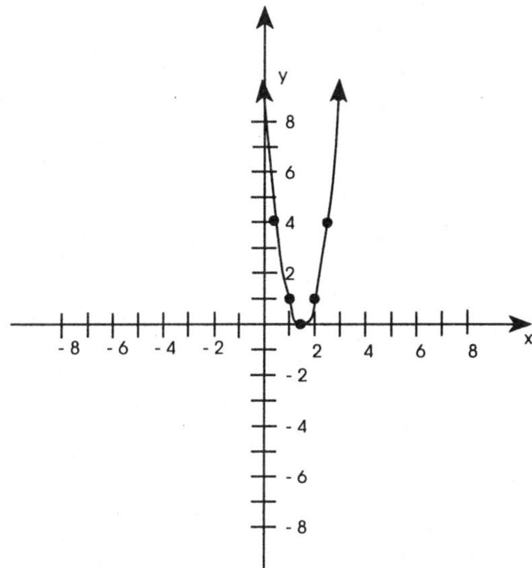

$$0 = (2x - 3)^2$$

$$0 = 2x - 3$$

$$3 = 2x$$

$$\frac{3}{2} = x$$

The x-intercept is $\frac{3}{2}$.

39. $y = x^2 - x - 1$

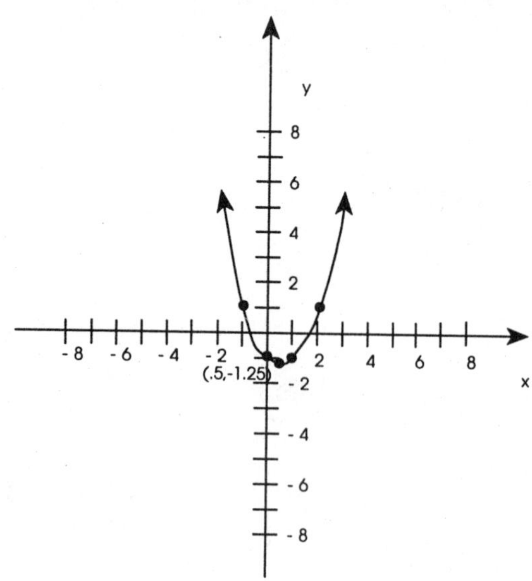

$$0 = x^2 - x - 1$$

$$a = 1,\ b = -1,\ c = -1$$

$$x = \frac{-b \pm \sqrt{b^2 - 4ac}}{2a}$$

$$x = \frac{-(-1) \pm \sqrt{(-1)^2 - 4(1)(-1)}}{2(1)}$$

$$x = \frac{1 \pm \sqrt{5}}{2}$$

The x-intercepts are $\frac{1+\sqrt{5}}{2} \approx 1.6$ and $\frac{1-\sqrt{5}}{2} \approx -.6$

41. Let x = length of the third side of the triangle.

By the Pythagorean Theorem,

$$8^2 + x^2 = 17^2$$
$$64 + x^2 = 289$$
$$x^2 = 225$$
$$x = \pm 15$$

We reject $x = -15$, since length cannot be negative. So $x = 15$. Since the length of the third side is 15, the perimeter of the triangle is $8 + 15 + 17 = 40$.

•43. Let x = Kim's average speed on the way there (in mph)

$x - 30$ = Kim's average speed on the way back (in mph)

$$\frac{280}{x}+\frac{280}{x-30}=11$$

$$x(x-30)\left(\frac{280}{x}+\frac{280}{x-30}\right)=x(x-30)(11)$$

$$\frac{\cancel{x}(x-30)}{1}\cdot\frac{280}{\cancel{x}}+x\cancel{(x-3)}\left(\frac{280}{\cancel{x-30}}\right)=11x(x-30)$$

$$280(x-30)+280x=11x^2-330x$$

$$280x-8400+280x=11x^2-330x$$

$$560x-8400=11x^2-330x$$

$$0=11x^2-890x+8400$$

$$0=(x-70)(11x-120)$$

$$0=x-70 \text{ or } 0=11x-120$$

$$70=x \text{ or } 120=11x$$

$$70=x \text{ or } \frac{120}{11}=x$$

$$70=x \text{ or } 10.9=x$$

We reject x = 10.9, since this value would make Kim's average speed negative on the way back. So x = 70.

Therefore, Kim's average speed on the way there is 70 mph.

Cumulative Practice Test: Chapters 10–11

1. $\sqrt{72}=\sqrt{36\cdot 2}$
$=\sqrt{36}\sqrt{2}$
$=6\sqrt{2}$

3. $3\sqrt{20}-4\sqrt{45}$
$=3\sqrt{4\cdot 5}-4\sqrt{9\cdot 5}$
$=3\sqrt{4}\sqrt{5}-4\sqrt{9}\sqrt{5}$
$=3(2\sqrt{5})-4(3\sqrt{5})$
$=6\sqrt{5}-12\sqrt{5}$
$=-6\sqrt{5}$

5. $\frac{8}{\sqrt{5}}=\frac{8\sqrt{5}}{\sqrt{5}\sqrt{5}}=\frac{8\sqrt{5}}{5}$

7. $\dfrac{24}{4-\sqrt{7}} = \dfrac{24\left(4+\sqrt{7}\right)}{\left(4-\sqrt{7}\right)\left(4+\sqrt{7}\right)}$

$= \dfrac{24\left(4+\sqrt{7}\right)}{16-7}$

$= \dfrac{\overset{8}{\cancel{24}}\left(4+\sqrt{7}\right)}{\cancel{9}\,3}$

$= \dfrac{8\left(4+\sqrt{7}\right)}{3}$

9. $\left(5\sqrt{t}+\sqrt{3}\right)\left(2\sqrt{t}-4\sqrt{3}\right)$

$= \left(5\sqrt{t}\right)\left(2\sqrt{t}\right)-\left(5\sqrt{t}\right)\left(4\sqrt{3}\right)+\sqrt{3}\left(2\sqrt{t}\right)-\sqrt{3}\left(4\sqrt{3}\right)$

$= 10t - 20\sqrt{3t} + 2\sqrt{3t} - 12$

$= 10t - 18\sqrt{3t} - 12$

11. $(x+3)(x-5) = 9$

$x^2 - 5x + 3x - 15 = 9$

$x^2 - 2x - 15 = 9$

$x^2 - 2x - 24 = 0$

$(x-8)(x+6) = 0$

$x - 8 = 0$ or $x + 6 = 0$

$x = 8$ or $x = -6$

13. $\dfrac{x}{2x-5} = \dfrac{x+4}{2x}$

$\dfrac{(\cancel{2x-5})(2x)}{1}\left(\dfrac{x}{\cancel{2x-5}}\right) = \dfrac{(2x-5)(\cancel{2x})}{1}\left(\dfrac{x+4}{\cancel{2x}}\right)$

$(2x)(x) = (2x-5)(x+4)$

$2x^2 = 2x^2 + 8x - 5x - 20$

$2x^2 = 2x^2 + 3x - 20$

$0 = 3x - 20$

$20 = 3x$

$\dfrac{20}{3} = x$

15. $2x^2 + 1 = 5x$

$2x^2 - 5x + 1 = 0$

$a = 2,\ b = -5,\ c = 1$

$$x = \frac{-b \pm \sqrt{b^2 - 4ac}}{2a}$$

$$x = \frac{-(-5) \pm \sqrt{(-5)^2 - 4(2)(1)}}{2(2)}$$

$$x = \frac{5 \pm \sqrt{17}}{4}$$

17.
$$(2x+5)^2 = 4x(x-3)$$
$$4x^2 + 20x + 25 = 4x^2 - 12x$$
$$20x + 25 = -12x$$
$$25 = -32x$$
$$-\frac{25}{32} = x$$

19. Let x = length of the rectangle (in inches)

By the Pythagorean Theorem,

$$x^2 + 5^2 = 10^2$$
$$x^2 + 25 = 100$$
$$x^2 = 75$$
$$x = \pm\sqrt{75}$$
$$x = \pm 5\sqrt{3}$$

We reject $x = -5\sqrt{3}$, since length cannot be negative. So $x = 5\sqrt{3}$. Then the area of the rectangle is (length) $\times$ (width) $= \left(5\sqrt{3}\right)(5) = 25\sqrt{3}$ sq. in.

21. $y = x^2 + 4x - 5$

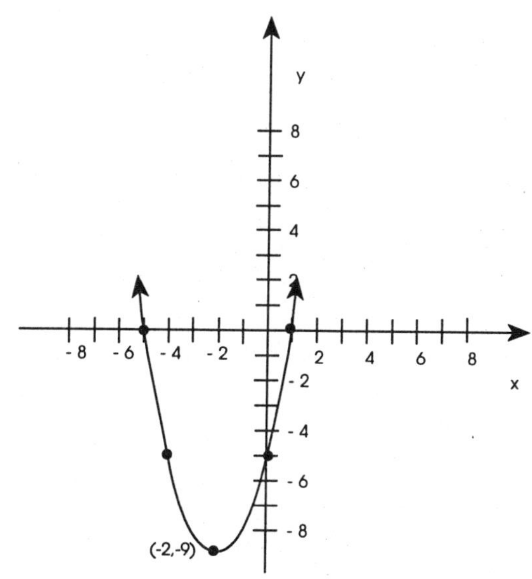

$0 = x^2 + 4x - 5$
$0 = (x + 5)(x - 1)$

$0 = x + 5$ or $0 = x - 1$

$-5 = x$ or $1 = x$

The x-intercepts are –5 and 1.

APPENDIX A
REVIEW OF ARITHMETIC

Exercises A1

1. $\frac{6}{9}=\frac{2}{3}=\frac{60}{90}$

5. $\frac{33}{44}=\frac{3}{4}=\frac{66}{88}$

9. $1=\frac{7}{7}=\frac{12}{12}$

13. $\frac{6}{42}=\frac{6\cdot 1}{6\cdot 7}=\frac{1}{7}$

17. $\frac{20}{49}$ cannot be reduced.

21. $\frac{90}{15}=\frac{15\cdot 6}{15\cdot 1}=\frac{6}{1}=6$

25. $\frac{81}{100}$ cannot be reduced.

29. $\frac{15}{7}=\frac{15\cdot 3}{7\cdot 3}=\frac{45}{21}$ So ? = 21.

33. $\frac{1}{18}=\frac{1\cdot 6}{18\cdot 6}=\frac{6}{108}$ So ? = 6

37. $\frac{54}{60}=\frac{\cancel{6}\cdot 9}{\cancel{6}\cdot 10}=\frac{9}{10}$ hit the bull's-eye

$\frac{6}{60}=\frac{\cancel{6}\cdot 1}{\cancel{6}\cdot 10}=\frac{1}{10}$ missed the bull's eye

39. $\frac{24}{28}=\frac{\cancel{4}\cdot 6}{\cancel{4}\cdot 7}=\frac{6}{7}$ was correct

$\frac{4}{28}=\frac{\cancel{4}\cdot 1}{\cancel{4}\cdot 7}=\frac{1}{7}$ was incorrect

Exercises A2

1. $\frac{2}{3}\cdot\frac{5}{7}=\frac{2\cdot 5}{3\cdot 7}=\frac{10}{21}$

5. $\frac{\cancel{8}^{2}}{_{3}\cancel{9}}\cdot\frac{\cancel{3}}{\cancel{4}}=\frac{2}{3}$

9. $\frac{5}{6}\div\frac{4}{15}=\frac{5}{\cancel{6}_{2}}\cdot\frac{\cancel{15}^{5}}{4}=\frac{25}{8}$

13. $\frac{15}{18}\div\frac{24}{25}=\frac{\cancel{15}^{5}}{18}\cdot\frac{25}{\cancel{24}_{8}}=\frac{125}{144}$

15. $\frac{\cancel{3}}{\cancel{5}}\cdot\frac{\cancel{5}}{\cancel{3}}=\frac{1}{1}=1$

17. $\frac{3}{5}\div\frac{5}{3}=\frac{3}{5}\cdot\frac{3}{5}=\frac{9}{25}$

21. $18\div\frac{3}{2}=\frac{\cancel{18}^{6}}{1}\cdot\frac{2}{\cancel{3}}=\frac{12}{1}=12$

25. $\frac{\cancel{28}^{7}}{\cancel{45}_{5}}\cdot\frac{\cancel{63}^{7}}{\cancel{40}_{10}}=\frac{49}{50}$

29. $\frac{12}{25}\cdot\frac{5}{2}\div\frac{6}{5}=$

$$\left(\frac{\overset{6}{\cancel{12}}}{\underset{5}{\cancel{25}}}\cdot\frac{\cancel{5}}{\cancel{2}}\right)\div\frac{6}{5}=$$

$$\frac{6}{5}\div\frac{6}{5}=$$

$$\frac{\cancel{6}}{\cancel{5}}\cdot\frac{\cancel{5}}{\cancel{6}}=\frac{1}{1}=1$$

31. The reciprocal of $\frac{5}{9}$ is $\frac{9}{5}$.

33. $\frac{4}{7}$ is the reciprocal of $\frac{7}{4}$.

35. $$\frac{4}{\cancel{5}}\cdot\frac{\overset{4,500}{\cancel{22,500}}}{1}=\frac{18,000}{1}$$

$$=18,000$$

Allison earns $18,000.

39. $$\frac{1}{\cancel{4}}\cdot\frac{\overset{75}{\cancel{300}}}{1}=\frac{75}{1}=75$$

$$\frac{3}{\cancel{5}}\cdot\frac{\overset{60}{\cancel{300}}}{1}=\frac{180}{1}=180$$

So 75 hours are for assembly and 180 hours are for testing. Therefore, 300 – (75 + 180) = 300 – 255 = 45 hours are left for packing.

Exercises A3

1. $$\frac{2}{7}+\frac{3}{7}=\frac{2+3}{7}=\frac{5}{7}$$

5. $$\frac{5}{12}+\frac{1}{12}=\frac{5+1}{12}=\frac{6}{12}$$

$$=\frac{\cancel{6}\cdot 1}{\cancel{6}\cdot 2}=\frac{1}{2}$$

9. $$\frac{1}{2}+\frac{1}{4}=\frac{1\cdot 2}{2\cdot 2}+\frac{1}{4}$$

$$=\frac{2}{4}+\frac{1}{4}=\frac{2+1}{4}=\frac{3}{4}$$

11. $$\frac{5}{6}-\frac{3}{8}=\frac{5\cdot 4}{6\cdot 4}-\frac{3\cdot 3}{8\cdot 3}$$

$$=\frac{20}{24}-\frac{9}{24}=\frac{20-9}{24}=\frac{11}{24}$$

15. $\frac{7}{12} + \frac{1}{3} = \frac{7}{12} + \frac{1 \cdot 4}{3 \cdot 4}$

$= \frac{7}{12} + \frac{4}{12} = \frac{7+4}{12} = \frac{11}{12}$

17. $\frac{1}{2} + \frac{1}{3} + \frac{1}{4} = \frac{1 \cdot 6}{2 \cdot 6} + \frac{1 \cdot 4}{3 \cdot 4} + \frac{1 \cdot 3}{4 \cdot 3}$

$= \frac{6}{12} + \frac{4}{12} + \frac{3}{12}$

$= \frac{6+4+3}{12} = \frac{13}{12}$

21. $\frac{3}{28} + \frac{2}{35} = \frac{3 \cdot 5}{28 \cdot 5} + \frac{2 \cdot 4}{35 \cdot 4}$

$= \frac{15}{140} + \frac{8}{140}$

$= \frac{15+8}{140} = \frac{23}{140}$

25. $5 \div \frac{3}{5} = \frac{5}{1} \cdot \frac{5}{3} = \frac{25}{3}$

29. $12\frac{4}{5} = 12 + \frac{4}{5}$

$= \frac{12}{1} + \frac{4}{5}$

$= \frac{12 \cdot 5}{1 \cdot 5} + \frac{4}{5}$

$= \frac{60}{5} + \frac{4}{5}$

$= \frac{60+4}{5} = \frac{64}{5}$

33. $\frac{43}{6} = \frac{42+1}{6}$

$= \frac{42}{6} + \frac{1}{6}$

$= 7 + \frac{1}{6} = 7\frac{1}{6}$

37. $\frac{5}{12} = \frac{5 \cdot 5}{12 \cdot 5} = \frac{25}{60}$

$\frac{7}{15} = \frac{7 \cdot 4}{15 \cdot 4} = \frac{28}{60}$

$\frac{9}{20} = \frac{9 \cdot 3}{20 \cdot 3} = \frac{27}{60}$

Therefore, $\frac{5}{12}$ is the smallest of the three fractions.

41. $240 \div 22\frac{1}{2} = \frac{240}{1} \div \frac{45}{2}$

$= \frac{\overset{16}{\cancel{240}}}{1} \cdot \frac{2}{\underset{3}{\cancel{45}}}$

$= \frac{32}{3}$ or $10\frac{2}{3}$

On a 240-mile trip, the car uses $10\frac{2}{3}$ gallons of gasoline.

45. $27\frac{1}{2} \div 4\frac{1}{2} = \frac{55}{2} \div \frac{9}{2}$

$= \frac{55}{\cancel{2}} \cdot \frac{\cancel{2}}{9}$

$= \frac{55}{9}$ or $6\frac{1}{9}$

Raul's rate of speed is $6\frac{1}{9}$ miles per hour.

Exercises A4

1.
```
   4.7
   3.5
 +21.7
  29.9
```

5.
```
  21.620
   4.100
 +57.236
  82.956
```

9.
```
  9.27
 -7.85
  1.42
```

15.
```
   13.05
  x 2.63
    3915
   7830
  2610
 34.3215
```

19.
```
     7.92
  5)39.60
    35
     46
     45
      10
      10
       0
```

23. $.16\overline{)3.28}$ becomes $16\overline{)328.0}$ = 20.5

```
     20.5
16)328.0
   32
    80
    80
     0
```

27.
```
   .032
 x  .05
 .00160
```

33. $.015\overline{)3.900}$ becomes $15\overline{)3900}$ = 260

```
      260
15)3900
   30
    90
    90
     00
     00
      0
```

37. $.12\overline{)102}$ becomes $12\overline{)10200}$ = 850

```
      850
12)10200
   96
    60
    60
     00
     00
      0
```

The customer used 850 kilowatt hours.

41.
```
   $18.65
  × 35.4
     7460
    9325
   5595
 $660.210
```

The mixture used for this process costs $660.21.

Exercises A5

3. .78 = 78%

7. 9% = .09

11. 28% = .28

15. 2 = 200%

19. .007 = .7%

21. 62.4% = .624

25. 30% of 70 = (.30)(70) = 21

27. 7.2% of 35 = (.072)(35) = 2.52

29. .8% of 5 = (.008)(5) = .04

33. $\frac{5}{8} = .625 = 62.5\%$

35. $\frac{8}{20} = \frac{2}{5} = .4 = 40\%$

39. $\frac{24.6}{.30}$ becomes $.30\overline{)24.6}$, which becomes $3\overline{)246}$

$$\begin{array}{r} 82 \\ 3\overline{)246} \\ \underline{24} \\ 6 \\ \underline{6} \\ 0 \end{array}$$

43. 6% of 12,500 = (.06)(12,500)
= 750

So in 5 years, the population of the town will be 12,500 +750 = 13,250.

47. 52% of 8,600 = (.52)(8,600)
= 4,472

48% of 9,250 = (.48)(9,250)
= 4,440

Therefore, more votes were cast in this town in 1984.

APPENDIX B
USING A HAND-HELD CALCULATOR

Exercises

1. 861.55

5. 65.5

9. 4,847.04

13. 30.45

17. 27.62

21 .06

25. (186,000)(60) = 11,160,000 miles

27. 15% of \$12 = (.15)(\$12) = \$1.80

    The wholesaler marks up the price to \$12 + \$1.80 = \$13.80

    12% of \$13.80 = (.12)(\$13.80) = \$1.66

    So the consumer pays \$13.80 + \$1.66 = \$15.46

SUPPLEMENTARY CHAPTER ON GEOMETRY

Exercises G1

1. $\angle AEB$ and $\angle BED$

   $\angle EAD$ and $\angle DAB$

3. (a) 60° (b) 30° (c) 45° (d) 72° (e) 1°

5. Let x = the angle

   180 – x = its supplement

   $$\begin{aligned} x &= 3(180 - x) - 28 \\ x &= 540 - 3x - 28 \\ x &= 512 - 3x \\ 4x &= 512 \\ x &= 128 \end{aligned}$$

   The angle is 128°

7. Let x = the angle

   90 – x = its complement

   $$\begin{aligned} x &= 2(90 - x) + 12 \\ x &= 180 - 2x + 12 \\ x &= 192 - 2x \\ 3x &= 192 \\ x &= 64 \end{aligned}$$

   The angle is 64°

9. $$\begin{aligned} x + (3x + 10) &= 90 \\ 4x + 10 &= 90 \\ 4x &= 80 \\ x &= 20 \end{aligned}$$

11. $$\begin{aligned} x + 90 + (3x + 10) &= 180 \\ 4x + 100 &= 180 \\ 4x &= 80 \\ x &= 20 \end{aligned}$$

13. $\angle 1 = 145°$ (supplement of $\angle 4$)

    $\angle 2 = 35°$ ($\angle 2$ and $\angle 4$ are vertical angles)

    $\angle 3 = 145°$ (supplement of $\angle 4$)

15. $\angle 1 = 65°$ ($\angle 1 + 90 + 25 = 180$)

    $\angle 2 = 115°$ (supplement of $\angle 1$)

    $\angle 3 = 65°$ ($\angle 1$ and $\angle 3$ are vertical angles)

17. $\angle 1 = \angle 3 = \angle 6 = \angle 8 = 55°$

    $\angle 2 = \angle 5 = \angle 7 = 125°$

19. $$\begin{aligned} 2x + (x + 12) &= 180 \\ 3x + 12 &= 180 \\ 3x &= 168 \\ x &= 56 \end{aligned}$$

21. $\angle 1 = \angle 4 = \angle 5 = \angle 7 = \angle 9 = \angle 11 = \angle 14 = \angle 16 = 140°$

$\angle 3 = \angle 6 = \angle 8 = \angle 10 = \angle 12 = \angle 13 = \angle 15 = 40°$

23. $\angle 1 = 50°$, $\angle 2 = 130°$, $\angle 3 = 50°$, $\angle 4 = 130°$

Exercises G2

1. $\angle C = 180 - (34 + 23) = 180 - 57 = 123°$

3. Since AC = BC, $\angle A = \angle B$. Let x be the number of degrees in $\angle A$ and in $\angle B$.

$$\begin{aligned} x + x + 40 &= 180 \\ 2x + 40 &= 180 \\ 2x &= 140 \\ x &= 70 \end{aligned}$$

So $\angle A = 70°$ and $\angle B = 70°$

5. $$\begin{aligned} x + 4x + (2x - 16) &= 180 \\ 7x - 16 &= 180 \\ 7x &= 196 \\ x &= 28 \end{aligned}$$

7. $$\begin{aligned} x + (2x + 20) + (2x + 20) &= 180 \\ 5x + 40 &= 180 \\ 5x &= 140 \\ x &= 28 \end{aligned}$$

9. $$\begin{aligned} x + 23 + 90 &= 180 \\ x + 113 &= 180 \\ x &= 67 \end{aligned}$$

11. $$\begin{aligned} x + (2x + 12) + 90 &= 180 \\ 3x + 102 &= 180 \\ 3x &= 78 \\ x &= 26 \end{aligned}$$

13. $\frac{|\overline{DE}|}{15} = \frac{8}{5}$, so $|\overline{DE}| = 24$

$\frac{|\overline{EF}|}{15} = \frac{4}{5}$, so $|\overline{EF}| = 12$

15. $\frac{|\overline{AC}|}{24} = \frac{10}{12}$, so $|\overline{AC}| = 20$

17. $\frac{|\overline{EA}|}{|\overline{EC}|} = \frac{|\overline{DB}|}{|\overline{DC}|}$

$\frac{|\overline{EA}|}{18} = \frac{6}{10}$, so $|\overline{EA}| = 10.8$

19. $\left|\overline{AB}\right|^2 = 3^2 + 4^2$

$\left|\overline{AB}\right|^2 = 9 + 16$

$\left|\overline{AB}\right|^2 = 25$

$\left|\overline{AB}\right| = 5$

21. $\left|\overline{AC}\right|^2 + 4^2 = 9^2$

$\left|\overline{AC}\right|^2 + 16 = 81$

$\left|\overline{AC}\right|^2 = 65$

$\left|\overline{AC}\right| = \sqrt{65}$

23. Let x = $\left|\overline{AC}\right| = \left|\overline{BC}\right|$

$x^2 + x^2 = 8^2$

$2x^2 = 64$

$x^2 = 32$

$x = \sqrt{32} = 4\sqrt{2}$

So $\left|\overline{AC}\right| = 4\sqrt{2}$ and $\left|\overline{BC}\right| = 4\sqrt{2}$

25. $\dfrac{\left|\overline{ED}\right|}{\left|\overline{DC}\right|} = \dfrac{\left|\overline{BA}\right|}{\left|\overline{AC}\right|}$

$\dfrac{\left|\overline{ED}\right|}{8} = \dfrac{6}{4}$, so $\left|\overline{ED}\right| = 12$

27. $\left|\overline{AC}\right|^2 = 4^2 + 8^2$

$\left|\overline{AC}\right|^2 = 16 + 64$

$\left|\overline{AC}\right|^2 = 80$

$\left|\overline{AC}\right| = \sqrt{80} = 4\sqrt{5}$

29. Let x = distance between top of ladder and ground (in feet)

$x^2 + 10^2 = 30^2$

$x^2 + 100 = 900$

$x^2 = 800$

$x = \sqrt{800} = 20\sqrt{2}$

The top of the ladder is $20\sqrt{2}$ (approximately 28.3) feet above the ground.

31. $x + 63 + 90 = 180$

$x + 153 = 180$

$x = 27$

33.

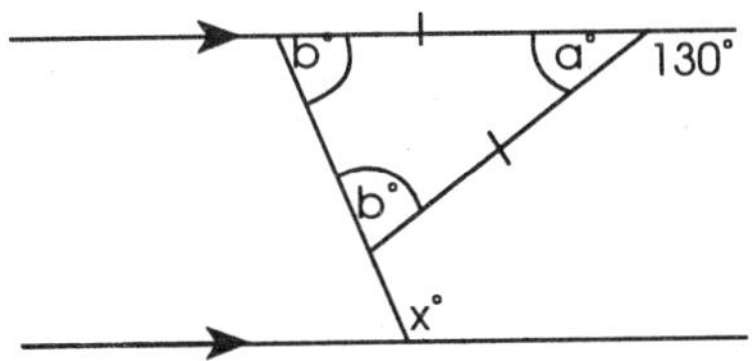

Referring to the sketch, a = 50, since it is the supplement of 130. Then

$$\begin{aligned} 50 + b + b &= 180 \\ 50 + 2b &= 180 \\ 2b &= 130 \\ b &= 65 \end{aligned}$$

Since the lines are parallel,

$$\begin{aligned} x + b &= 180 \\ x + 65 &= 180 \\ x &= 115 \end{aligned}$$

35.

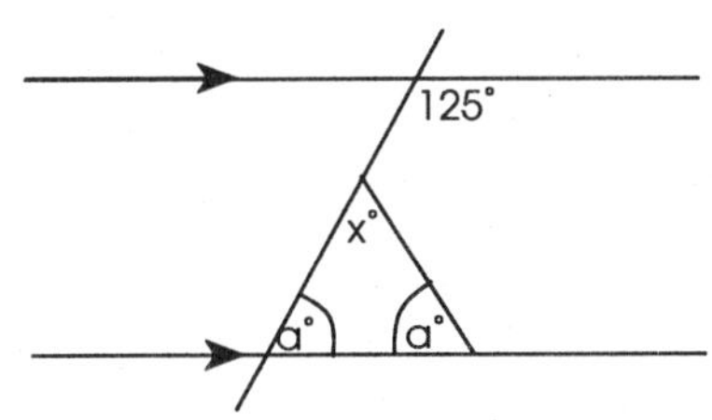

Referring to the sketch, a + 125 = 180 (since the lines are parallel.) So a = 55. Then

$$\begin{aligned} 55 + 55 + x &= 180 \\ 110 + x &= 180 \\ x &= 70 \end{aligned}$$

37.

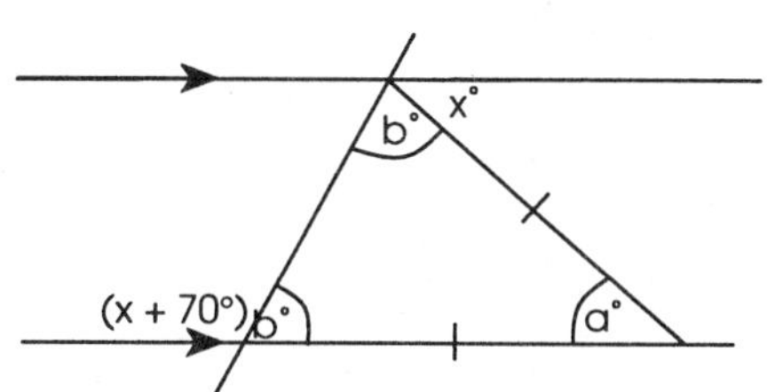

Referring to the sketch, a = x (since the lines are parallel) and b + x = x + 70 (for the same reason)

From the second equation, we get b = 70. Then a + 70 + 70 = 180, so a = 40. Therefore, x = 40

Exercises G3

1. $\angle C = 360 - (110 + 130 + 40) = 360 - 280 = 80°$

3. $\angle C = 60°$ (opposite angles in a parallelogram are equal)
   $\angle B = 120°$ (adjacent angles in a parallelogram are supplementary)
   $\angle D = 120°$ (opposite angles in a parallelogram are equal)

5. $$\begin{aligned} 2x + 3 &= x + 8 \\ x + 3 &= 8 \\ x &= 5 \end{aligned}$$

7. $$\begin{aligned} 5x - 18 + 6x &= 180 \\ 11x - 18 &= 180 \\ 11x &= 198 \\ x &= 18 \end{aligned}$$

9. Let x = length of the diagonal (in cm)

   $$\begin{aligned} x^2 &= 5^2 + 8^2 \\ x^2 &= 25 + 64 \\ x^2 &= 89 \\ x &= \sqrt{89} \end{aligned}$$
   The length of the diagonal is $\sqrt{89}$ (approximately 9.4) cm.

11. Let x = length of the diagonal (in inches)

   $$\begin{aligned} x^2 &= 4^2 + 4^2 \\ x^2 &= 16 + 16 \\ x^2 &= 32 \\ x &= \sqrt{32} = 4\sqrt{2} \end{aligned}$$
   The length of the diagonal is $4\sqrt{2}$ (approximately 5.7) in.

13. Let x = length of the side (in cm)

   $$\begin{aligned} x^2 + x^2 &= 5^2 \\ 2x^2 &= 25 \\ x^2 &= \frac{25}{2} \\ x &= \sqrt{\frac{25}{2}} = \frac{\sqrt{25}}{\sqrt{2}} = \frac{5}{\sqrt{2}} = \frac{5\sqrt{2}}{2} \end{aligned}$$
   The length of the side is $\frac{5\sqrt{2}}{2}$ (approximately 3.5) cm.

15. p = 2(5) + 2(8) = 10 + 16 = 26 ft.

   A = (5)(8) = 40 sq. ft.

17. p = 2(6) + 2(12) = 12 + 24 = 36 in.

   A = (6)(12) = 72 sq. in.

19. $p = 4(3) = 12$ in.

$A = 3^2 = 9$ sq. in.

21. $A = (8)(6) = 48$

23. $A = (10)(6) = 60$

25. $A = \frac{1}{2}(8)(6) = 24$

27. $A = \frac{1}{2}(12)(4) = 24$

29. $A = \frac{1}{2}(6)(4) = 12$

31. $A = \frac{1}{2}(5+12)(6) = 51$

33. $|\overline{AD}|^2 = 5^2 + 12^2$

$|\overline{AD}|^2 = 25 + 144$

$|\overline{AD}|^2 = 169$

$|\overline{AD}| = 13$

Then $p = 2(13) + 2(15) = 56$

$A = (15)(12) = 180$

35. Let x = length of the hypotenuse

$x^2 = 5^2 + 8^2$

$x^2 = 25 + 64$

$x^2 = 89$

$x = \sqrt{89}$

Then $p = 5 + 8 + \sqrt{89} = 13 + \sqrt{89}$ (approximately 22.4)

$A = \frac{1}{2}(5)(8) = 20$

37. Let x = length of the missing leg.

$x^2 + 7^2 = 9^2$

$x^2 + 49 = 81$

$x^2 = 32$

$x = \sqrt{32} = 4\sqrt{2}$

Then $p = 7 + 9 + 4\sqrt{2} = 16 + 4\sqrt{2}$ (approximately 21.7)

$A = \frac{1}{2}(4\sqrt{2})(7) = 14\sqrt{2}$ (approximately 19.8)

39. Let x = height of the rectangle (in cm.)

$$x^2 + 8^2 = 12^2$$
$$x^2 + 64 = 144$$
$$x^2 = 80$$
$$x = \sqrt{80} = 4\sqrt{5}$$

Then $p = 2(4\sqrt{5}) + 2(8) = 8\sqrt{5} + 16$ (approximately 33.9) cm

$A = (8)(4\sqrt{5}) = 32\sqrt{5}$ (approximately 71.6) sq. cm.

41. Let x = side of the square (in mm)

$$x^2 + x^2 = 10^2$$
$$2x^2 = 100$$
$$x^2 = 50$$
$$x = \sqrt{50} = 5\sqrt{2}$$

Then $p = 4(5\sqrt{2}) = 20\sqrt{2}$ (approximately 28.3) mm

$A = (5\sqrt{2})^2 = 50$ sq. mm.

43. Drop a perpendicular from D to AB. Call its foot E. Then $|\overline{EB}| = 5$, so $|\overline{AE}| = 3$.

Using the Pythagorean Theorem,

$|\overline{AD}|^2 = 3^2 + 4^2 = 9 + 16 = 25,$

so $|\overline{AD}| = 5.$

Then $p = 8 + 4 + 5 + 5 = 22$

$A = \frac{1}{2}(5 + 8)(4) = 26$

45. $p = 10 + 10 + 10 + 26 = 56$

Drop perpendiculars from D and C to AB, calling their feet E and F respectively.

Since $|\overline{EF}| = 10$ and $|\overline{AE}| = |\overline{FB}|$, it follows that $|\overline{AE}| = 8$. Then

$$|\overline{DE}|^2 + 8^2 = 10^2$$
$$|\overline{DE}|^2 + 64 = 100$$
$$|\overline{DE}|^2 = 36$$
$$|\overline{DE}| = 6$$

Then $A = \dfrac{1}{2}(10 + 26)(6) = 108$

47. Area of ABCDE = (Area of rectangle ABDE) – (Area of triangle BCD)

$$|\overline{BD}|^2 = 6^2 + 6^2$$
$$|\overline{BD}|^2 = 36 + 36$$
$$|\overline{BD}|^2 = 72$$
$$|\overline{BD}| = \sqrt{72} = 6\sqrt{2}$$

Then Area of rectangle ABDE = $9(6\sqrt{2}) = 54\sqrt{2}$

and Area of triangle BCD = $\frac{1}{2}(6)(6) = 18$.

So Area of ABCDE = $54\sqrt{2} - 18$ (approximately 58.4)

49. From the sketch, $|\overline{DG}| = 4$ and $|\overline{FC}| = 2$

Area of AEFG = (Area of rectangle ABCD)
− (Area of triangle ADG) − (Area of triangle GCF)
− (Area of triangle FBE)

Area of rectangle ABCD = (10)(8) = 80

Area of triangle ADG = $\frac{1}{2}(8)(4) = 16$

Area of GCF = $\frac{1}{2}(6)(2) = 6$

Area of triangle FBE = $\frac{1}{2}(6)(7) = 21$

So Area of AEFG = 80 − 16 − 6 − 21 = 37

51. From the sketch, $|\overline{DF}| = 6$ and $|\overline{DG}| = 9$.

Area of the figure = (Area of rectangle ABCD)
+ (Area of triangle FDG) + (Area of triangle ECG)

Area of rectangle ABCD = (12)(8) = 96

Area of triangle FDG = $\frac{1}{2}(6)(9) = 27$

Area of triangle ECG = $\frac{1}{2}(3)(6) = 9$

So Area of the figure = 96 + 27 + 9 = 132

53. From the sketch, $|\overline{DE}| = 9$.

Area of shaded region = (Area of rectangle ACDF)
− (Area of trapezoid BCDE) − (Area of triangle EFG)

Area of rectangle ACDF = (12)(6) = 72

Area of trapezoid BCDE = $\frac{1}{2}(4 + 9)(6) = 39$

Area of triangle EFG = $\frac{1}{2}(2)(3) = 3$

So Area of shaded region = 72 − 39 − 3 = 30

Exercises G4

1. $\widehat{AB} = 35°$

3. $$\begin{aligned} x + (2x - 30) &= 360 \\ 3x - 30 &= 360 \\ 3x &= 390 \\ x &= 130 \\ \text{So } \widehat{AB} &= 130° \end{aligned}$$

5. 6π inches

7. 8π feet

9. Length of arc $= \frac{30}{360}(18\pi) = \frac{3\pi}{2}$ inches

11. Length of arc AB $= \frac{65}{360}(12\pi) = \frac{13\pi}{6}$

Then perimeter of sector $= 6 + 6 + \frac{13\pi}{6} = 12 + \frac{13\pi}{6}$

13. Length of arc AB $= \frac{10}{360}(10\pi) = \frac{5\pi}{18}$

Then perimeter of sector $= 5 + 5 + \frac{5\pi}{18} = 10 + \frac{5\pi}{18}$

15. 9π sq. in.

17. 36π sq. ft.

19. Length of $\widehat{AB} = \frac{1}{2}(20\pi) = 10\pi$

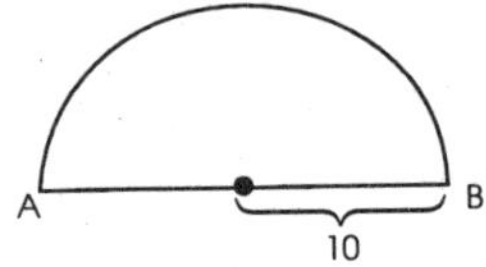

Length of $\overline{AB} = 20$

So perimeter of semicircle $= 10\pi + 20$ inches

Area of semicircle $= \frac{1}{2}(100\pi) = 50\pi$ sq. in.

21. $\frac{80}{360}(25\pi) = \frac{50\pi}{9}$ sq. in.

23. Radius of the semicircle = 3, so the area of the semicircle is $\frac{1}{2}(9\pi) = \frac{9}{2}\pi$

Since area of the rectangle is $(6)(8) = 48$, the area of the figure is $48 + \frac{9}{2}\pi$.

Length of semicircular arc $= \frac{1}{2}(6\pi) = 3\pi$

So the perimeter of the figure is $8 + 6 + 8 + 3\pi = 22 + 3\pi$

25. Area of shaded region = (Area of square ABCD) − (Area of circle)
Area of square ABCD $= 8^2 = 64$
Area of circle $= \pi(4)^2 = 16\pi$
So Area of shaded region $= (64 - 16\pi)$ sq. cm.

27. Radius of large semicircle $= 6 + 8 = 14$, so area of large semicircle $= \frac{1}{2}\left(\pi(14)^2\right) = 98\pi$

Radius of small semicircle = 6, so area of small semicircle $= \frac{1}{2}\left(\pi(6)^2\right) = 18\pi$

So area of the figure $= 98\pi - 18\pi = 80\pi$

29. $|\overline{AB}|^2 = 6^2 + 8^2$

$|\overline{AB}|^2 = 36 + 64$

$|\overline{AB}|^2 = 100$

$|\overline{AB}| = 10$

Then length of $\widehat{AB} = \frac{1}{2}(10\pi) = 5\pi$,

so the perimeter of the figure is $6 + 8 + 5\pi = 14 + 5\pi$

31. Perimeter $= |\overline{AE}| + |\overline{ED}| + |\overline{DC}| + |\overline{CB}| +$ length of $\widehat{AB} = 20 + 10 + 30 + 10 +$ length of $\widehat{AB}$

Since $|\overline{AB}| = |\overline{EB}| - |\overline{EA}| = 30 - 20 = 10$, $\widehat{AB}$ is a semicircle with diameter = 10.

So length of $\widehat{AB} = \frac{1}{2}\pi(10) = 5\pi$.

Then perimeter $= (70 + 5\pi)$cm

Area = (Area of rectangle BCDE) + (Area of semicircle) $= (30)(10) + \frac{1}{2}\pi(5)^2 = (300 + \frac{25\pi}{2})$ sq. cm.

Exercises G5

1. S.A. = 396 sq. cm. V = 504 cu. cm.

3. S.A. = 40π sq. ft. V = 32π cu. ft.

5. S.A. = 24π sq. m. V = 12π cu. m.

7. S.A. = 275.56 sq. ft. V = 762.38π cu. ft.

9. S.A. = 113.1 sq. ft. V = 113.1 cu. ft.

11. S.A. = 226.2 sq. cm. V = 251.3 cu. cm.

13. S.A. = 134.6 sq. m. V = 64.8 cu. m.

15. S.A. = 502.7 sq. ft. V = 599.5 cu. ft.

17. S.A. = 4,712.4 sq. mm. V = 29,452.4 cu. mm.

19. V = 144π cu. m.

21. $V = \frac{500\pi}{3}$ cu. cm.

23. Lateral S.A. = 60π sq. in.

25. 15 cu. yds.

27. $377.75

29. 859

31. $35.60

33. 790.3 sq. cm.

35. 11,196.6 cu. ft.

37. 360π cu. cm.

CHAPTER TESTS, CUMULATIVE REVIEWS and CUMULATIVE TESTS, AND PRACTICE FINAL

Chapter 1 Test A

1. Let A = {5, 10, 15, 20, . . . , 40} and let B = {x|x is an even integer between 13 and 31}. Answer parts (a) through (d) "true" or "false":

   (a) $24 \in A$

   (b) $24 \in B$

   (c) $12 \in B$

   (d) A has more elements than B

   (e) List the set $C = \{x | x \in A \text{ and } x \in B\}$.

In each of the following problems, compute the given expression.

2. $-8 + 1 - 2 - 4 + 5$

3. $|2 - 7| + |7 - 2|$

4. $2 + (-5) - (-3) - 4$

5. $\dfrac{(-2)(-5)(-3)}{-4-6}$

6. $\dfrac{(+1)-(-2)(-3)}{(+1)(-2)-(-3)}$

7. $(3 + 2)(5 - 9)$

8. $3 + 2 \cdot 5 - 9$

9. $-9[3 + 2(5 - 9)]$

10. $7 - 2[7 + 2(7 - 2)]$

11. Factor the number 72 into its prime factors.

Chapter 1 Test B

1. Let A = {3, 6, 9, 12, . . . , 27} and let B = {x|x is a prime less than 27}. Answer parts (a) through (d) "true" or "false":

   (a) $21 \in A$

   (b) $21 \in B$

   (c) $2 \in B$

   (d) A has fewer elements than B

   (e) List the set $C = \{x | x \in A \text{ and } x \in B\}$

In each of the following problems, compute the given expression.

2. $6 - 3 - 7 + 1 - 2$

3. $|3 - 5| - |3 + 5|$

4. $-5 - (-1) + 6 + (-2)$

5. $\dfrac{(-4)(-3)(+2)}{5-(-1)}$

6. $\dfrac{-1-(-3)(-5)}{(-1)(-3)-(-5)}$

7. $(2-4\cdot 3)-6$

8. $2-(4\cdot 3-6)$

9. $5[(2-4)(3-6)]$

10. $1+3[1-3(1+3)]$

11. Factor the number 60 into its prime factors.

Chapter 2 Test A

In problems 1–4, evaluate each of the given expressions.

1. $(-2)^6$

2. -2^6

3. $(-8+5-(-2))^7$

4. $3-5(3-5)^3$

In problems 5–8, evaluate the given expression for x = –1, y = 4, and z = –2.

5. $x+y-z$

6. $x(y-z)$

7. x^2y-xz^3

8. $|xy-z|+z$

In problems 9–13, perform the indicated operations and simplify as completely as possible.

9. $x^4-2xy^3+x^3y-3xy^3+4x^4-6x^3y$

10. $-5x^3y^2(-2xy)(xy^4)$

11. $x(xy-3y^2)+3y(xy+x)-xy(x+1)$

12. $4x^2y^2(x+2y^3)-8xy^3(xy^2)$

13. $3+[x-2(x+1)]$

14. If we let n stand for the "number," translate each of the following phrases:

    (a) three less than four times a number

    (b) seven more than three times a number is equal to one less than the number

15. Tin plates cost $4 each and copper plates cost $7 each. A customer bought a certain number of tin plates and one less than three times that many copper plates.

    (a) How many tin plates did the customer buy?

    (b) How much does a copper plate cost?

    (c) How many copper plates did the customer buy?

    (d) How much does a tin plate cost?

    (e) How much did the customer spend on the copper plates?

    (f) How much did the customer spend on the tin plates?

    (g) How much did the customer spend altogether?

Chapter 2 Test B

In problems 1–4, evaluate each of the given expressions.

1. $(-4)^3$

2. -4^3

3. $(5-8-(-1))^4$

4. $1-2(1-2)^7$

In problems 5–8, evaluate the given expression for x = 2, y = – 3, and z = –1.

5. $2x - y + z$

6. $2(x - y) + z$

7. $x^3 - xyz^2$

8. $|x + y| + |xy|z$

In problems 9–13, perform the indicated operations and simplify as completely as possible.

9. $2x^2y^2 - xy - 3x^3 + 5xy - x^3 + 4x^2y^2$

10. $3xy^4(-x^3y^2)(-2xy^3)$

11. $y(x^3y + 2x) - x(xy + x^2y^2) - 2xy(1 - x)$

12. $5x^3y^3(y - 4x) + 2xy^2(10x^3y)$

13. $4 - [x + 3(x - 2)]$

14. If we let n stand for the "number," translate each of the following phrases:

    (a) four times three less than a number

    (b) six more than five times a number is equal to ten more than the number

15. Walnuts cost $5 a pound and cashews cost $6 a pound. Mrs. O'Brien bought a certain number of pounds of walnuts and one more than four times that many pounds of cashews.

    (a) How much does a pound of cashews cost?

    (b) How many pounds of walnuts did Mrs. O'Brien buy?

    (c) How much does a pound of walnuts cost?

    (d) How many pounds of cashews did Mrs. O'Brien buy?

    (e) How much did Mrs. O'Brien spend on walnuts?

    (f) How much did Mrs. O'Brien spend on cashews?

    (g) How much did Mrs. O'Brien spend altogether?

Chapter 3 Test A

1. Determine whether the given equation is conditional, an identity, or a contradiction.

    (a) $2x - 4(2x - 1) = -6x + 4$

    (b) $2x - 4(2x - 1) = 6x + 4$

    (c) $2x - 4(2x - 1) = -6x - 4$

2. Determine whether or not the given value is a solution to the equation or inequality.

(a) $x(x - 2) - (x + 1) = -2 - x$; $x = -1$

(b) $s(s + 5) + 2 = (s + 1)(s + 4) - 2$; $s = 2$

(c) $5 + 4(u + 3) > 9$; $u = -2$

3. Solve each of the following equations or inequalities.

(a) $4 + 2a = -2a + 16$

(b) $3(2z + 1) - (2 - z) = 4(z - 1) - 1$

(c) $t^2 - t(1 - t) = 2t(t - 2) + 12$

(d) Solve and sketch the solution set on a number line:

$5 - 3(x + 1) \le -4$

(e) Solve and sketch the solution set on a number line:

$-7 < 1 - 4x < 13$

4. One number is 7 more than 4 times another number. Find the numbers if their sum is 32.

5. A purse contains only dimes and quarters. If there are 16 coins altogether, and their total value is $2.50, how many of each type of coin are there in the purse?

6. A man leaves his home and travels at the rate of 40 mph. One hour later, his wife leaves and travels along the same road at the rate of 60 mph. How far from their home will she catch up to her husband?

Chapter 3 Test B

1. Determine whether the given equation is conditional, an identity, or a contradiction.

(a) $4x - 3(1 - x) = x - 3$

(b) $4x - 3(1 - x) = 7x + 3$

(c) $4x - 3(1 - x) = 7x - 3$

2. Determine whether or not the given value is a solution to the equation or inequality.

(a) $(x + 3)(x - 1) - (3 - x) = 2x$; $x = 2$

(b) $b(b - 4) + 3 = b^2 + 7b$; $b = -1$

(c) $-2 - 5(t - 1) \le 13$; $t = -2$

3. Solve each of the following equations or inequalities.

(a) $-2 + 5n = -2n + 5$

(b) $3(3 - y) - 4(y + 4) = 2 - 4y$

(c) $4t^2 + 10 = 2t(2t + 1)$

(d) Solve and sketch the solution set on a number line:

$2 + 3(4 - x) < 2$

(e) Solve and sketch the solution set on a number line:

$-4 \leq 2 - 3x \leq -1$

4. Admission to a play costs \$3 for a child's ticket and \$5 for an adult's ticket. If 100 people attend the play and \$380 is collected, how many children's tickets were sold?

5. Two sides of a triangle are equal in length. The third side is three less than the sum of the other two. If the perimeter of the triangle is 25 inches, find the lengths of its sides.

6. If 11 more than 3 times a number is 1 less than 5 times the number, what is the number?

Cumulative Review: Chapters 1–3

In exercises 1–24. perform the indicated operations and simplify as completely as possible.

1. $-2 - 3 - 4$
2. $(-2)(-3)(-4)$
3. $(8 - 2 \cdot 3) - 1$
4. $8 - (2 \cdot 3 - 1)$
5. $(-3)^4$
6. -3^4
7. $a^2a^4a^6$
8. $3bb^5b^2 + 2b^3b^4b$
9. $3r^3s^5 - 5r^5s^3 + r^5s^3 - r^3s^5$
10. $6y^2 + 3y - 4 - 5y^2 - 6y + 2$
11. $3x^2(2y - 5z)$
12. $3x^2(2y)(-5z)$
13. $-t^3(u^2)(-6t)$
14. $-t^3(u^2 - 6t)$
15. $8(p + 2q) - 4(p - 4q)$
16. $6(3c^5 - 2d^4) + 4(3d^4 - 2c^5)$
17. $v^2w^3(5vw + 6v^3w^2) - v^3w^2(4w^2 - 3v^2w^3)$
18. $2xz^3(3yz - x^2y) + 3yz^2(x^3z - 2xz^2)$
19. $5a^2b^2 - 3ab - (5ab - 3a^2b^2)$
20. $r(s - 2t) - (rs - 2t)$
21. $h + 6\{h - 5(h + 4)\}$
22. $n - [n^2 - n(n^2 - n)]$
23. $u(u - 2v) + 2v(u - 2v) - (u^2 - 4v^2)$
24. $c^3 - c(c^2 - cd + d^2) - d(c^2 - cd + d^2)$

n exercises 25–32, evaluate the given expression for $x = 4$, $y = -2$, and $z = -3$.

25. $x - y^2$

26. $(x - y)^2$

27. $x^2 - y^2$

28. $xy + (x + y)z$

29. $|xy + z - x(y + z)|$

30. $|xy + z| - |x(y + z)|$

31. $(y+1)^3 + (z+1)^2$

32. $(x - 4)(4y^4)(3z^3)$

In exercises 33–44, solve the equation or inequality. In the case of an inequality, sketch the solution set on a number line.

33. $4x + 1 = 9x - 9$

34. $8 - 3s = 3s - 10$

35. $3(b + 5) + 2(2b - 1) = -1$

36. $4(3a + 2) - 3(2a - 1) = 11$

37. $x(6x + 5) - 3x(2x - 1) = -40$

38. $y - 3(y - 3) > 3$

39. $3(2z - 5) + 2(3z + 5) \leq 7z$

40. $4(k + 2) + 2(k + 4) = 5(k + 1) - 3(k + 3)$

41. $-2 \leq 3(c - 2) + 4 < 7$

42. $1 < 1 + 3(4 - 2y) < 7$

43. $3\{x - 3(x - 3)\} = -x - 3$

44. $x - 3\{x - 2(x - 3)\} = 2$

Solve each of the following problems algebraically. Be sure to clearly label what the variable represents.

45. One number is 3 more than 2 times another. If their difference is 11, find the numbers.

46. The length of a rectangle is 2 less than 5 times its width. If the perimeter is 11 times the width, find the dimensions of the rectangle.

47. Sara and David live 51 miles apart. Sara can pedal her bicycle at the rate of 6 mph, while David can pedal his bicycle at the rate of 7 mph. If Sara leaves her house at 9:00 a.m. and David leaves his house two hours later, when will they meet?

48. A florist sells red geranium plants for \$2 apiece and white geranium plants for \$3 apiece. If he sells 50 plants and earns \$113, how many white geranium plants did he sell?

Cumulative Test: Chapters 1–3

1. Evaluate each of the following:

   (a) $-2^3 + 3(-3)^2$

   (b) $(-3 + 5 - 6)^3$

2. Evaluate each of the following for $a = 2, b = -3$:

(a) $(b - a)^2 - ab^2$

(b) $|3a + 2b| + |2a + 3b|$

3. Perform the indicated operations and simplify as completely as possible.

(a) $3x^3 - 8x^2 + 5x - x^3 + 4x^2 + 2x$

(b) $3(2x + y) - 5(x - y)$

(c) $2ab(a^3 + ab) - 3a^2(a^2b - 2b^2)$

(d) $5(m - 2n) - 3(m - 4n) - 2(n + m)$

(e) $3u^2v^4(u^3 + v^2) - 3uv(2u^3v)(2uv^2)$

(f) $3 - 2b(3 - 2b(3 - 2b))$

4. Solve each of the following equations or inequalities.

(a) $4 - 5x = 10 - 7x$

(b) $3(z - 4) + 4(z - 3) = 11$

(c) $2 + 3(4 - c) > -10$

(d) $4(s + 1) - 3(s - 1) = 2(s + 2) - (s - 3)$

(e) $2(t - 3) + 3(t - 2) = 12 - t$

(f) $3t + 7(5 - t) = 4(8 - t)$

5. Solve the following inequalities and sketch the solution set on a number line.

(a) $5 - 2(1 + x) < -7$

(b) $-3 < 5x + 2 \leq 2$

6. Solve each of the following problems algebraically.

(a) One number is 6 more than 4 times another number. If their difference is 36, find the numbers.

(b) Cities A and B are 2200 miles apart. A train leaves city A at 10:00 a.m. and travels toward city B at the rate of 200 mph. A second train leaves city B at 11:00 a.m. and travels toward city A at the rate of 300 mph. At what time will the two trains pass one another?

(c) A charity car wash collects $156 by washing 40 cars. If it costs $3 to wash a compact car and $5 to wash a full-sized car, how many full-sized cars were washed?

Chapter 4 Test A

1. Reduce each of the following to lowest terms:

(a) $\dfrac{12}{-21}$

(b) $\dfrac{y^9}{y^4}$

(c) $\dfrac{-12x^8}{-8x^7}$

(d) $\dfrac{-28a^3b^4}{8a^6b}$

2. Perform the indicated operations and express your answer in lowest terms.

(a) $\dfrac{3x^2y}{4xy^3} \cdot \dfrac{6x^2y^2}{xy}$

(b) $\dfrac{3x^2y}{4xy^3} \div \dfrac{6x^2y^2}{xy}$

(c) $\dfrac{5s^2t}{8s^3} \cdot \dfrac{-4st^3}{15t^2} \cdot \dfrac{-6}{t^2}$

(d) $\dfrac{3a^4b^3}{7a^5} \div 6b^2$

(e) $\dfrac{c}{4} + \dfrac{c}{4}$

(f) $\dfrac{c}{4} \cdot \dfrac{c}{4}$

(g) $\dfrac{5}{3x^2} + \dfrac{1}{3x^2}$

(h) $\dfrac{u}{2} - \dfrac{2u}{5}$

(i) $\dfrac{3}{4y} + \dfrac{2}{3y}$

(j) $\dfrac{2}{5xy^3} - \dfrac{5}{2x^2y^2}$

(k) $\dfrac{x^2 - 3x + 4}{6x^2} - \dfrac{x^2 + 3x + 4}{6x^2}$

3. Solve each of the following equations or inequalities.

(a) $\dfrac{5x}{6} - \dfrac{x}{4} = 7$

(b) $\dfrac{2x - 1}{3} + \dfrac{x + 1}{8} \le -1$

(c) $\dfrac{5 - y}{5} - \dfrac{4 - y}{4} = \dfrac{1}{10}$

(d) $.01z + .1z = 1.32$

Solve each of the following problems algebraically.

4. If there are 39.37 inches in a meter, how many meters are there in 100 inches?

5. When $\frac{3}{5}$ of a number is subtracted from 5 more than the number, the result is one less than the number. Find the number.

6. A bank contains pennies, nickels, and dimes. There are 15 more pennies than nickels, and 3 times as many dimes as pennies. If the value of the coins is \$6.45, how many of each type of coin are in the bank?

7. An amount of money is invested at 7% and \$1000 more than that amount is invested at 10%. If the total interest on the two investments is \$950, how much was invested at each rate?

8. The length of a rectangle is 15 cm more than $\frac{5}{6}$ of its width. If the perimeter is 10 cm less than 7 times the width, find the dimensions of the rectangle.

9. How many liters of a 24% salt solution must be mixed with 40 liters of a 30% salt solution in order to produce a 26% salt solution?

Chapter 4 Test B

1. Reduce each of the following to lowest terms:

(a) $\dfrac{-15}{-40}$ (b) $\dfrac{z^{11}}{z^3}$

(c) $\dfrac{15x^9}{-9x^6}$ (d) $\dfrac{-30p^5q^2}{12p^4q^3}$

2. Perform the indicated operations and express your answer in lowest terms.

(a) $\dfrac{2xy^3}{3x^2y^2}\cdot\dfrac{x^3y^2}{12xy}$ (b) $\dfrac{2xy^3}{3x^2y^2}\div\dfrac{x^3y^2}{12xy}$

(c) $\dfrac{3u^3v}{-7uv^3}\cdot\dfrac{14v}{15u}\cdot\dfrac{-5v^2}{2uv}$ (d) $\dfrac{5c^2d^4}{6d^3}\div 10c^3$

(e) $\dfrac{m}{6}+\dfrac{m}{6}$ (f) $\dfrac{m}{6}\cdot\dfrac{m}{6}$

(g) $\dfrac{13}{5x^3}-\dfrac{3}{5x^3}$ (h) $\dfrac{v}{6}+\dfrac{4v}{9}$

(i) $\dfrac{4}{5r}-\dfrac{1}{4r}$ (j) $\dfrac{3}{4a^2b}+\dfrac{5}{3ab^3}$

(k) $\dfrac{x^2+8x-7}{2x^3}+\dfrac{7-4x-x^2}{2x^3}$

3. Solve each of the following equations or inequalities.

(a) $\dfrac{3x}{8}+\dfrac{x}{6}=13$ (b) $\dfrac{x-3}{4}-\dfrac{3x+1}{5}>-2$

(c) $\dfrac{6-y}{6}+\dfrac{5-y}{5}=\dfrac{8}{15}$ (d) $.2t + .04t = 14.4$

Solve each of the following problems algebraically.

4. If there are 35.31 cubic feet in a cubic meter, how many cubic meters are there in 200 cubic feet?

5. When two times a number is added to $\dfrac{2}{3}$ of one more than the number, the result is one less than three times the number. Find the number.

6. A man has 60 coins in his pocket, consisting of nickels, dimes, and quarters. He has three times as many dimes as quarters and the total value of his coins is $6.50. How many of each type of coin are in the man's pocket?

7. Two people leave from the same point and travel in opposite directions. One leaves at 1:00 p.m. and drives at the rate of 50 kph. The other leaves 15 minutes later and drives at the rate of 60 kph. At what time will they be 205 kilometers apart?

8. A woman invests a sum of money at 8% and twice as much as 9%. If her annual interest from the two investments is $390, how much was invested at each rate?

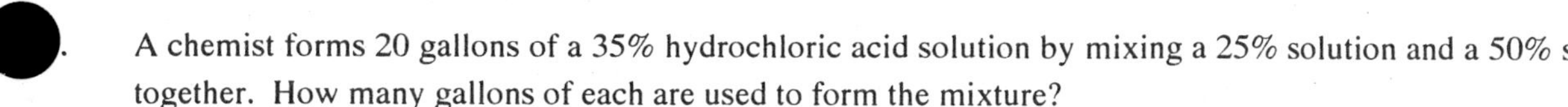

A chemist forms 20 gallons of a 35% hydrochloric acid solution by mixing a 25% solution and a 50% solution together. How many gallons of each are used to form the mixture?

Chapter 5 Test A

1. Determine whether or not the point (2, –3) satisfies the equation $2x + 3y = 5(5y + 7x)$.

2. Find the missing coordinates for the ordered pair of the given equation.

   (a) $4x + y = 9$ (3,)

   (b) $x + 2y = 8$ (, – 2)

   (c) $2x - 3y = 12$ (0,)

3. Sketch the graphs of each of the following in a rectangular coordinate system using the intercept method.

   (a) $x + 3y = 6$

   (b) $2x - 5y = 10$

   (c) $3x + 4y = 0$

   (d) $x = -1$

   (e) $y = 5$

4. Find the slope of the line passing through the points (4, – 3) and (– 3, 4).

5. Write an equation of the line with slope = $-\frac{2}{5}$ which passes through the point (6, – 1).

6. Write an equation of the line passing through the points (4, – 1) and (0, 7).

7. Write equations of the horizontal and vertical lines which pass through the point (– 2, 3).

8. What is the slope of a line whose equation is $2x + 3y + 4 = 0$?

9. Write an equation of the line passing through the point (– 1, 2) which is parallel to the line whose equation is $y = 3x + 7$.

10. Write an equation of the line whose graph appears in the following figure.

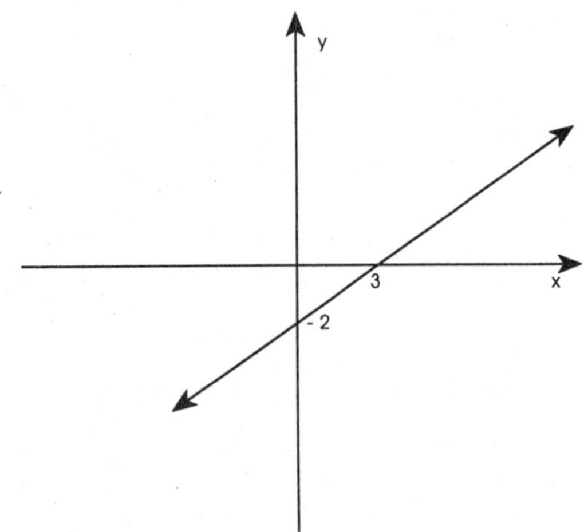

Chapter 5 Test B

1. Determine whether or not the point $(-3, 1)$ satisfies the equation $3(x + 4y) = 3y + 2x$.

2. Find the missing coordinates for the ordered pair of the given equation.

   (a) $5x - y = 7$ $(, 3)$

   (b) $x - 3y = -2$ $(-11,)$

   (c) $4x + 5y = 40$ $(, 0)$

3. Sketch the graphs of each of the following in a rectangular coordinate system using the intercept method.

   (a) $3x + 2y = -6$

   (b) $4x - 3y = 12$

   (c) $x + 3y = 0$

   (d) $x = 5$

   (e) $y = -1$

4. Find the slope of the line passing through the points $(-5, 2)$ and $(-2, 5)$.

5. Write an equation of the line with slope $= -\frac{3}{4}$ which passes through the point $(1, -4)$.

6. Write an equation of the line passing through the points $(3, -2)$ and $(0, 10)$.

7. Write equations of the horizontal and vertical lines which pass through the point $(4, -2)$.

8. What is the slope of a line whose equation is $3x - 2y + 8 = 0$?

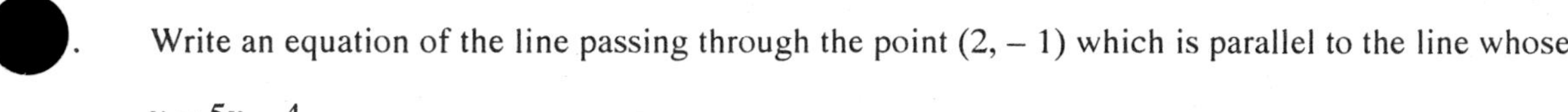

Write an equation of the line passing through the point (2, – 1) which is parallel to the line whose equation is $y = 5x - 4$.

10. Write an equation of the line whose graph appears in the following figure.

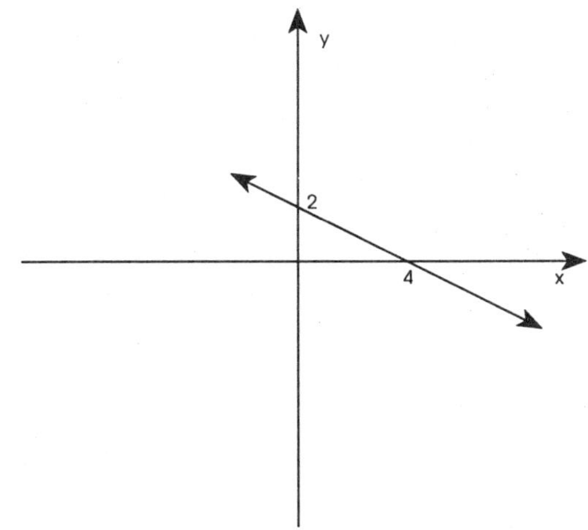

Chapter 6 Test A

1. Solve the following system of equations by the graphical method:

$$\begin{cases} x + 3y = 6 \\ 2x - y = 5 \end{cases}$$

Solve each of the following systems of equations algebraically.

2. $\begin{cases} 2x + 3y = 5 \\ x - 2y = 6 \end{cases}$

3. $\begin{cases} 4x - 5y = -10 \\ 5x + 6y = 12 \end{cases}$

4. $\begin{cases} x + \dfrac{3}{4}y = 1 \\ \dfrac{4}{3}x + y = 2 \end{cases}$

Solve the following problem algebraically.

5. Five pounds of plums and four pounds of peaches cost \$5.00. Two pounds of plums and three pounds of peaches cost \$2.70. Find the cost of a pound of plums and a pound of peaches.

Chapter 6 Test B

1. Solve the following system of equations by the graphical method:

$$\begin{cases} 3x - y = 3 \\ x + 2y = 8 \end{cases}$$

Solve each of the following systems of equations algebraically.

2. $\begin{cases} 5x - 6y = -9 \\ 4x + 3y = -15 \end{cases}$

3. $\begin{cases} 3x + 4y = 9 \\ 4x - 3y = 12 \end{cases}$

4. $\begin{cases} x - \frac{5}{2}y = 2 \\ \frac{2}{5}x - y = 1 \end{cases}$

Solve the following problem algebraically.

5. Three packages of veal and four packages of chicken cost $19. Five packages of veal and two packages of chicken cost $20. Find the cost of a package of veal and a package of chicken.

Cumulative Review: Chapters 4–6

In exercises 1–4, reduce the fraction to lowest terms.

1. $\dfrac{56}{-35}$

2. $\dfrac{6x^3}{16x^8}$

3. $\dfrac{45a^3b^5}{25a^4b^4}$

4. $\dfrac{2c + 3c + 4c}{5c^2 + 6c^2 + 7c^2}$

In exercises 5–20, perform the indicated operations and simplify as completely as possible.

5. $\dfrac{8x^2}{27} \div \dfrac{x}{6}$

6. $\dfrac{8x^2}{27} \cdot \dfrac{x}{6}$

7. $\dfrac{8x^2}{27} - \dfrac{x}{6}$

8. $\dfrac{8}{3y^3} - \dfrac{2}{3y^3}$

9. $\dfrac{3u^2 + 1}{4u^2} + \dfrac{u^2 - 1}{4u^2}$

10. $\dfrac{2x + 5}{7x} + \dfrac{x - 3}{7x} - \dfrac{3x + 2}{7x}$

11. $\dfrac{10a^4b^5}{21b^3c^2} \div \dfrac{15a^2c^4}{14b^4}$

12. $\dfrac{10a^4b^5}{21b^3c^2} \cdot \dfrac{15a^2c^4}{14b^4}$

13. $\dfrac{8}{5d} + \dfrac{5}{3d}$

14. $\dfrac{3}{4k} - \dfrac{5}{6\ell}$

15. $\dfrac{5}{8u^3v^2} + \dfrac{3}{10u^2v^3}$

16. $\dfrac{5}{8u^3v^2} \cdot \dfrac{3}{10u^2v^3}$

17. $(\dfrac{5}{2y} - \dfrac{1}{y}) \div \dfrac{4}{y}$

18. $\dfrac{5}{2y} - (\dfrac{1}{y} \div \dfrac{4}{y})$

19. $3x - \dfrac{2}{x} + \dfrac{4}{x^3}$

20. $\dfrac{3}{2p^3q} + \dfrac{5}{4p^2p^2} - \dfrac{7}{6pq^3}$

In exercises 21–28, solve the given equation.

21. $\frac{x}{4}+\frac{x}{6}=\frac{2x+9}{12}$

22. $\frac{t+1}{5}-\frac{2t}{15}=\frac{t-1}{3}$

23. $\frac{w}{2}-\frac{w}{3}=\frac{w}{5}$

24. $\frac{z}{4}+\frac{z}{12}=\frac{z}{3}$

25. $\frac{1-3y}{4}+\frac{4-3y}{6}=\frac{9y+7}{3}$

26. $\frac{x}{6}-\frac{1}{5}=\frac{x}{8}$

27. $.05\ (x + 1) + .5\ (x - 1) = .65$

28. $\frac{2}{3}(2u-3)-\frac{3}{4}(3u+2)=2$

In exercises 29–36, sketch the graph of the given equation in a rectangular coordinate system. Label the intercepts.

29. $y = -x + 4$

30. $2x - 7y = 14$

31. $5y + 4x = 10$

32. $6y - 9x = 18$

33. $y - 4 = 0$

34. $x + 4 = 0$

35. $y + 4x = 0$

36. $x - 4y = 0$

In exercises 37–40, find the slope of the line passing through the given pair of points.

37. (5, –3) and (3, –7)

38. (1, –8) and (–5, 7)

39. (3, 11) and (3, –11)

40. (3, 11) and (–3, 11)

In exercises 41–44, write an equation of the line with the given slope which passes through the given point.

41. $m = -5$, (1, –4)

42. $m = \frac{2}{3}$, (0, –6)

43. $m = 0$, (–8, 2)

44. m is undefined, (6, 7)

45. Write an equation of the line passing through the points (4, – 2) and (– 3, 5).

46. Write an equation of the line whose x-intercept is 6 and whose y-intercept is 5.

In exercises 47–48, solve the given system of equations graphically.

47. $\begin{cases} 3x-2y=6 \\ 2x+\ y=-10 \end{cases}$

48. $\begin{cases} 3x+2y=10 \\ 2x+3y=10 \end{cases}$

In exercises 49–56, solve each system of equations algebraically.

49. $\begin{cases} 3x-2y=\ \ 6 \\ 2x+\ y=-10 \end{cases}$

50. $\begin{cases} 3x+2y=10 \\ 2x+3y=10 \end{cases}$

51. $\begin{cases} 6x+4y=1 \\ 3x-8y=8 \end{cases}$

52. $\begin{cases} 8x+3y=19 \\ 9x-8y=10 \end{cases}$

53. $\begin{cases} 4x = 3 - y \\ 3y = -4 + x \end{cases}$

54. $\begin{cases} 5x - y = 5 \\ 4x + \dfrac{y}{5} = 1 \end{cases}$

55. $\begin{cases} 2x - 5y = 10 \\ 10y = 4x - 20 \end{cases}$

56. $\begin{cases} 3x - y = 9 \\ x - \dfrac{1}{3}y = 1 \end{cases}$

Solve each of the problems algebraically. Be sure to label what the variables represent.

57. The sum of four consecutive odd numbers is 14 more than twice the largest one. Find the four numbers.

58. The ratio of married couples to single people in an apartment house is 9 to 5. If there are 135 married couples living in the apartment, how many single people live there?

59. A typist charges $1.50 more for typing a page of technical material than she does for typing a page of regular material. If a report contains 18 pages of regular material and 12 pages of technical material and if the total bill is $108, what is the charge for typing a page of regular material?

60. 100 liters of a 30% chlorine solution is to be formed by mixing a 20% chlorine solution, a 25% chlorine solution, and a 40% chlorine solution. If there is to be twice as much of the 25% solution in the mixture as the 20% solution, how many liters of each solution must be mixed?

61. A wallet contains only five and ten dollar bills. If there are 23 bills altogether and their total value is $200, how many of each type of bill are in the wallet?

Cumulative Test: Chapters 4–6

In problems 1–6, perform the indicated operations and simplify as completely as possible. Final answers should be reduced to lowest terms.

1. $\dfrac{8p^4q^3}{3r^2} \cdot \dfrac{9r}{4pq^3}$

2. $\dfrac{12a^2b^3}{5c^5} \div 30a^3b^2c$

3. $\dfrac{4y}{3ab} + \dfrac{7y}{3ab} - \dfrac{2y}{3ab}$

4. $\dfrac{2w+5}{5w} - \dfrac{5-3w}{5w}$

5. $\dfrac{3}{4c^3d} - \dfrac{5}{6cd^3}$

6. $\left(8 \cdot \dfrac{2}{x^2}\right) \div \dfrac{12}{x^4}$

7. Solve for h: $\dfrac{h}{4} + \dfrac{h}{10} = 7$

8. Solve for v: $\dfrac{4v+5}{2} - \dfrac{2v+1}{6} = v + 1$

In problems 9–10, sketch the graph of the equation in a rectangular coordinate system. Label the intercepts.

9. $3x - 4y = 12$

10. $4x + 3y = 12$

Problems 11–13 refer to ℓ, the straight line that passes through the points (1, –3) and (–3, 5).

11. Find the slope of ℓ.

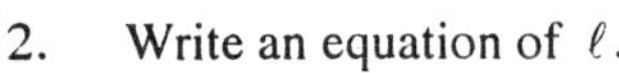

12. Write an equation of ℓ.

13. Find the y-intercept of ℓ.

In problems 14–15, solve the system of equations algebraically.

14. $\begin{cases} 5x - 6y = 13 \\ 3x + 2y = -9 \end{cases}$

15. $\begin{cases} x - \dfrac{5}{3}y = -3 \\ -6x + 10y = 15 \end{cases}$

Solve each of the following problems algebraically. Be sure to clearly label what your variable represents.

16. If there are 5,280 feet in one mile, how many miles are there in 20,000 feet?

17. Two consecutive odd numbers have the property that the sum of $\frac{2}{5}$ of the smaller and $\frac{2}{3}$ of the larger is equal to one more than the larger. Find the numbers.

18. There are 200 orchestra seats and 120 balcony seats in a theater. If an orchestra seat costs \$6 more than a balcony seat, and if a sold-out performance brings in \$4080, find the cost of a balcony seat.

19. Bill invests some money at 8% and four times as much money at 13%. The total annual interest he receives on these investments is \$100 more than he would have received if he had invested all of his money at 10%. How much did Bill invest at each rate?

Chapter 7 Test A

1. Evaluate $2^{-3} - 3^{-2}4^{0}$

In problems 2–5, simplify as completely as possible. Express final answers with positive exponents only.

2. $x^{-2}x^{-3}$

3. $(x^{-2})^{-3}$

4. $\dfrac{(x^2y)^3(xy^2)^2}{x^4y^8}$

5. $\dfrac{(-2x^{-2}y^{-3})^3}{2(x^{-1}y^{-2})^5}$

6. Given the polynomial $3x^6 + 5x^3 - x^2 + 7x + 1$:

   (a) How many terms are there?

   (b) What is the coefficient of the second degree term? the first degree term?

   (c) What is the degree of the polynomial?

In problems 7–13, perform the indicated operations and simplify as completely as possible.

7. $2x^3y^2(5xy)(-3x^2y^3)$

8. $2x^3y^2(5xy - 3x^2y^3)$

9. $3y^2(x - y) - xy(3y + 2x) + y(2x^2 + 5y^2)$

10. $(2x + 1)(5x^2 - 3x + 4)$

11. Add $x^3 + 8x^2 - 6x + 1$ and $2x - 5x^2 + 3$

12. Subtract $x^2 - 8x + 9$ from $x^2 + 8x - 9$

13. $(z - 2)^2 - (z - 3)^2$

14. Write in scientific notation:

(a) 258,000

(b) .0258

15. Compute using scientific notation:

$$\frac{(.0032)(600,000)}{.08}$$

Chapter 7 Test B

1. Evaluate $3^0 4^{-1} + 5^{-2}$

In problems 2–5, simplify as completely as possible. Express final answers with positive integers only.

2. $x^{-1}x^{-5}$

3. $(x^{-1})^{-5}$

4. $\frac{(xy^3)^3(x^3y^2)^2}{x^{10}y^{11}}$

5. $\frac{(-3x^{-3}y^{-4})^2}{-3(x^{-2}y^{-1})^4}$

6. Given the polynomial $4x^7 + 3x^4 - 5x^3 - x + 2$:

(a) How many terms are there?

(b) What is the coefficient of the fourth degree term? the third degree term?

(c) What is the degree of the polynomial?

In problems 7–13, perform the indicated operations and simplify as completely as possible.

7. $4x^2y^4(2x^3y)(-3xy^3)$

8. $4x^2y^4(2x^3y - 3xy^3)$

9. $2x^2y(1 - xy) - y(x^3 + 2x^2) + 2x^3(y + y^2)$

10. $(2x - 1)(3x^2 + 5x - 2)$

11. Add $2x^3 - 6x^2 + 5x - 3$ and $4 - 8x + 3x^2$

12. Subtract $x^2 - 6x - 4$ from $x^2 + 6x + 4$

13. $(1 - y)^2 - (2 - y)^2$

14. Write in scientific notation:

(a) 7,950,000

(b) .000795

15. Compute using scientific notation:

$$\frac{.00075}{(.03)(5,000,000)}$$

Chapter 8 Test A

In problems 1–11, factor as completely as possible. If not factorable, say so.

1. $2x^4 - 8x^3 + 6x^2$

2. $3x^3y^2 + 12x^2y^3 - 9x^2y^2$

3. $x^2 + 13x + 36$

4. $x^2 + xy - 12y^2$

5. $4x^5 - 16x^3$

6. $7x^2 + 14x + 14$

7. $x^2 - 16x + 64$

8. $2x^2 - 9x + 4$

9. $2x^2 - 4x + 9$

10. $8x^2 + 13x - 6$

11. $x^2y^2 + 6xy + 9$

12. Divide: $\dfrac{40a^2b^3 + 30a^3b^4 - 24a^2b^2}{20a^3b^3}$

13. Divide: $\dfrac{3x^3 + 5x^2 - 7}{x + 2}$

Chapter 8 Test B

In problems 1–11, factor as completely as possible. If not factorable, say so.

1. $5x^4 + 15x^3 + 10x^2$

2. $7x^3y^4 - 28x^4y^3 - 14x^3y^3$

3. $x^2 - 11x + 24$

4. $x^2 - 7xy - 18y^2$

5. $3x^4 - 27x^2$

6. $4x^2 + 12x + 16$

7. $x^2 + 18x + 81$

8. $3x^2 + 4x - 15$

9. $3x^2 - 4x + 15$

10. $9x^2 - 14x - 8$

11. $x^2y^2 - 10xy + 25$

12. Divide: $\dfrac{54p^4q^3 - 24p^5q^4 + 27p^3q^3}{18p^4q^4}$

13. Divide: $\dfrac{2x^3 - 3x^2 + 1}{x - 3}$

Chapter 9 Test A

1. Reduce to lowest terms: $\dfrac{3x^2 + 9x}{x^2 - 9x}$

2. Reduce to lowest terms: $\frac{3x^2 + 9x}{x^2 - 9}$

In problems 3–7, perform the indicated operations and simplify as completely as possible.

3. $\frac{4}{x-2} - \frac{2}{x-4}$

4. $\frac{x^2}{4x^2 - 8x} \cdot \frac{2x^2 - 8}{(x+2)^2}$

5. $\frac{6}{x^2 - 4x} + \frac{3}{2x}$

6. $\frac{x^2 - 3x - 4}{x^2 - x - 6} \div \frac{x^2 - 4x}{x^2 + 2x}$

7. $\frac{2x-3}{x^2 - 2x - 3} + \frac{5x-4}{x^2 - 2x - 3} - \frac{3x+5}{x^2 - 2x - 3}$

8. Solve for x: $\frac{4}{x+2} + \frac{2}{x} = \frac{14}{3x}$

9. Solve for y: $2xy + z = 1 - \frac{2y}{3}$

10. Solve for b: $\frac{2b+8}{b+3} + 3 = \frac{2}{b+3}$

Solve each of the following problems algebraically.

11. What number must be added to the numerator and the denominator of the fraction $\frac{3}{5}$ so that the resulting fraction is equal to $\frac{5}{3}$?

12. George can build a bookcase in 6 hours. If he works together with his assistant Frank, the two men can build the bookcase in 4 hours. How long would it take Frank to build the bookcase if he works alone?

Chapter 9 Test B

1. Reduce to lowest terms: $\frac{x^2 - 16x}{2x^2 - 8x}$

2. Reduce to lowest terms: $\frac{x^2 - 16}{2x^2 - 8x}$

In problems 3–7, perform the indicated operations and simplify as completely as possible.

3. $\frac{2}{x+5} + \frac{1}{x+2}$

4. $\frac{3x^3 + 9x^2}{3x} \cdot \frac{(x-3)^2}{3x^2 - 27}$

5. $\frac{6}{x^2 + 9x} - \frac{2}{3x}$

6. $\frac{x^2 + 4x - 5}{x^2 + x - 12} + \frac{x^2 + 5x}{x^2 - 3x}$

7. $\frac{6x-5}{x^2 - 5x + 4} - \frac{2x+3}{x^2 - 5x + 4} - \frac{x+4}{x^2 - 5x + 4}$

8. Solve for x: $\frac{1}{x} + \frac{3}{x-3} = \frac{17}{2x}$

9. Solve for q: $3pq - 2r = 4 + \frac{1}{4}q$

10. Solve for a: $\frac{4a-7}{4-a} + 1 = \frac{5+a}{4-a}$

Solve each of the following problems algebraically.

11. The denominator of a fraction is one more than three times the numerator. If 2 is added to both the numerator and denominator, the resulting fraction is equal to $\frac{2}{5}$. Find the original fraction.

12. Julia can run 4 times as fast as she can walk. If it takes her $3\frac{1}{4}$ hours to complete a trip in which she walks 5 miles and runs 6 miles, find her walking speed.

Cumulative Review: Chapters 7–9

In exercises 1–14, perform the indicated operations and simplify as completely as possible.

1. $(x + 2)(x - 7)$

2. $(x + 2)(x^2 + x - 7)$

3. $(x^2 + x + 2)(x^2 + x - 7)$

4. $5t(t + 1)(t - 4)$

5. $(6r - 5s)(3r + 2s)$

6. $3m(m - 2) + 2m(m - 3)$

7. $(x + 4)(x + 16) - (x - 8)^2$

8. $(2x + 1)(x - 8) + 2(x + 2)^2$

9. $(m + 3)(3m - 1)(3m + 1)$

10. $(x - 2)(x + 5)(x + 2)(x - 5)$

11. Subtract $5z^2 + 4z - 8$ from $3z^2 + 6z - 1$

12. Add $y^2 - 8xy + 2x^2$ to the product of x + 3y and 2x – y

13. (a) What is the degree of the polynomial $7x^6 - 5x^4 + 3x^2 - 1$?

(b) What is the coefficient of the fourth degree term?

14. Write the polynomial $2x - 3x^2 - 1 + 4x^4$ in complete standard form.

In exercises 15–20, divide the given polynomials. Use long division where necessary.

15. $\frac{2x^3 - 18x^4}{6x^2}$

16. $\frac{18s^3t^2 + 24s^2t^2 - 12s^2t^3}{8s^3t^3}$

17. $\frac{a^2 - 4a + 7}{a + 3}$

18. $\frac{3y^3 + y^2 - 12y - 4}{y + 2}$

19. $\dfrac{12t^3 + 10t^2 - 9}{2t - 1}$

20. $\dfrac{b^4 + 3b + 2}{b - 1}$

In exercises 21–28, simplify the expression as completely as possible. Final answers should be expressed with positive exponents only.

21. $2^1 + 2^0 + 2^{-1}$

22. $(3-1)^{-2} - (4-1)^{-2}$

23. $\dfrac{x^4 x^5}{(x^4)^5}$

24. $\dfrac{y^{-10}}{y^{-5}}$

25. $\dfrac{(-3x^4)^3}{-9(x^7)^2}$

26. $\dfrac{(a^{-1}b^{-2})^{-3}}{(b^{-3}a^{-2})^{-1}}$

27. $\dfrac{(2u^3v^{-3})^{-2}}{(3u^{-2}v^2)^{-3}}$

28. $\left(\dfrac{4mn^{-2}}{3m^{-3}n}\right)^{-1}$

29. Write in scientific notation: 123,000,000

30. Write in scientific notation: .00000975

31. Evaluate: $\dfrac{(3 \times 10^{-5})(2.4 \times 10^4)}{.6 \times 10^2}$

32. Evaluate: $\dfrac{(.00015)(48,000,000)}{(.144)(80,000)}$

In exercises 33–52, factor the polynomial as completely as possible. If the polynomial is not factorable, say so.

33. $x^2 - 8x$

34. $x^2 - 8x + 12$

35. $x^2 - 8x - 9$

36. $x^2 - 8x + 9$

37. $4a^3b^2 - 6a^2b + 8ab^3$

38. $8r^4s^6 - 2r^6s^4$

39. $18v^2 - 32$

40. $18v^2 + 32$

41. $t^5 - 49t^3$

42. $3s^2 - 7s - 10$

43. $6t^2 - 13t + 6$

44. $6t^2 - 37t + 6$

45. $6t^2 - 12t + 6$

46. $6t^2 - 13t - 6$

47. $56 - x - x^2$

48. $x^2 - 29xy + 100y^2$

49. $x^2 + 100y^2$

50. $x^2 + ax + bx + ab$

51. $x^2 + 3x - 2xy - 6y$

52. $81x^4 - 16y^4$

In exercises 53–54, simplify the fraction.

53. $\dfrac{y^2 + 2y}{y^2 - 4}$

54. $\dfrac{s^2 - 3s - 4}{s^2 + 5s + 4}$

In exercises 55–63, perform the indicated operations and simplify as completely as possible.

55. $\dfrac{2}{5x} - \dfrac{1}{3x + 2}$

56. $\dfrac{4}{3cd^2} + \dfrac{3}{4c^2d}$

57. $\dfrac{x^2 - 1}{x - 1} \cdot \dfrac{x + 1}{x^2 + 1}$

58. $\dfrac{m^2n - mn^2}{m^2 - n^2} \div \dfrac{2mn}{m^2 + 2mn + n^2}$

59. $\dfrac{4}{x^2 + 4x + 3} - \dfrac{2}{x^2 + 3x + 2}$

60. $\dfrac{1}{a - 2} - \dfrac{1}{2a} - \dfrac{2}{3a - 6}$

61. $\dfrac{\frac{s}{3} - \frac{3}{s}}{\frac{s^2 + 6s + 9}{6s}}$

62. $\dfrac{3t}{t - 1} - \dfrac{t^2 + 2t}{t^2 - t}$

63. $\dfrac{5r + 4}{2r + 8} + \dfrac{3r + 7}{2r + 8} - \dfrac{8r + 3}{2r + 8}$

In exercises 64–71, solve the given equation. If it contains more than one variable, solve for the indicated variable.

64. $\dfrac{5}{x} - \dfrac{1}{4} = \dfrac{7}{2x}$

65. $\dfrac{2}{4a + 1} + \dfrac{5}{12a + 3} = \dfrac{1}{9}$

66. $3r + 2s = 8 - r - 2s$ for s

67. $\dfrac{3}{4}u - 8v + 2 = \dfrac{2}{3}u + v + 1$ for u

68. $\dfrac{3z - 2}{z + 5} - 2 = \dfrac{2z - 7}{z + 5}$

69. $\dfrac{1}{2}(x - 1) + \dfrac{1}{6}(x + 1) = \dfrac{1}{3}(x - 1) + \dfrac{1}{4}(x + 1)$

70. $\dfrac{6p + 1}{3p + 2} = \dfrac{2p - 3}{p}$

71. $\dfrac{6p + 1}{3p - 1} = \dfrac{2p + 1}{p}$

Solve each of the following problems algebraically. Be sure to label what the variable represents.

72. Tom can complete an assignment in 3 hours, Dick in 4 hours, and Harry in 12 hours. If the three work together, how long will it take for them to complete the assignment?

73. A man walks a distance of 4 kilometers at a certain rate and runs back to his starting point at three times his walking speed. If the round trip takes 2 hours, what is the man's running speed?

74. The denominator of a fraction is 5 more than twice its numerator. If 7 is subtracted from both the numerator and the denominator, the resulting fraction is equal to $-\frac{1}{2}$. Find the original fraction.

Cumulative Test: Chapters 7–9

In problems 1–6, perform the indicated operations and simplify as completely as possible. Final answers should be reduced to lowest terms, and should be expressed with positive exponents only.

1. $(3s - 4t)(s^3t + 2s^2t^2 - 3st^3)$

2. $\dfrac{12a^3b^{-4}}{9a^{-2}b^{-1}}$

3. $(2x + 1)(x - 2) + (x - 3)(3x + 5)$

4. $(2n - 1)^2 - 2(n - 1)^2$

5. $\dfrac{(4x^{-2})^{-2}}{(2x^{-3})^{-3}}$

6. $\dfrac{16k^3\ell^5 - 10k^5\ell^3}{4k^4\ell^4}$

7. Use long division to find the quotient and remainder: $\dfrac{x^3 + 3x^2 - x + 3}{x + 1}$

8. Use long division to find the quotient and remainder: $\dfrac{18 - 11x^2 + x^4}{x + 3}$

9. Write the following in scientific notation:

   (a) .000000531

   (b) 531,000,000

10. Compute using scientific notation:

$$\frac{(.00009)(1,600,000)}{(.0045)(320,000)}$$

11. Subtract $x^3 + 3x^2 - 4x + 1$ from the sum of $x^3 - 2x + 2$ and $5x^2 - x - 4$

In problems 12–19, factor the given polynomial as completely as possible.

12. $x^2 - 5x + 6$

13. $x^2 - 5x - 6$

14. $2s^4t^6 - 8s^6t^4$

15. $10x^2 + 30x - 70$

16. $3y^2 - y - 14$

17. $12z^2 + 13z - 25$

18. $c(2c + 1) - 3(2c + 1)$

19. $x^3 + xy^2 - 2x^2 - 2y^2$

In problems 20–21, reduce to lowest terms.

20. $\dfrac{x^3 - 4x}{x^2 + 4x + 4}$

21. $\dfrac{2a^2 - 7a + 6}{a^2 - 5a + 6}$

In problems 22–26, perform the indicated operations and simplify as completely as possible.

22. $\dfrac{6}{x+3}+\dfrac{4}{x-2}$

23. $\dfrac{3u^3-12u}{u^2+3u-10}\cdot\dfrac{u^2+6u+5}{6u^3+12u^2+6u}$

24. $\dfrac{9}{2x^2+5x+2}+\dfrac{3}{2x^2+x}$

25. $\dfrac{r^2+8r+15}{r^2+8r+16}\div\dfrac{r^2+6r+9}{r^2+6r+8}$

26. $\dfrac{\frac{1}{a}+\frac{1}{b}}{\frac{4}{a^2}-\frac{4}{b^2}}$

In problems 27–29, solve the given equation.

27. $\dfrac{8}{2t-3}=\dfrac{5}{6t-9}+\dfrac{19}{15}$

28. Solve for w: $\dfrac{3}{4}w+5v=kw-\dfrac{1}{2}v+4$

29. $\dfrac{2x+3}{x+7}-\dfrac{2}{3}=\dfrac{x-4}{x+7}$

30. The distance between towns A and B is 60 miles. Adam drives from A to B at a certain speed and drives 10 mph faster when he returns from B to A. If the entire trip took $3\frac{1}{2}$ hours, find Adam's driving speed each way.

Chapter 10 Test A

Perform the indicated operations. Make sure your final answer is in simplest radical form. Assume all variables appearing under radical signs are non-negative.

1. $\sqrt{16x^{10}y^{12}}$

2. $\sqrt{32}+4\sqrt{72}-\sqrt{200}$

3. $\sqrt{45x^5}-x^2\sqrt{20x}$

4. $\dfrac{\sqrt{180x^8y^7}}{\sqrt{12x^2y^3}}$

5. $\sqrt{48x^5y^{11}}-2x^2y^3\sqrt{12xy^5}$

6. $\dfrac{6y^3}{\sqrt{8y}}$

7. $(\sqrt{x}+2\sqrt{3})(\sqrt{4x}-\sqrt{27})$

8. $(\sqrt{a}+\sqrt{6})^2-(\sqrt{a+6})^2$

9. $\dfrac{7}{3+\sqrt{2}}$

10. $\dfrac{2x-8y}{\sqrt{x}-2\sqrt{y}}$

11. Decide whether or not $1+\sqrt{3}$ is a solution to the equation $x^2-2=2x$

12. Solve for x: $\sqrt{4x}-1=5$

Chapter 10 Test B

Perform the indicated operations. Make sure your final answer is in simplest radical form. Assume all variables appearing under radical signs are non-negative.

1. $\sqrt{36x^{14}y^8}$

2. $\sqrt{50} - 3\sqrt{98} + \sqrt{128}$

3. $\sqrt{75x^5} - x\sqrt{27x^3}$

4. $\dfrac{\sqrt{120x^5y^{12}}}{\sqrt{12xy^4}}$

5. $\sqrt{72x^9y^5} - 2x^3y\sqrt{18x^3y^3}$

6. $\dfrac{8z^4}{\sqrt{12z}}$

7. $(\sqrt{x} - 3\sqrt{2})(\sqrt{9x} + \sqrt{8})$

8. $(\sqrt{b+7})^2 - (\sqrt{b} + \sqrt{7})^2$

9. $\dfrac{6}{4 - \sqrt{10}}$

10. $\dfrac{3x - 27y}{\sqrt{x} + 3\sqrt{y}}$

11. Decide whether or not $1 - \sqrt{3}$ is a solution to the equation $x^2 = 2x + 2$.

12. Solve for x: $\sqrt{3x} + 1 = 7$

Chapter 11 Test A

Solve each of the following equations. Choose any method you like.

1. $(x + 1)(x - 1) = 8$

2. $2x^2 - 4x - 1 = 0$

3. $3x^2 = 8x + 3$

4. $(2x - 1)^2 = 4$

5. $(2x - 1)^2 = 4x$

6. $3x^2 - 6x + 1 = -x^2 + 2x + 1$

7 $\dfrac{x+3}{x} = \dfrac{x}{x-1}$

8. $(x^2 + 1)(x + 1) = (x^2 - 2)(x - 2)$

9. $\dfrac{1}{x} + \dfrac{1}{x-1} = \dfrac{3}{2}$

10. Solve the following problem by completing the square, and check your answer by using the quadratic formula: $2x^2 - 16x = -5$.

11. Sketch the graph of $y = x^2 + 4x - 5$. Label the intercepts.

Chapter 11 Test B

Solve each of the following equations. Choose any method you like.

1. $(x - 2)(x + 2) = 21$

2. $2x^2 + 4x + 1 = 0$

3. $5x^2 + 3x = 2$

4. $(3x + 1)^2 = 16$

5. $(3x + 1)^2 = 16x$

6. $2x^2 + 5x - 2 = x^2 - x + 1$

7. $\frac{2x - 3}{x} = \frac{4x}{2x + 1}$

8. $(x^2 + 2)(x - 1) = (x^2 - 2)(x + 1)$

9. $\frac{1}{x + 1} + \frac{1}{x} = \frac{3}{2}$

10. Solve the following problem by completing the square, and check your answer by using the quadratic formula: $3x^2 + 18x = -2$.

11. Sketch the graph of $y = x^2 - 4x - 5$. Label the intercepts.

Cumulative Review: Chapters 10–11

In exercises 1–8, simplify the expression as completely as possible. Fractions should be reduced to lowest terms.

1. $\sqrt{4a^{20}b^{30}}$

2. $\sqrt{\frac{12w^8z^{11}}{108wz^7}}$

3. $\frac{8}{\sqrt{11}}$

4. $\frac{8}{\sqrt{11} - 3}$

5. $\frac{12}{\sqrt{14} + \sqrt{10}}$

6. $\frac{21x}{\sqrt{7x}}$

7. $\sqrt{192}$

8. $\sqrt{84u^6v^7}$

In exercises 9–17, perform the indicated operations and simplify as completely as possible.

9. $\sqrt{24rs} + \sqrt{54rs}$

10. $\sqrt{3}(4\sqrt{5} + \sqrt{10}) - 2\sqrt{3}(\sqrt{10} + \sqrt{20})$

11. $4\sqrt{\frac{5}{2}} - \sqrt{40}$

12. $(3\sqrt{a} - \sqrt{4b})(\sqrt{4a} + 3\sqrt{b})$

13. $2\sqrt{50p^8q^5} - p^2q^2\sqrt{98p^4q}$

14. $\frac{5}{\sqrt{6}} + \frac{7}{\sqrt{96}}$

15. $(2\sqrt{6} + \sqrt{7})(2\sqrt{6} - \sqrt{7})$

16. $(\sqrt{x + 5})^4$

17. $(\frac{8}{\sqrt{13} - 3} - \frac{26}{\sqrt{13}})^2$

In exercises 18–33, solve the given equation.

18. $x^2 + 4x = 0$

19. $x^2 + 4x = 21$

20. $1 - \frac{5}{x} = \frac{24}{x^2}$

21. $\frac{x+4}{x+1} = \frac{x-3}{x-2}$

22. $x^2 + 5x = 4$

23. $4x^2 - 7x + 5 = x^2 - 2x + 3$

24. $(x-2)(x+3) + 0$

25. $(x-2)(x+3) = 6$

26. $(2x+1)(x-2) = 18$

27. $(2x+1)(x-2) = (x+2)(2x-3)$

28. $(a-3)^2 = 4$

29. $(a-4)^2 = 3$

30. $\sqrt{x+2} = 5$

31. $\sqrt{x} + 2 = 5$

32. $7 - \sqrt{4s} = 11$

33. $2\sqrt{y-3} + 5 = 9$

34. Solve by completing the square: $x^2 + 4x + 2 = 0$

35. Solve by completing the square: $2x^2 - 6x + 1 = 0$

In exercises 36–39, sketch the graph of the given equation. Label the intercepts. Round to the nearest tenth where necessary.

36. $y = x^2 - 8x + 7$

37. $y = (3x+4)^2$

38. $y = 16 - 9x^2$

39. $y = -x^2 + x + 3$

40. The length of a rectangle is 2 less than three times its width. If the area of the rectangle is 40, find its dimensions.

41. Consider the right triangle shown below:

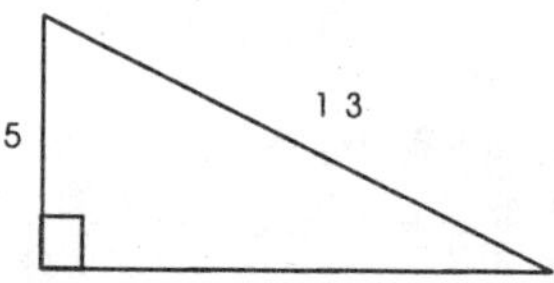

(a) Find its perimeter

(b) Find its area

42. Find the length of the side of a square if its diagonal is 8 inches.

43. A man jogs for 10 miles and then walks back to his starting point along the same route. His jogging speed is 5 mph faster than his walking speed. If the round trip takes him 3 hours, find his walking speed.

44. The sum of three times a number and twice the reciprocal of the number is 7. Find the number(s).

Cumulative Test: Chapters 10–11

In problems 1–10, perform the indicated operations and simplify as completely as possible. Assume all variables appearing under radical signs are non-negative.

1. $\sqrt{128}$

2. $\sqrt{27r^3s^5t^8}$

3. $3\sqrt{28} + 2\sqrt{63}$

4. $5\sqrt{40b^3c^8} - 3bc^2\sqrt{90bc^4}$

5. $\dfrac{45}{\sqrt{10}}$

6. $\sqrt{16x^{16}}$

7. $\dfrac{45}{\sqrt{10} - 1}$

8. $(3\sqrt{2} + 1)^2$

9. $(\sqrt{16w} - \sqrt{24})(\sqrt{9w} + \sqrt{96})$

10. $\dfrac{\sqrt{11}}{\sqrt{7} - \sqrt{3}} - \dfrac{\sqrt{3}}{\sqrt{11} + \sqrt{7}}$

11. $(x - 1)(x + 4) = 50$

12. $(x + 5)^2 = 5$

13. $\dfrac{2x}{4x - 1} = \dfrac{x + 2}{2x}$

14. $\sqrt{3x + 1} + 1 = 3$

15. $3(x^2 + 1) = 10x$

16. $\dfrac{r^2}{4} + \dfrac{1}{3} = 2$

17. $(2x + 1)^2 = 4x(x + 1)$

18. $(2x + 1)^2 = 4(x + 1)$

19. Find the area of a square with a diagonal of 6 cm.

20. A motorboat travels 28 miles at a certain speed before the motor fails. The driver then has to row an additional 6 miles to reach the shore, at a speed that is 12 mph less than the motorboat's speed. Find the rowing speed if the entire trip takes 5 hours.

In problems 21–22, sketch the graph of the given equation. Label the intercepts. Round to the nearest tenth when necessary.

21. $y = -x^2 - 6x - 8$

22. $y = x^2 - 3$

Practice Final Exam

In problems 1–4, evaluate the given expression.

1. $\dfrac{(-1)(-2) - (-3)(-4)}{-5}$

2. $|1 - 2[3 - 4(5 - 6)]|$

3. $x^2 - xy - |x - y - z|$ when $x = -3, y = 3,$ and $z = 2$

4. $(2^{-1} + 3^{-1})^{-1}$

In problems 5–12, perform the indicated and operations and simplify as completely as possible. (Where applicable, express your answers with positive exponents only in simplest radical form.)

5. $x^2(2x - 3y^2) + 2x(xy^2) - 3(x^3 - 2x(xy))$

6. Subtract the sum of $x - 1$ and $2x^2 - x + 2$ from the product of $x - 1$ and $2x^2 - x + 2$.

7. $\frac{1}{8y} - \frac{3x^2}{5y^2} \cdot \frac{y}{6x}$

8. $\frac{(x^3y^2)^{-1}(x^{-2}y^3)^{-2}}{(xy^{-2})^{-3}}$

9. $\frac{x^2 - 4x - 5}{x^2 - 5x + 4} \div \frac{x^2 + 5x + 4}{x^2 + 4x - 5}$

10. $\frac{1}{x^2 - x} + \frac{1}{x^2 + x}$

11. $4x\sqrt{24x} - \frac{2}{5}\sqrt{150x^3}$

12. $\frac{3}{3 - \sqrt{6}} - \frac{3}{\sqrt{6} + \sqrt{3}}$

In problems 13–14, solve the given inequality and sketch its solution set on a number line.

13. $-11 \le 2(4 - x) - 5(1 + x) < 24$

14. $\frac{x + 1}{2} - \frac{x + 2}{3} + \frac{x + 3}{4} > 1$

In problems 15 - 20, solve the given equation.

15. $3(p + 2) + p(p + 3) = p(p - 2) + 2(p - 3)$

16. $\frac{3x - 2}{4} + 2 = \frac{2x - 3}{5}$

17. $\frac{3}{2x + 1} + \frac{2}{3x} = \frac{2}{x}$

18. $(3a + 1)(a - 1) = (2a - 1)(a + 1)$

19. $y^2 - 1 = \frac{5}{6}y$

20. $5x^2 + 6x - 1 = 0$

21. Compute using scientific notation: $\frac{(7,200,000)(.0048)}{(64,000)(.000003)}$

In problems 22–23, factor as completely as possible.

22. $4x^4 + 18x^3 - 36x^2$

23. $x^3 - xy^2 + 2x^2y - 2y^3$

24. Find the quotient and the remainder: $\frac{2x^3 - 1}{x + 1}$

25. Use the intercept method to sketch the graph of $5x - 6y = 15$.

26. Write an equation of the line whose x-intercept is 2 and whose y-intercept is –8.

27. Solve the following system algebraically: $\begin{cases} 7x + 12y = 3 \\ 5x - 3y = 6 \end{cases}$

Solve each of the following problems algebraically.

28. My father's age is seven more than three times my age. If the sum of our ages is five less than five times my age, how old am I?

29. Red poker chips are worth 2 cents, blue poker chips are worth 3 cents, and white poker chips are worth 5 cents. There are 150 chips in a pot, with twice the number of white chips as blue chips. If the total value of the pot is $5.80, how many red chips are contained in the pot?

30. Mrs. Brown invests some money at 6% and the rest at 9%, yielding her an annual interest of $306. If she had invested the same amounts in the opposite way, her annual interest would have increased to $324. How much money has Mrs. Brown invested in all?

31. Sam needs three more hours to plow a field than Hugo needs. Together, they can plow the field in two hours. How long would it take Hugo to plow the field if he works alone?

32. Find the two numbers whose sum is 1 and whose product is $-\frac{3}{4}$.

ANSWERS TO CHAPTER TESTS, CUMULATIVE REVIEWS AND TESTS, AND PRACTICE FINAL

Chapter 1: Test A

1. (a) F (b) T (c) F (d) F
   (e) {20, 30}

2. –8 3. 10 4. –4 5. 3

6. –5 7. –20 8. 4 9. 45

10. –27 11. $2 \cdot 2 \cdot 2 \cdot 3 \cdot 3$

Chapter 1: Test B

1. (a) T (b) F (c) T (d) F
   (e) {3}

2. –5 3. –6 4. 0 5. 4

6. –2 7. –16 8. –4 9. 30

10. –32 11. $2 \cdot 2 \cdot 3 \cdot 5$

Chapter 2: Test A

1. 64 2. –64 3. –1 4. 43

5. 5 6. –6 7. –4 8. 0

9. $5x^4 - 5xy^3 - 5x^3y$ 10. $10x^5y^7$

11. $2xy$ 12. $4x^3y^2$

13. $-x + 1$

14. (a) $4n - 3$ (b) $3n + 7 = n - 1$

15. (a) n (b) 7 dollars (c) $3n - 1$ (d) 4 dollars
    (e) $7(3n - 1)$ dollars (f) $4n$ dollars (g) $25n - 7$ dollars

Chapter 2: Test B

1. –64 2. –64 3. 16 4. 3

5. 6 6. 9 7. 14 8. –5

9. $6x^2y^2 + 4xy - 4x^3$ 10. $6x^5y^9$

11. x^2y

12. $5x^3y^4$

13. $-4x + 10$

14. (a) $4(n - 3)$ (b) $5n + 6 = n + 10$

15. (a) 6 dollars (b) n (c) 5 dollars (d) $4n + 1$

(e) 5n dollars (f) $6(4n + 1)$ dollars (g) $29n + 6$ dollars

Chapter 3: Test A

1. (a) identity (b) conditional (c) contradiction
2. (a) no (b) yes (c) no
3. (a) 3 (b) -2 (c) 4

   (d) $x \geq 2$ (e) $-3 < x < 2$

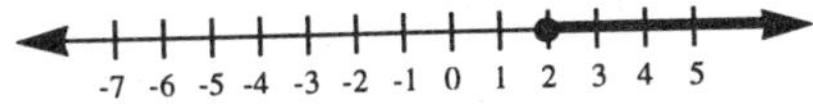

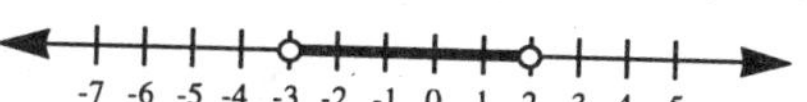

4. 5 and 27

5. 10 dimes and 6 quarters

6. 120 miles

Chapter 3: Test B

1. (a) conditional (b) contradiction (c) identity
2. (a) yes (b) no (c) yes
3. (a) 1 (b) -3 (c) 5

   (d) $x > 4$ (e) $1 \leq x \leq 2$

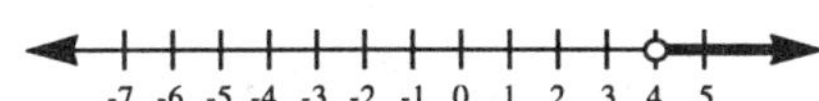

4. 60 children's tickets

5. 7 inches, 7 inches, 11 inches

6. 6

Cumulative Review: Chapters 1–3

1. -9 2. -24 3. 1 4. 3

5. 81 6. -81 7. a^{12} 8. $5b^8$

9. $2r^3s^5 - 4r^5s^3$

10. $y^2 - 3y - 2$

11. $6x^2y - 15x^2z$

12. $-30x^2yz$

13. $6t^4u^2$

14. $-t^3u^2 + 6t^4$

15. $4p + 32q$

16. $10c^5$

17. $v^3w^4 + 9v^5w^5$

18. x^3yz^3

19. $8a^2b^2 - 8ab$

20. $-2rt + 2t$

21. $-23h - 120$

22. $n - 2n^2 + n^3$

23. 0
24. $-d^3$
25. 0
26. 36

27. 12
28. –14
29. 9
30. –9

31. 3
32. 0
33. 2
34. 3

35. –2
36. 0
37. –5

38. $y < 3$ 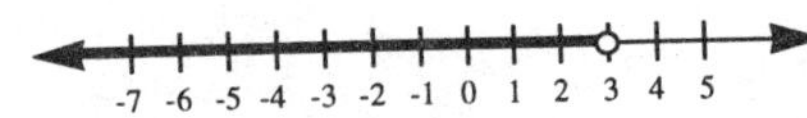

39. $z \le 1$ 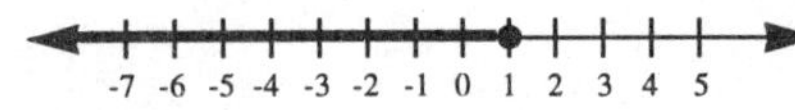

40. –5

41. $0 \le c < 3$

42. $1 < y < 2$ 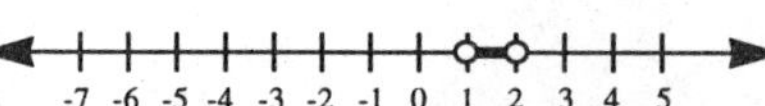

43. 6
44. 5
45. 8 and 19
46. width = 4, length = 18

47. 2 p.m.
48. 13 white geraniums

Cumulative Test: Chapters 1–3

1. (a) 19 (b) –64

2. (a) 7 (b) 5

3. (a) $2x^3 - 4x^2 + 7x$ (b) $x + 8y$ (c) $-a^4b + 8a^2b^2$ (d) 0

   (e) $3u^2v^6 - 9u^5v^4$ (f) $3 - 6b + 12b^2 - 8b^3$

4. (a) 3 (b) 5 (c) $c < 8$ (d) identity

(e) 4 (f) contradiction

5. (a) $x > 5$ (b) $-1 < x \le 0$

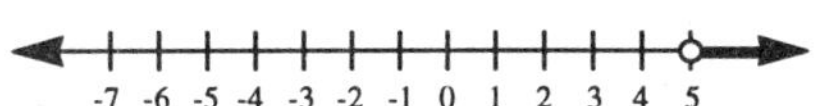

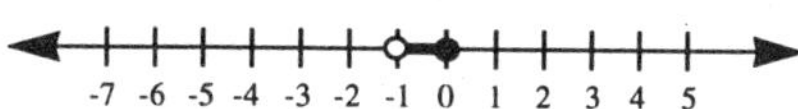

6. (a) 46 and 10 (b) 3 p.m. (c) 18 full-sized cars

Chapter 4: Test A

1. (a) $-\frac{4}{7}$ (b) y^5 (c) $\frac{3x}{2}$ (d) $-\frac{7b^3}{a^3}$

2. (a) $\frac{9x^2}{2y}$ (b) $\frac{1}{8y^3}$ (c) 1 (d) $\frac{b}{14a}$

   (e) $\frac{c}{2}$ (f) $\frac{c^2}{16}$ (g) $\frac{2}{x^2}$ (h) $\frac{u}{10}$

   (i) $\frac{17}{12y}$ (j) $\frac{4x - 25y}{10x^2y^3}$ (k) $-\frac{1}{x}$

3. (a) 12 (b) $x \le -1$ (c) 2 (d) 12

4. 2.54 meters

5. 10

6. 5 nickels, 20 pennies, 60 dimes

7. \$5000 at 7%, \$6000 at 10%

8. width = 12 cm, length = 25 cm

9. 80 liters

Chapter 4: Test B

1. (a) $\frac{3}{8}$ (b) z^8 (c) $-\frac{5x^3}{3}$ (d) $-\frac{5p}{2q}$

2. (a) $\frac{xy^2}{18}$ (b) $\frac{8}{x^3}$ (c) 1 (d) $\frac{d}{12c}$

   (e) $\frac{m}{3}$ (f) $\frac{m^2}{36}$ (g) $\frac{2}{x^3}$ (h) $\frac{11v}{18}$

   (i) $\frac{11}{20r}$ (j) $\frac{9b^2 + 20a}{12a^2b^3}$ (k) $\frac{2}{x^2}$

3. (a) 24 (b) $x < 3$ (c) 4 (d) 60

4. 5.66 cubic meters

5. 5

6. 10 quarters, 20 nickels, 30 dimes

7. 3 p.m.

8. $1500 at 8%, $3000 at 9%

9. 12 gallons of 25% solution, 8 gallons of 50% solution

Chapter 5: Test A

1. yes

2. (a) –3 (b) 12 (c) –4

3. (a)

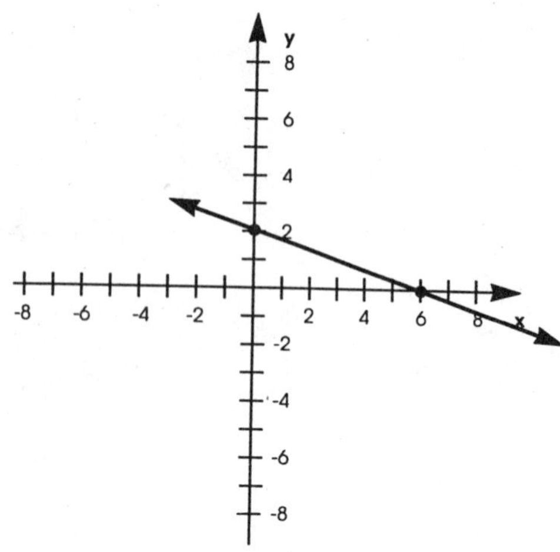

(b)

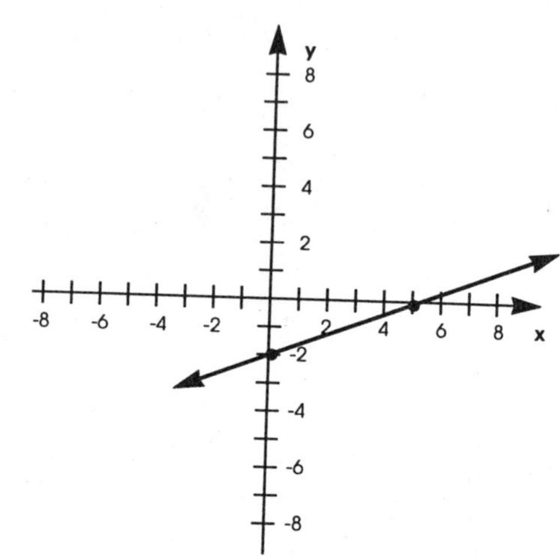

(c)

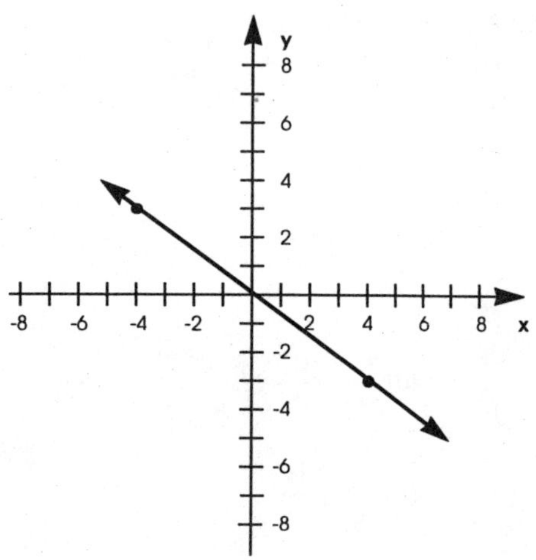

(d)

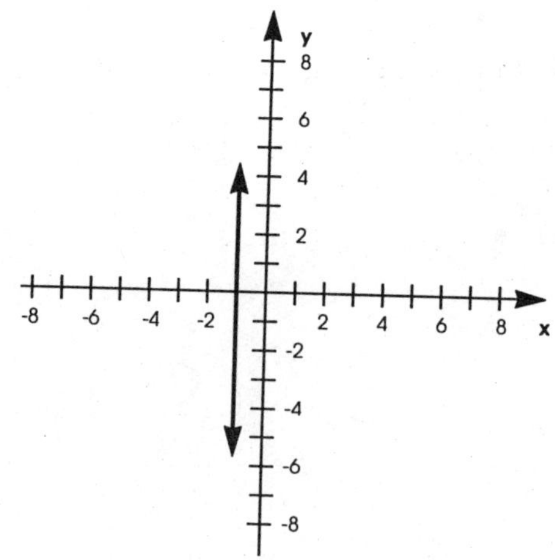

(e)

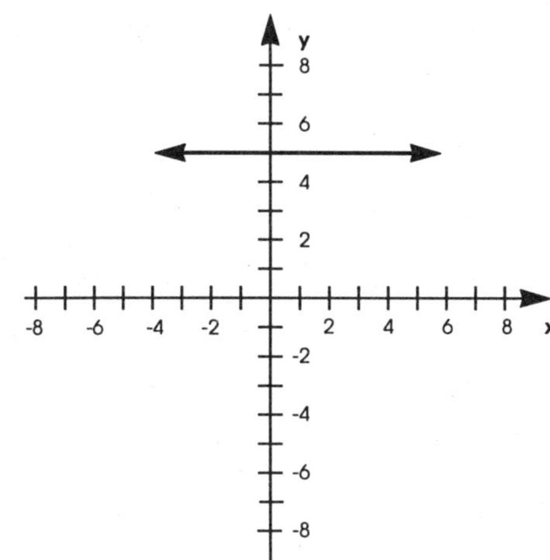

4. -1

5. $y + 1 = -\frac{2}{5}(x - 6)$ or $y = -\frac{2}{5}x + \frac{7}{5}$

6. $y = -2x + 7$

7. $y = 3$; $x = -2$

8. $-\frac{2}{3}$

9. $y = 3x + 5$

10. $y = \frac{2}{3}x - 2$

Chapter 5: Test B

1. no

2. (a) 2 (b –3 (c) 10

3. (a)

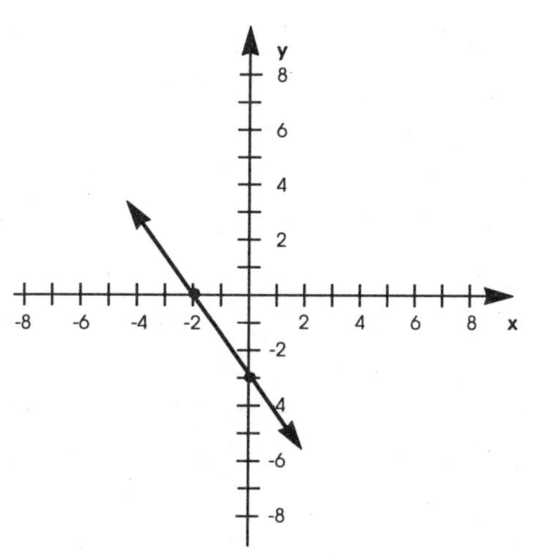

(b)

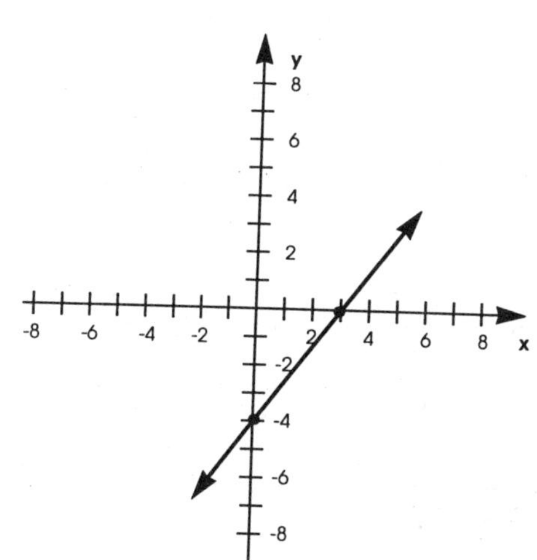

(c)

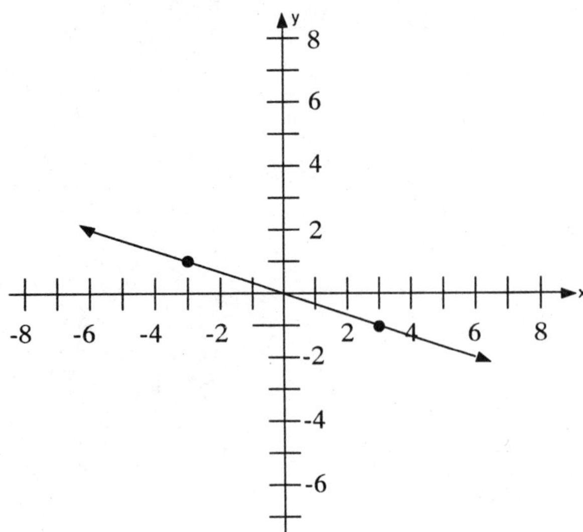

(d)

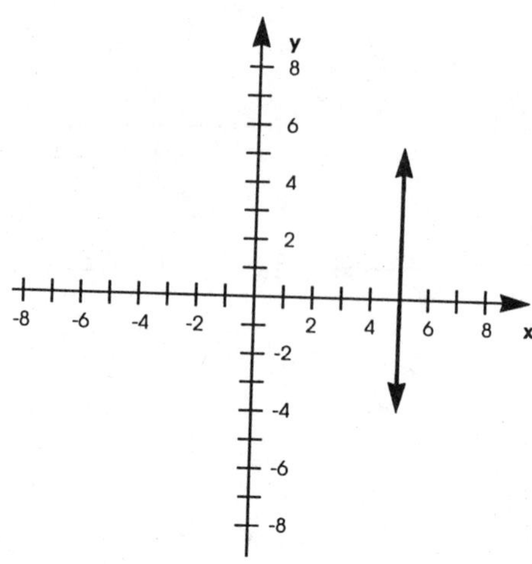

(e)

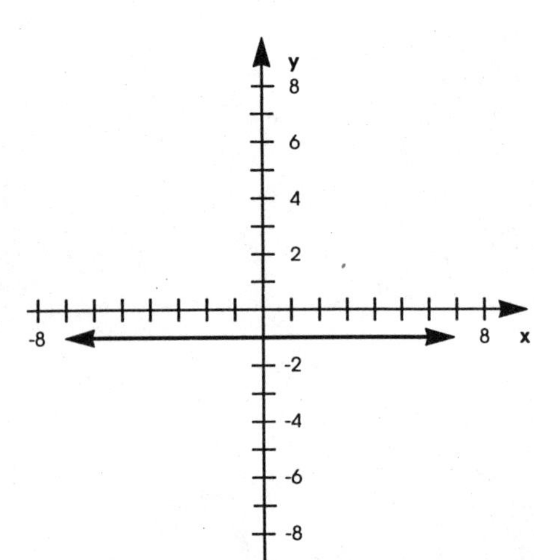

4. 1

5. $y + 4 = -\frac{3}{4}(x - 1)$ or $y = -\frac{3}{4}x - \frac{13}{4}$

6. $y = -4x + 10$ 7. $y = -2;\ x = 4$ 8. $\frac{3}{2}$ 9. $y = 5x - 11$

10. $y = -\frac{1}{2}x + 2$

Chapter 6: Test A

1.

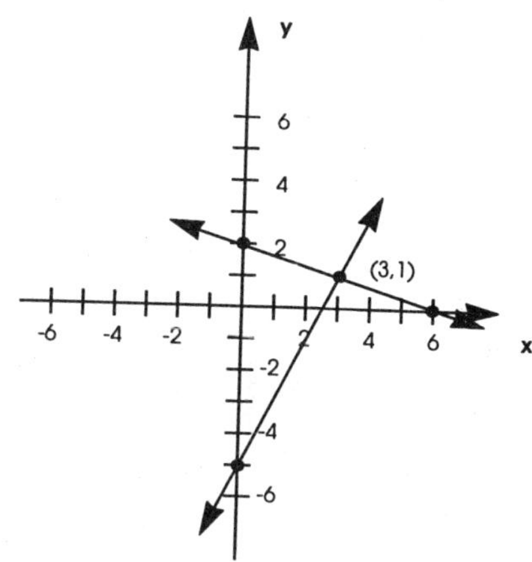

2. (4, –1)

3. (0, 2)

4. no solution

5. plums: \$.60/pound; peaches: \$.50/pound

Chapter 6: Test B

1.

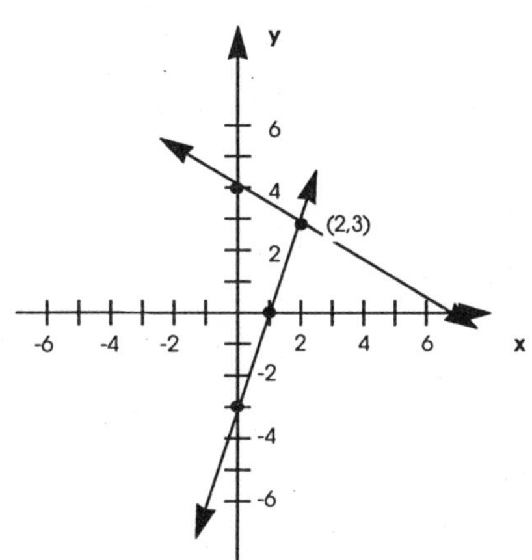

2. (–3, –1)

3. (3, 0)

4. no solution

5. veal: \$3/package; chicken: \$2.50/package

Cumulative Review: Chapters 4–6

1. $-\frac{8}{5}$	2. $\frac{3}{8x^5}$	3. $\frac{9b}{5a}$	4. $\frac{1}{2c}$
5. $\frac{16x}{9}$	6. $\frac{4x^3}{81}$	7. $\frac{16x^2 - 9x}{54}$	8. $\frac{2}{y^3}$
9. 1	10. 0	11. $\frac{4a^2b^6}{9c^6}$	12. $\frac{25a^6c^2}{49b^2}$
13. $\frac{49}{15d}$	14. $\frac{9\ell - 10k}{12k\ell}$	15. $\frac{25v + 12u}{40u^3v^3}$	16. $\frac{3}{16u^5v^5}$
17. $\frac{3}{8}$	18. $\frac{10 - y}{4y}$	19. $\frac{3x^4 - 2x^2 + 4}{x^3}$	20. $\frac{18q^2 + 15pq - 14p^2}{12p^3q^3}$
21. 3	22. 2	23. 0	24. identity
25. $-\frac{1}{3}$	26. $\frac{24}{5}$	27. 2	28. –6

29.

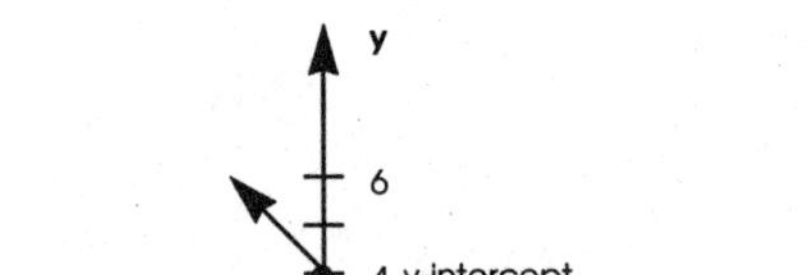

30.

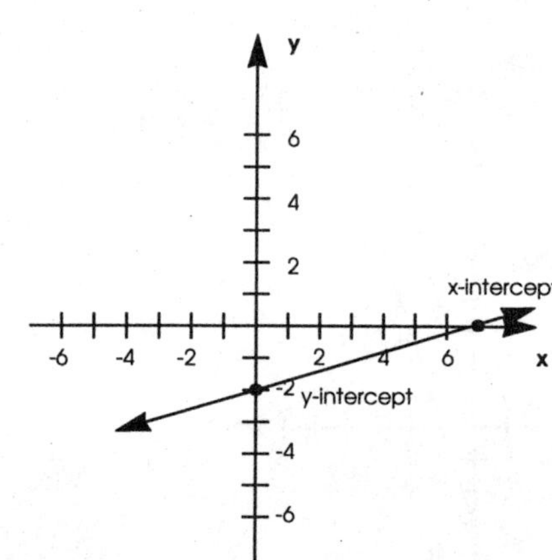

31.

y
6
4
y-intercept
2
x-intercept
-6
-4
-2
2
4
6
x
-2
-4
-6

32.

y
6
4
y-intercept
2
x-intercept
-6
-4
-2
2
4
6
x
-2
-4
-6

33.

y
6
4
y-intercept
2
-6
-4
-2
2
4
6
x
-2
-4
-6

34.

y
6
4
2
x-intercept
-6
-4
-2
2
4
6
x
-2
-4
-6

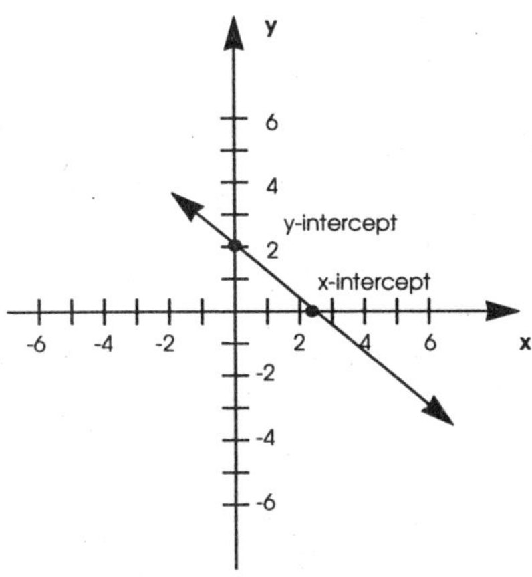

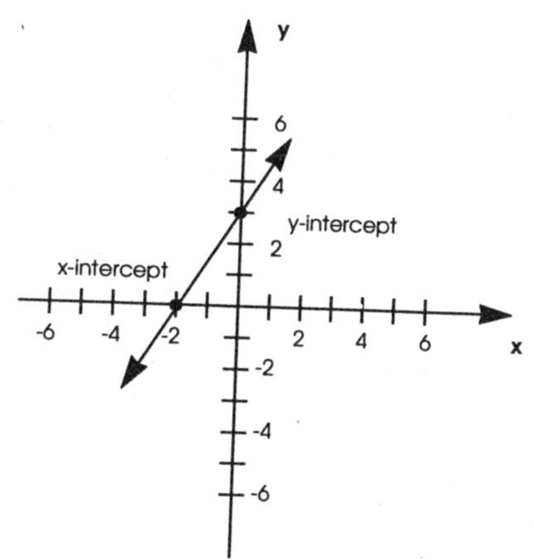

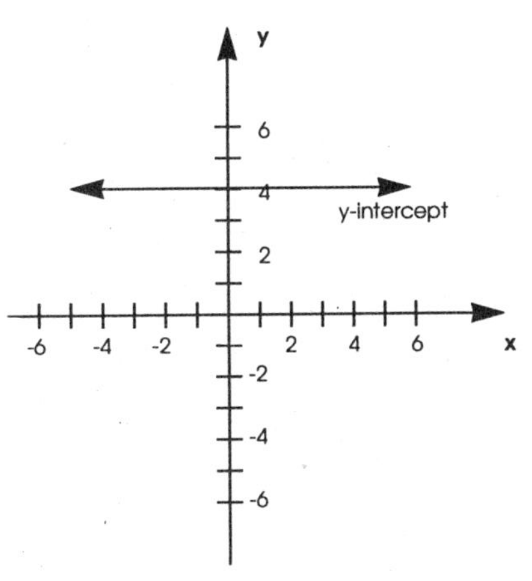

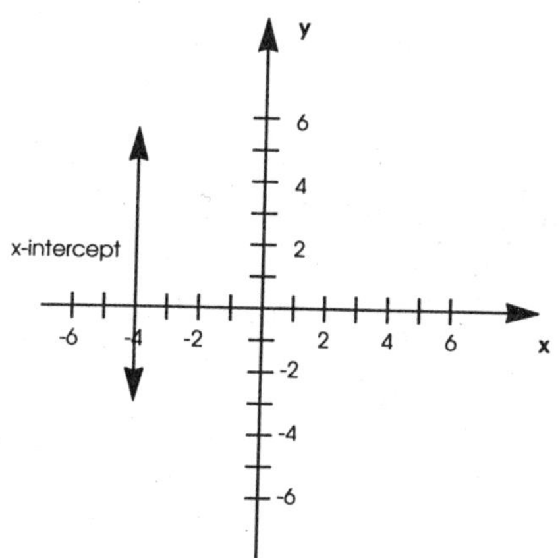

35.

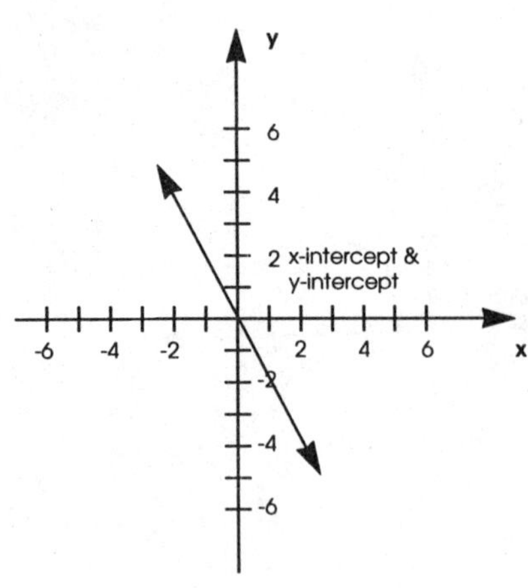

36.

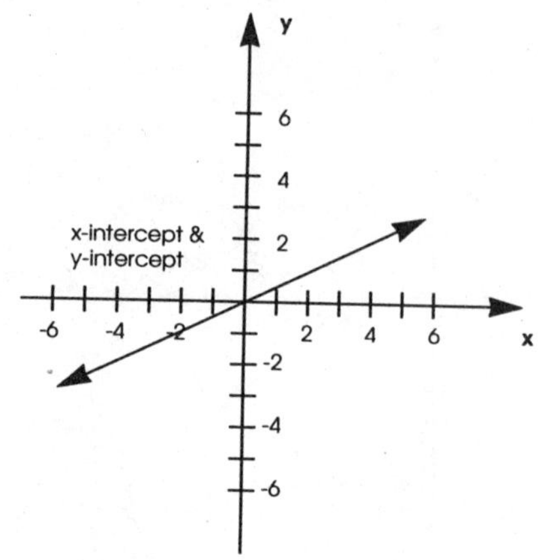

37. 2

38. $-\frac{5}{2}$

39. undefined

40. 0

41. $y + 4 = -5(x - 1)$ or $y = -5x + 1$

42. $y = \frac{2}{3}x - 6$

43. $y = 2$

44. $x = 6$

45. $y + 2 = -1(x - 4)$ or $y = -x + 2$

46. $y = -\frac{5}{6}x + 5$

47.

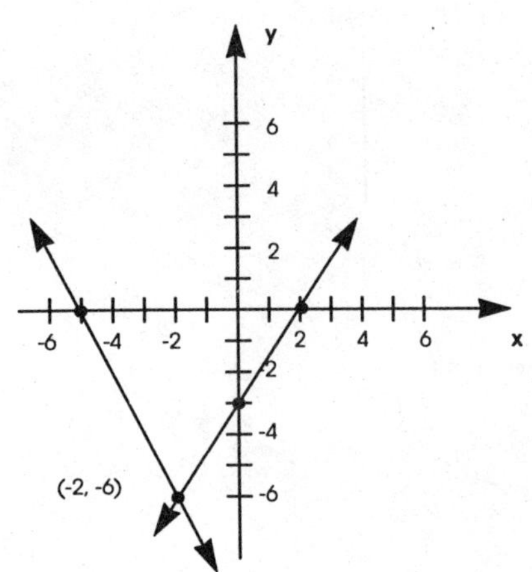

48.

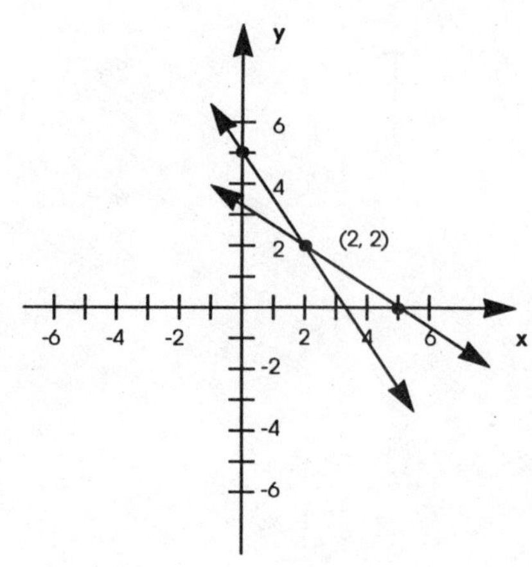

49. $(-2, -6)$ 50. $(2, 2)$ 51. $(\frac{2}{3}, -\frac{3}{4})$ 52. $(2, 1)$

53. $(1, -1)$ 54. $(\frac{2}{5}, -3)$

55. infinitely many solutions:

$\{(x, y) | 2x - 5y = 10\}$

56. no solution 57. 7, 9, 11, 13 58. 75 single people 59. \$3

60. 20 liters of 20% solution, 40 liters of 25% solution, 40 liters of 40% solution

61. 6 \$5 bills, 17 \$10 bills

Cumulative Test: Chapters 4–6

1. $\frac{6p^3}{r}$ 2. $\frac{2b}{25ac^6}$ 3. $\frac{3y}{ab}$ 4. 1

5. $\frac{9d^2 - 10c^2}{12c^3d^3}$ 6. $\frac{4x^2}{3}$ 7. 20 8. -2

9.

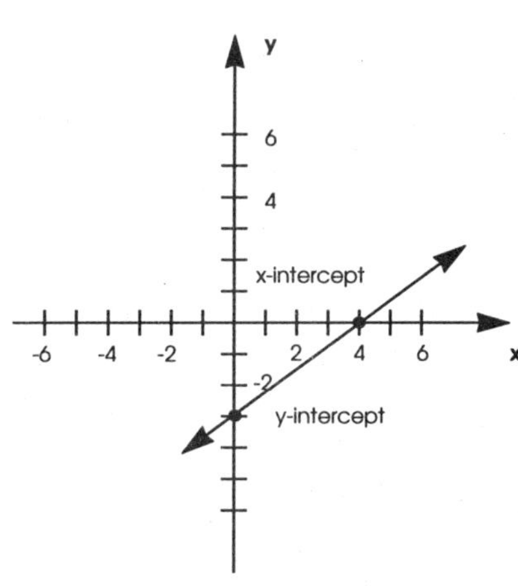

10.

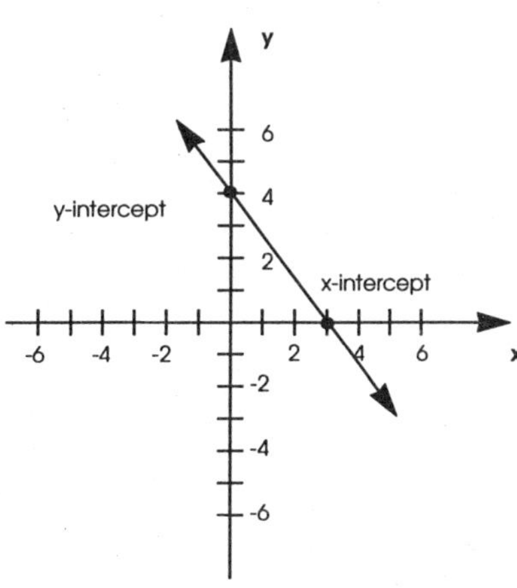

11. -2 12. $y - 5 = -2(x + 3)$ or $y = -2x - 1$

13. -1 14. $(-1, -3)$ 15. no solution 16. 3.79 miles

17. 25 and 27 18. \$9 19. \$1000 at 8%, \$4000 at 13%

Chapter 7: Test A

1. $\frac{1}{72}$ 2. $\frac{1}{x^5}$ 3. x^6 4. $\frac{x^4}{y}$

5. $-\frac{4y}{x}$

6. (a) 5 (b) −1; 7 (c) 6

7. $-30x^6y^6$ 8. $10x^4y^3 - 6x^5y^5$ 9. $2y^3$ 10. $10x^3 - x^2 + 5x + 4$

11. $x^3 + 3x^2 - 4x + 4$ 12. $16x - 18$ 13. $2z - 5$

14. (a) 2.58×10^5 (b) 2.58×10^{-2}

15. 2.4×10^5

Chapter 7: Test B

1. $\frac{29}{100}$ 2. $\frac{1}{x^6}$ 3. x^5 4. $\frac{y^2}{x}$

5. $-\frac{3x^2}{y^4}$

6. (a) 5 (b) 3; −5 (c) 7

7. $-24x^6y^8$ 8. $8x^5y^5 - 12x^3y^7$ 9. x^3y

10. $6x^3 + 7x^2 - 9x + 2$ 11. $2x^3 - 3x^2 - 3x + 1$ 12. $12x + 8$

13. $2y - 3$

14. (a) 7.95×10^6 (b) 7.95×10^{-4}

15. 5×10^{-9}

Chapter 8: Test A

1. $2x^2(x-1)(x-3)$ 2. $3x^2y^2(x+4y-3)$

3. $(x + 4)(x + 9)$ 4. $(x - 3y)(x + 4y)$

5. $4x^3(x-2)(x+2)$ 6. $7(x^2 + 2x + 2)$

7. $(x - 8)(x - 8)$ 8. $(2x - 1)(x - 4)$

9. not factorable 10. $(8x - 3)(x + 2)$

11. $(xy + 3)(xy + 3)$

12. $\frac{2}{a} + \frac{3b}{2} - \frac{6}{5ab}$

13. quotient: $3x^2 - x + 2$

remainder: -11

Chapter 8: Test B

1. $5x^2(x+1)(x+2)$

2. $7x^3y^3(y-4x-2)$

3. $(x-3)(x-8)$

4. $(x-9y)(x+2y)$

5. $3x^2(x-3)(x+3)$

6. $4(x^2+3x+4)$

7. $(x+9)(x+9)$

8. $(3x-5)(x+3)$

9. not factorable

10. $(9x+4)(x-2)$

11. $(xy-5)(xy-5)$

12. $\frac{3}{q} - \frac{4p}{3} + \frac{3}{2pq}$

13. quotient: $2x^2 + 3x + 9$;

remainder: 28

Chapter 9: Test A

1. $\frac{3(x+3)}{x-9}$

2. $\frac{3x}{x-3}$

3. $\frac{2(x-6)}{(x-2)(x-4)}$

4. $\frac{x}{2(x+2)}$

5. $\frac{3}{2(x-4)}$

6. $\frac{x+1}{x-3}$

7. $\frac{4}{x+1}$

8. 4

9. $y = \frac{3-3z}{6x+2}$

10. no solution

11. -8

12. 12 hours

Chapter 9: Test B

1. $\frac{x-16}{2(x-4)}$

2. $\frac{x+4}{2x}$

3. $\frac{3(x+3)}{(x+5)(x+2)}$

4. $\frac{x(x-3)}{3}$

5. $-\frac{2}{3(x+9)}$

6. $\frac{x-1}{x+4}$

7. $\frac{3}{x-1}$

8. 5

9. $q = \frac{8r+16}{12p-1}$

10. no solution

11. $\frac{4}{13}$

12. 2 mph

Cumulative Review: Chapters 7–9

1. $x^2 - 5x - 14$

2. $x^3 + 3x^2 - 5x - 14$

3. $x^4 + 2x^3 - 4x^2 - 5x - 14$

4. $5t^3 - 15t^2 - 20t$

5. $18r^2 - 3rs - 10s^2$

6. $5m^2 - 12m$

7. $36x$

8. $4x^2 - 7x$

9. $9m^3 + 27m^2 - m - 3$

10. $x^4 - 29x^2 + 100$

11. $-2z^2 + 2z + 7$

12. $4x^2 - 3xy - 2y^2$

13. (a) 6

(b) -5

14. $4x^4 + 0x^3 - 3x^2 + 2x - 1$

15. $\frac{x}{3} - 3x^2$

16. $\frac{9}{4t} + \frac{3}{st} - \frac{3}{2s}$

17. quotient: $a - 7$

remainder: 28

18. quotient: $3y^2 - 5y - 2$

remainder: 0

19. quotient: $6t^2 + 8t + 4$

remainder: -5

20. quotient: $b^3 + b^2 + b + 4$

remainder: 6

21. $\frac{7}{2}$

22. $\frac{5}{36}$

23. $\frac{1}{x^{11}}$

24. $\frac{1}{y^5}$

25. $\frac{3}{x^2}$

26. ab^3

27. $\frac{27v^{12}}{4u^{12}}$

28. $\frac{3n^3}{4m^4}$

29. 1.23×10^8

30. 9.75×10^{-6}

31. 1.2×10^{-2} or .012

32. 6.25×10^{-1} or .625

33. $x(x - 8)$

34. $(x - 2)(x - 6)$

35. $(x - 9)(x + 1)$

36. not factorable

37. $2ab(2a^2b - 3a + 4b^2)$

38. $2r^4s^4(2s - r)(2s + r)$

39. $2(3v - 4)(3v + 4)$

40. $2(9v^2 + 16)$

41. $t^3(t - 7)(t + 7)$

42. $(3s - 10)(s + 1)$

43. $(2t - 3)(3t - 2)$

44. $(6t - 1)(t - 6)$

45. $6(t - 1)(t - 1)$

46. not factorable

47. $-(x + 8)(x - 7)$

48. $(x - 25y)(x - 4y)$

49. not factorable

50. $(x + a)(x + b)$

51. $(x - 2y)(x + 3)$

52. $(9x^2 + 4y^2)(3x + 2y)(3x - 2y)$

53. $\frac{y}{y - 2}$

54. $\frac{s - 4}{s + 4}$

55. $\frac{x+4}{5x(3x+2)}$

56. $\frac{16c+9d}{12c^2d^2}$

57. $\frac{x^2+2x+1}{x^2+1}$

58. $\frac{m+n}{2}$

59. $\frac{2}{(x+2)(x+3)}$

60. $\frac{-a+6}{6a(a-2)}$

61. $\frac{2(s-3)}{s+3}$

62. 2

63. $\frac{4}{r+4}$

64. 6

65. 8

66. $s = 2 - r$

67. $u = 12(9v - 1)$

68. no solution

69. 3

70. –1

71. no solution

72. $1\frac{1}{2}$ hours

73. 8 kph

74. $\frac{4}{13}$

Cumulative Test: Chapters 7–9

1. $3s^4t + 2s^3t^2 - 17s^2t^3 + 12st^4$

2. $\frac{4a^5}{3b^3}$

3. $5x^2 - 7x - 17$

4. $2n^2 - 1$

5. $\frac{1}{2x^5}$

6. $\frac{4\ell}{k} - \frac{5k}{2\ell}$

7. quotient: $x^2 + 2x - 3$

   remainder: 6

8. quotient: $x^3 - 3x^2 - 2x + 6$

   remainder: 0

9. (a) 5.31×10^{-7} (b) 5.31×10^8

10. 1×10^{-1} or .1

11. $2x^2 + x - 3$

12. $(x-2)(x-3)$

13. $(x-6)(x+1)$

14. $2s^4t^4(t-2s)(t+2s)$

15. $10(x^2+3x-7)$

16. $(3y-7)(y+2)$

17. $(12z+25)(z-1)$

18. $(2c+1)(c-3)$

19. $(x^2+y^2)(x-2)$

20. $\frac{x(x-2)}{x+2}$

21. $\frac{2a-3}{a-3}$

22. $\frac{10x}{(x+3)(x-2)}$

23. $\frac{u+2}{2(u+1)}$

24. $\frac{6}{x(x+2)}$

25. $\frac{(r+5)(r+2)}{(r+4)(r+3)}$

26. $\frac{ab}{4(b-a)}$

27. 4

28. $w=\frac{22v-16}{4k-3}$

29. no solution

30. 30 mph from A to B
40 mph from B to A

Chapter 10: Test A

1. $4x^5y^6$

2. $18\sqrt{2}$

3. $x^2\sqrt{5x}$

4. $\sqrt{15}x^3y^2$

5. 0

6. $\frac{3y^2\sqrt{2y}}{2}$

7. $2x+\sqrt{3x}-18$

8. $2\sqrt{6a}$

9. $3-\sqrt{2}$

10. $2(\sqrt{x}+2\sqrt{y})$

11. yes

12. 9

Chapter 10: Test B

1. $6x^7y^4$

2. $-8\sqrt{2}$

3. $2x^2\sqrt{3x}$

4. $\sqrt{10}x^2y^4$

5. 0

6. $\frac{4z^3\sqrt{3z}}{3}$

7. $3x - 7\sqrt{2x} - 12$

8. $-2\sqrt{7b}$

9. $4 + \sqrt{10}$

10. $3(\sqrt{x} - 3\sqrt{y})$

11. yes

12. 12

Chapter 11: Test A

1. ± 3

2. $\frac{2 \pm \sqrt{6}}{2}$

3. $3, -\frac{1}{3}$

4. $\frac{3}{2}, -\frac{1}{2}$

5. $\frac{2 \pm \sqrt{3}}{2}$

6. $0, 2$

7. $\frac{3}{2}$

8. $\frac{-1 \pm \sqrt{5}}{2}$

9. $2, \frac{1}{3}$

10. $\frac{8 \pm 3\sqrt{6}}{2}$

11.

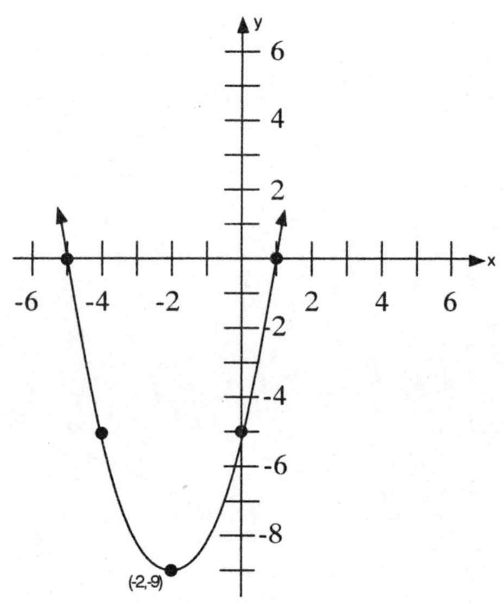

The x-intercepts are –5 and 1.

Chapter 11: Test B

1. ± 5

2. $\frac{-2 \pm \sqrt{2}}{2}$

3. $\frac{2}{5}, -1$

4. $1, -\frac{5}{3}$

5. $\frac{1}{9}, 1$

6. $-3 \pm 2\sqrt{3}$

7. $-\frac{3}{4}$

8. 0, 2

9. $1, -\frac{2}{3}$

10. $\frac{-9 \pm 5\sqrt{3}}{3}$

11.

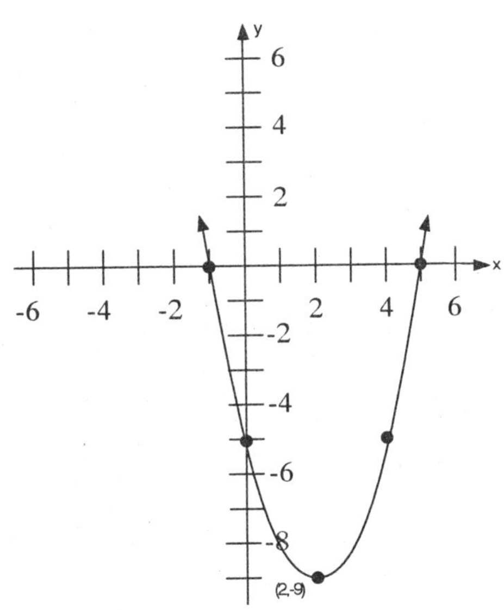

The x-intercepts are –1 and 5.

Cumulative Review: Chapters 10–11

1. $2a^{10}b^{15}$

2. $\frac{w^3z^2}{3}\sqrt{w}$

3. $\frac{8\sqrt{11}}{11}$

4. $4\left(\sqrt{11}+3\right)$

5. $3\left(\sqrt{14}-\sqrt{10}\right)$

6. $3\sqrt{7x}$

7. $8\sqrt{3}$

8. $2u^3v^3\sqrt{21v}$

9. $5\sqrt{6rs}$

10. $-\sqrt{30}$

11. 0

12. $6a + 5\sqrt{ab} - 6b$

13. $3p^4q^2\sqrt{2q}$

14. $\frac{9\sqrt{6}}{8}$

15. 17

16. $x^2 + 10x + 25$

17. 36

18. 0, – 4

19. – 7, 3

20. – 3, 8

21. $\frac{5}{4}$

22. $\frac{-5 \pm \sqrt{41}}{2}$

23. $\frac{2}{3}$, 1

24. – 3, 2

25. – 4, 3

26. $-\frac{5}{2}$, 4

27. 1

28. 1, 5

29. $4 \pm \sqrt{3}$

30. 23

31. 9

32. no solution

33. 7

34. $-2 \pm \sqrt{2}$

35. $\frac{3 \pm \sqrt{7}}{2}$

36.

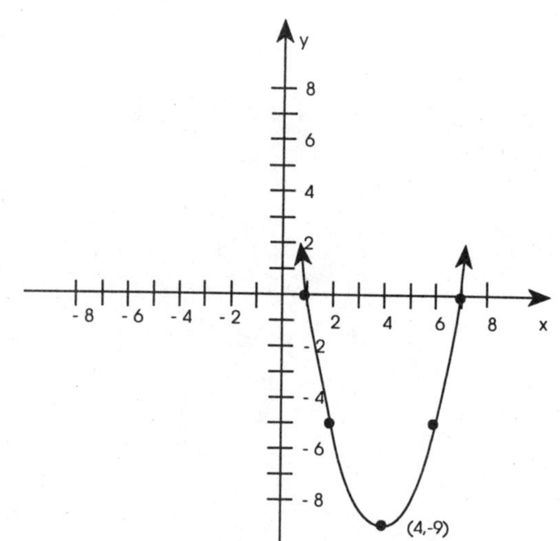

The x-intercepts are 1 and 7.

37.

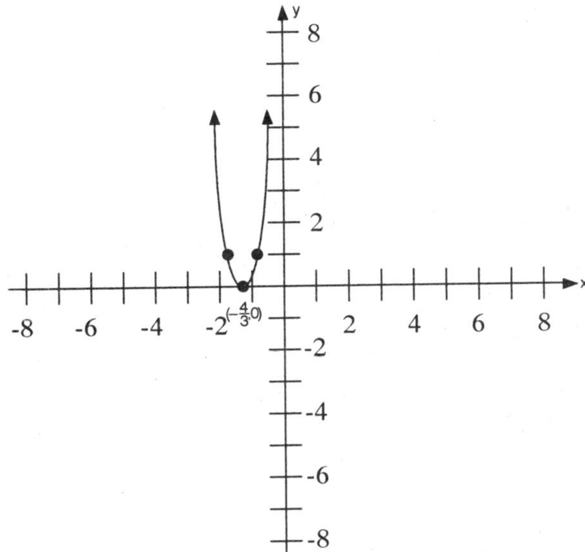

The x-intercept is $-\frac{4}{3}$.

38.

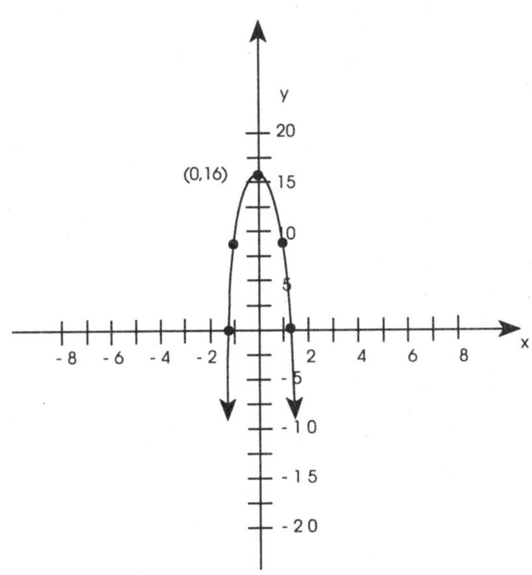

The x-intercepts are $-\frac{4}{3}$ and $\frac{4}{3}$.

39.

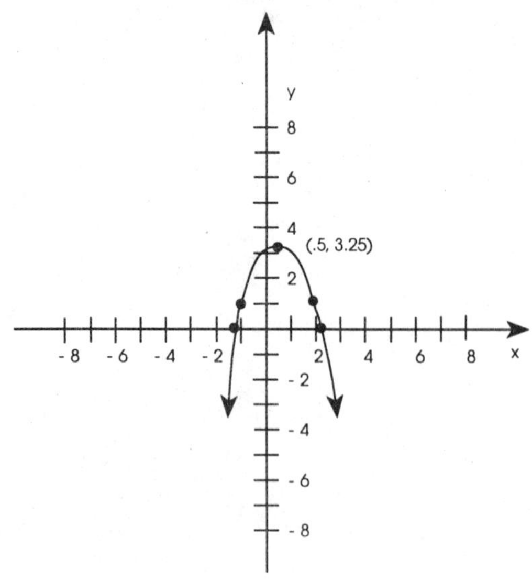

The x-intercepts are $\frac{1-\sqrt{13}}{2} \approx -1.3$ and $\frac{1+\sqrt{13}}{2} \approx 2.3$.

40. width = 4, length = 10

41. (a) 30 (b) 30

42. $4\sqrt{2}$ inches

43. 5 mph

44. $\frac{1}{3}$, 2

Cumulative Test: Chapters 10–11

1. $8\sqrt{2}$
2. $3rs^2t^4\sqrt{3rs}$
3. $12\sqrt{7}$
4. $bc^4\sqrt{10b}$
5. $\frac{9\sqrt{10}}{2}$
6. $4x^8$
7. $5\left(\sqrt{10}+1\right)$
8. $19+6\sqrt{2}$
9. $12w+10\sqrt{6w}-48$
10. $\frac{\sqrt{7}\left(\sqrt{11}+\sqrt{3}\right)}{4}$
11. -9, 6
12. $-5\pm\sqrt{5}$
13. $\frac{2}{7}$
14. 1
15. $\frac{1}{3}$, 3
16. $\frac{\pm 2\sqrt{15}}{3}$
17. no solution
18. $\pm\frac{\sqrt{3}}{2}$

19. 18 sq. cm 20. 14 mph

21.

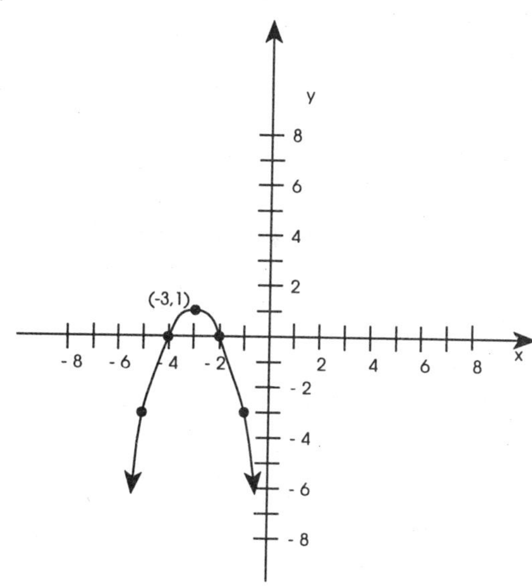

The x-intercepts are –4 and –2.

22.

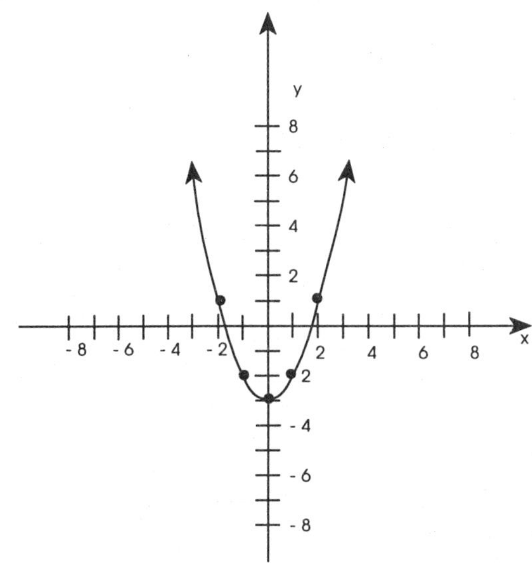

The x-intercepts are $-\sqrt{3}$ and $\sqrt{3}$.

Practice Final Exam

1. 2
2. 13
3. 10
4. $\frac{6}{5}$
5. $-x^3 - x^2y^2 + 6x^2y$
6. $2x^3 - 5x^2 + 3x - 3$

7. $\frac{5-4x}{40y}$

8. $\frac{x^4}{y^{14}}$

9. $\frac{(x-5)(x+5)}{(x-4)(x+4)}$

10. $\frac{2}{x^2-1}$

11. $6x\sqrt{6x}$

12. $3+\sqrt{3}$

13. $-3 < x \le 2$

14. $x > 1$

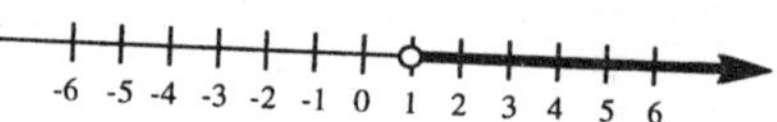

15. -2

16. -6

17. 4

18. $0, 3$

19. $-\frac{2}{3}, \frac{3}{2}$

20. $\frac{-3 \pm \sqrt{14}}{5}$

21. 1.8×10^5

22. $2x^2(2x-3)(x+6)$

23. $(x + 2y)(x - y)(x + y)$

24. quotient: $2x^2 - 2x + 2$
remainder: -3

25.

26. $y = 4x - 8$

27. $(1, -\frac{1}{3})$

28. 12 years old

29. 30 red chips

30. \$4200

31. 3 hours

32. $-\frac{1}{2}$ and $\frac{3}{2}$